W0111960

Beginning Android 3D
Game Development

Robert Chin

Beginning Android 3D Game Development

Copyright © 2014 by Robert Chin

This work is subject to copyright. All rights are reserved by the Publisher, whether the whole or part of the material is concerned, specifically the rights of translation, reprinting, reuse of illustrations, recitation, broadcasting, reproduction on microfilms or in any other physical way, and transmission or information storage and retrieval, electronic adaptation, computer software, or by similar or dissimilar methodology now known or hereafter developed. Exempted from this legal reservation are brief excerpts in connection with reviews or scholarly analysis or material supplied specifically for the purpose of being entered and executed on a computer system, for exclusive use by the purchaser of the work. Duplication of this publication or parts thereof is permitted only under the provisions of the Copyright Law of the Publisher's location, in its current version, and permission for use must always be obtained from Springer. Permissions for use may be obtained through RightsLink at the Copyright Clearance Center. Violations are liable to prosecution under the respective Copyright Law.

IISBN-13 (pbk): 978-1-4302-6547-4

ISBN-13 (electronic): 978-1-4302-6548-1

Trademarked names, logos, and images may appear in this book. Rather than use a trademark symbol with every occurrence of a trademarked name, logo, or image, we use the names, logos, and images only in an editorial fashion and to the benefit of the trademark owner, with no intention of infringement of the trademark.

The images of the Android Robot (01 / Android Robot) are reproduced from work created and shared by Google and used according to terms described in the Creative Commons 3.0 Attribution License. Android and all Android and Google-based marks are trademarks or registered trademarks of Google, Inc., in the U.S. and other countries. Apress Media, L.L.C. is not affiliated with Google, Inc., and this book was written without endorsement from Google, Inc.

The use in this publication of trade names, trademarks, service marks, and similar terms, even if they are not identified as such, is not to be taken as an expression of opinion as to whether or not they are subject to proprietary rights.

While the advice and information in this book are believed to be true and accurate at the date of publication, neither the author nor the editors nor the publisher can accept any legal responsibility for any errors or omissions that may be made. The publisher makes no warranty, express or implied, with respect to the material contained herein.

President and Publisher: Paul Manning
Lead Editor: Steve Anglin
Technical Reviewer: Jim Graham
Developmental Editor: Anne Marie Walker
Editorial Board: Steve Anglin, Mark Beckner, Ewan Buckingham, Gary Cornell, Louise Corrigan, Jim DeWolf, Jonathan Gennick, Jonathan Hassell, Robert Hutchinson, Michelle Lowman, James Markham, Matthew Moodie, Jeff Olson, Jeffrey Pepper, Douglas Pundick, Ben Renow-Clarke, Dominic Shakeshaft, Gwenan Spearing, Matt Wade, Steve Weiss
Coordinating Editors: Anamika Panchoo, Christine Ricketts
Copy Editor: Michael G. Laraque
Compositor: SPi Global
Indexer: SPi Global
Artist: SPi Global
Cover Designer: Anna Ishchenko

Distributed to the book trade worldwide by Springer Science+Business Media New York, 233 Spring Street, 6th Floor, New York, NY 10013. Phone 1-800-SPRINGER, fax (201) 348-4505, e-mail orders-ny@springer-sbm.com, or visit www.springeronline.com. Apress Media, LLC is a California LLC and the sole member (owner) is Springer Science + Business Media Finance Inc (SSBM Finance Inc). SSBM Finance Inc is a Delaware corporation.

For information on translations, please e-mail rights@apress.com, or visit www.apress.com.

Apress and friends of ED books may be purchased in bulk for academic, corporate, or promotional use. eBook versions and licenses are also available for most titles. For more information, reference our Special Bulk Sales–eBook Licensing web page at www.apress.com/bulk-sales.

Any source code or other supplementary materials referenced by the author in this text is available to readers at www.apress.com. For detailed information about how to locate your book's source code, go to www.apress.com/source-code/.

Contents at a Glance

Contents

About the Author

Robert Chin holds a bachelor of science in computer engineering and is experienced in C/C++, Unreal Script, Java, DirectX, OpenGL, and OpenGL ES 2.0. He has written 3D games for the Windows and Android platforms. He is the author of *Beginning iOS 3D Unreal Games Development*, published by Apress, and was the technical reviewer for *UDK Game Development*, published by Course Technology, a part of CENGAGE Learning.

About the Technical Reviewer

Jim Graham received a bachelor of science in electronics with a specialty in telecommunications from Texas A&M University and graduated with his class (Class of '88) in 1989. He was published in the International Communications Association's 1988 issue of *ICA Communique* ("Fast Packet Switching: An Overview of Theory and Performance"). He has worked as an associate network engineer in the Network Design Group at Amoco Corporation in Chicago, Illinois; a senior network engineer at Tybrin Corporation in Fort Walton Beach, Florida; and as an intelligence systems analyst at both 16th Special Operations Wing Intelligence and HQ US Air Force Special Operations Command Intelligence at Hurlburt Field, Florida. He received a formal letter of commendation from the 16th Special Operations Wing Intelligence on December 18, 2001

Acknowledgments

I would like to thank developmental editor Anne Marie Walker for making my manuscript as smooth and informative as possible. I would like to thank Christine Ricketts and Anamika Panchoo, coordinating editors, for keeping me on track with deadlines and answering other important questions I had. I would also like to thank copy editor Michael G. Laraque for correcting formatting errors and making my manuscript more consistent and readable. Finally, I would like to thank Jim Graham, the technical reviewer, for making sure all the Android examples in this book worked and that the technical information was as accurate as possible.

Introduction

This book is meant to be a quick-start guide to developing 3D games for the Android platform using Java and OpenGL ES 2.0. Development will utilize the Eclipse Integrated Development Environment (IDE) with Android Development Tools (ADT) plug-ins installed. The goal is to cover key concepts and illustrate them, using concrete hands-on examples and case studies. A single book cannot cover every aspect of Android game development or Android software development in general. Thus, this book is not meant as a reference guide. The following is a summary of each chapter in this book.

Chapter 1: "Let's Meet the Android." In this chapter, I provide an overview of Android, an overview of the Android SDK, instructions on how to set up your computer for Android development, and a hands-on example involving a simple "Hello World" program for those unfamiliar with Android.

Chapter 2: "Java for Android." In this chapter, I offer an overview of the Java language, the basic Android Java program framework, and information on the basic Java OpenGL ES framework.

Chapter 3: "3D Math Review." In this chapter, 3D math, vectors, matrices, and vector and matrix operations are discussed.

Chapter 4: "3D Graphics Using OpenGL ES 2.0." In this chapter, I provide an overview of OpenGL ES 2.0 on Android, 3D meshes, lighting, materials, textures, saving persistent data, and creating a gravity grid using vertex and fragment shaders.

Chapter 5: "Motion and Collision." In this chapter, collision and Newtonian mechanics are covered.

Chapter 6: "Game Environment." In this chapter, sounds and the heads-up display are discussed.

Chapter 7: "Drone Grid Case Study: Creating the Player." In this chapter, I explain how to create a player, including elements associated with a player within our Drone Grid game, such as weapons, ammunition and player's HUD.

Chapter 8: "Drone Grid Case Study: Creating the Enemies." This chapter details how to create the enemies in our Drone Grid game. The enemies are arena objects and tanks. Arena objects are fairly simple in their behavior. Tanks are more complex enemy objects that will require the use of complex artificial intelligence, which I also cover.

Chapter 9: "Drone Grid Case Study: The User Interface." User interfaces for our Drone Grid game are discussed in this chapter, including the Main Menu System, the creation of the high score table, and the high score entry menu.

Chapter 10: "The Final Drone Grid Game." This chapter brings together everything from previous chapters into the final Drone Grid game. A final complete working game that integrates all the elements from previous chapters is presented. The final game will use elements discussed previously such as menus, heads up display, and enemy objects such as arena objects and tanks.

Chapter 11: "The Android Native Development Kit (NDK)." This chapter covers the Android Native Development Kit and discusses the Java Native Interface (JNI) in addition.

Chapter 12: "Publishing and Marketing Your Final Game." This chapter discusses how to publish and market your final Android game. It includes a list of Android marketplaces from which you can upload your game distribution file, a list of numerous ad networks that support Android, and a list of game sites that review Android games.

Let's Meet the Android

Android mobile phones dominate the mobile smartphone market, surpassing even Apple's iPhone. There are hundreds of millions of mobile phones using the Android operating system in over 190 countries around the world. Every day, a million new users begin using their Android phones to surf the Web, to e-mail friends, and to download apps and games. In fact, in the Google Play Store alone, there are 1.5 billion downloads per month of Android games and applications. If you include other web sites that offer Android games and apps for sale, such as Amazon Appstore for Android, then the number is even higher.

In this chapter, you will learn about the Android Software Development Kit (SDK). You will learn how to set up the Android development environment. You will also learn about the major components of this environment, such as Eclipse. We then go through the creation and deployment of a simple "Hello World" program for Android, to both a virtual Android emulator program and also a real Android device.

Overview of Android

The Android operating system is a widely used operating system available on mobile phones and tablets. It is even used on a video game console called the Ouya. Android phones range from expensive phones that require a contract to inexpensive prepaid phones that do not require any contract. Developing programs for the Android platform does not require any developer's fees, unlike Apple mobile devices, which require yearly fees in order to even be able to run your program on their devices. A good working prepaid no-contract Android phone that can develop 3D games using OpenGL ES 2.0 can be bought on Amazon.com for as little as $75–$100 with free shipping.

Overview of the Android SDK

This section discusses the Android SDK. Development system requirements and important individual pieces of the SDK, such as the SDK Manager, Android Virtual Device Manager, and the actual Android emulator will be covered.

Android Software Development Kit (SDK) Requirements

Android development can be done on a Windows PC, Mac OS machine, or a Linux machine. The exact operating system requirements are as follows:

Operating Systems:

- Windows XP (32-bit), Vista (32- or 64-bit), or Windows 7 (32- or 64-bit)
- Mac OS X 10.5.8 or later (x86 only)
- Linux (tested on Ubuntu Linux, Lucid Lynx)
    - GNU C Library (glibc) 2.7 or later is required.
    - On Ubuntu Linux, version 8.04 or later is required.
    - 64-bit distributions must be capable of running 32-bit applications.

Developing Android programs also requires installation of the Java Development Kit. Java Development Kit requirements are JDK 6 or later and are located at `www.oracle.com/technetwork/java/javase/downloads/index.html`.

If you are using a Mac, then Java may already be installed.

The Eclipse IDE program modified with the Android Development Tools (ADT) plug-in forms the basis for the Android development environment. The requirements for Eclipse are as follows:

- Eclipse 3.6.2 (Helios) or greater located at `http://eclipse.org`
- Eclipse JDT plug-in (included in most Eclipse IDE packages)
- Android Development Tools (ADT) plug-in for Eclipse located at `http://developer.android.com/tools/sdk/eclipse-adt.html`

Notes   Eclipse 3.5 (Galileo) is no longer supported with the latest version of ADT. For the latest information on Android development tools, go to `http://developer.android.com/tools/index.html`.

Android SDK Components Overview

The different components of the Android SDK are the Eclipse program, the Android SDK Manager, and the Android Virtual Device Manager and emulator. Let's look at each in more detail.

Eclipse with Android Development Tools Plug-in

The actual part of the Android SDK that you will spend most of your time dealing with is a program called Eclipse, which is customized specifically for use with Android through the ADT software plug-in. You will enter new code, create new classes, run programs on the Android emulator and on real devices from this program. On older, less capable computers, the emulator may run so slowly that the best option would be running the program on an actual Android device. Because we are dealing with CPU-intensive 3D games in this book, you should use an actual Android device to run the example projects (see Figure 1-1).

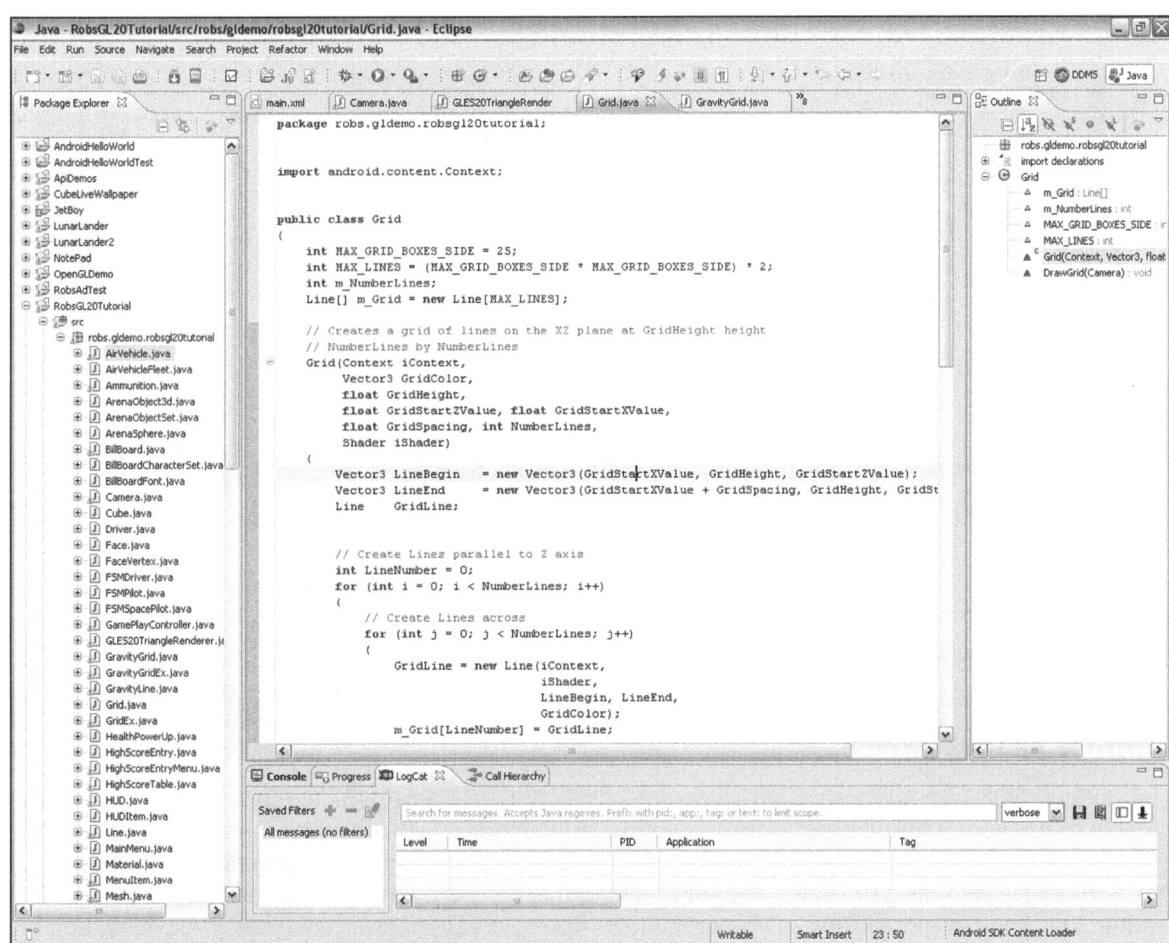

Figure 1-1. Eclipse with Android Development Tools plug-ins

Android SDK Manager

The Android SDK Manager allows you to download new Android platform versions and tools through its interface. Current tools and platform versions that are installed are also displayed. For example, in Figure 1-2, the Android 2.2 platform has already been installed and is ready for use for development. This means that you can compile your source code to target this platform.

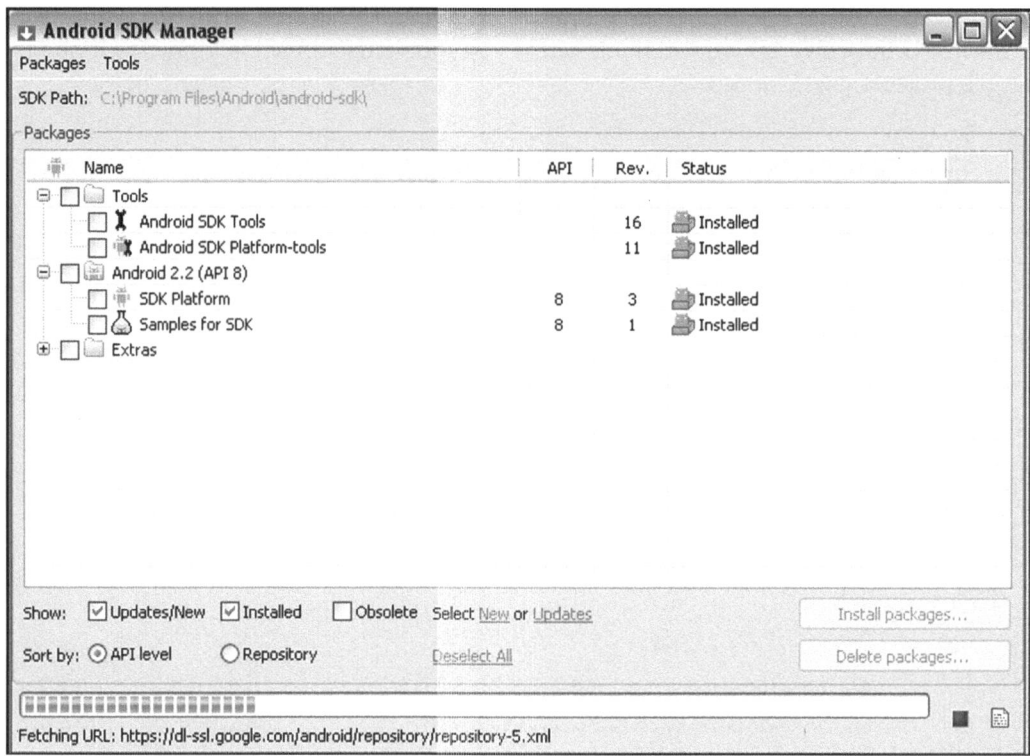

Figure 1-2. *The Android SDK Manager*

Android Virtual Device

The Android SDK also supports a virtual device emulator (see Figure 1-3). In many instances, you will be able to run your Android programs on a software emulator on your development system rather than an actual device. However, this works best for non-graphic intensive applications. Because this book deals with 3D games, we will not be using this software emulator but an actual Android device. The Android Virtual Device Manager allows you to create new virtual Android devices, edit existing Android devices, delete existing devices, and start up an existing virtual Android device. Figure 1-3 indicates that there is a valid virtual Android device named "Android22," which emulates the 2.2 version of the Android operating system (API Level 8) and simulates the ARM CPU type. The 2.2 version of the Android operating system is important because it is the first version that supports OpenGL ES 2.0, which we will be using in this book to develop our 3D graphics. OpenGL is the graphics system that allows the programmer to create 3D graphics on the Android platform. It is designed to be hardware-independent. That is, OpenGL graphics commands are designed to be the same across many different hardware platforms, such as PC, Mac, Android, etc. The OpenGL 2.0 version of OpenGL is the first version of OpenGL that includes programmable vertex and fragment shaders. OpenGL ES is a subset of regular OpenGL and contains fewer features.

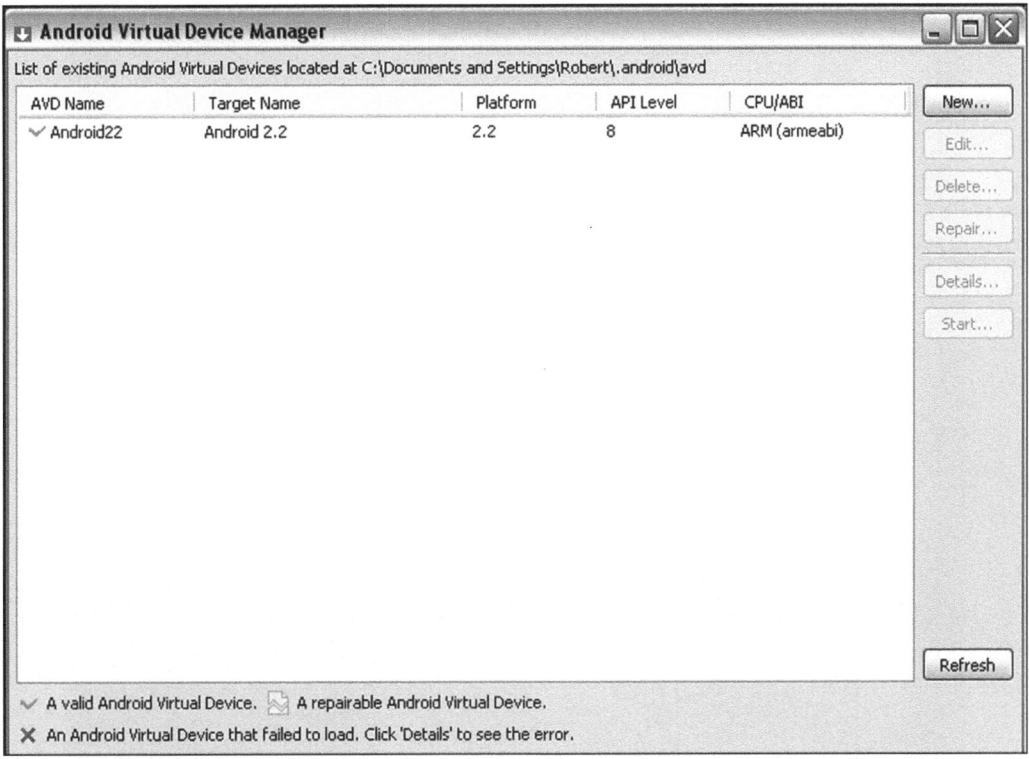

Figure 1-3. *The Android Virtual Device Manager*

Figure 1-4 depicts the actual emulator after it is launched. The emulator depicted is the one for version 2.2 of the Android operating system.

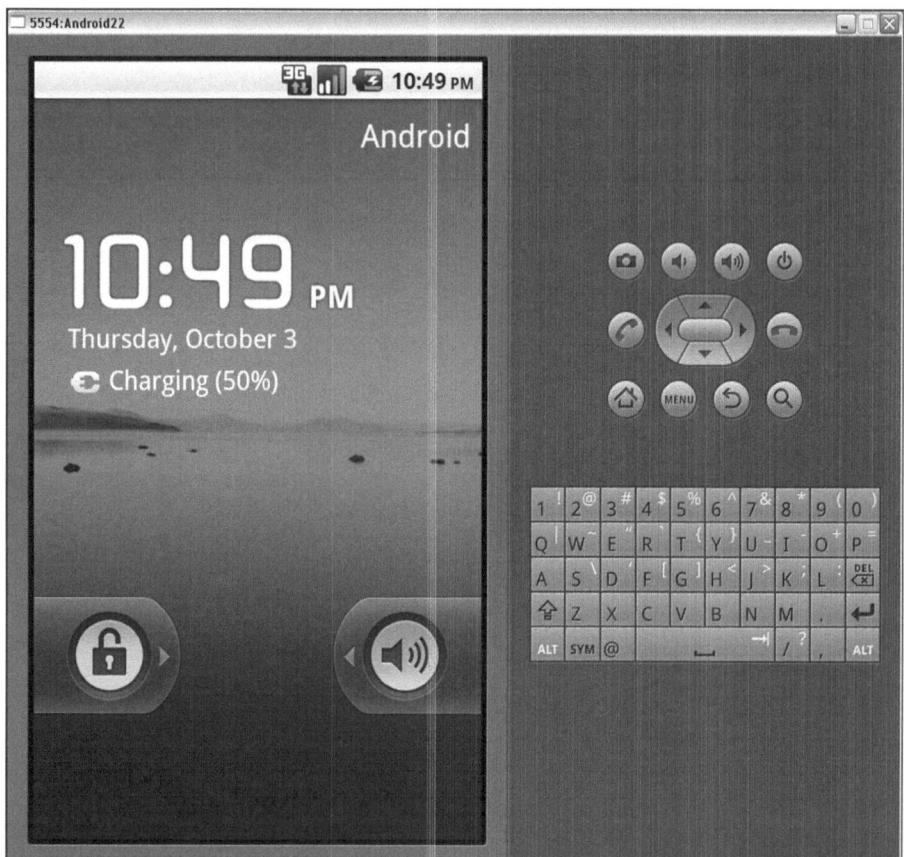

Figure 1-4. *The actual Android Virtual Device emulator*

How to Set Up for Development

First, you need to download and install the Java Development Kit Version 6 or greater. The Android development environment requires this as a prerequisite. After you verify that it is installed and working, then you will have to install the main components of the Android SDK.

The quickest and easiest way to do this is to download the ADT Bundle located at http://developer.android.com/sdk/index.html under the "Download for Other Platforms" section.

The ADT Bundle is a downloadable zip file that contains a special version of Eclipse with the Android Development Tools plug-in, the Android Virtual Device Manager, the SDK Manager and tools, as well as the latest Android platform, and the latest Android system image for the Android emulator. All you have to do to install this ADT Bundle is to create a new directory and unzip the file into it. You can use a free tool such as 7-Zip to uncompress the file. After doing this, you can execute the new ADT Integrated Development Environment by executing the `eclipse.exe` file located in the Eclipse directory under the main bundle directory.

> **Note**   7-Zip can be downloaded at `www.7-zip.org`.

Android Development Tools Integrated Development Environment (IDE) Overview

The Eclipse IDE consists of several important sections that I will discuss here. The important sections are the Package Explorer window, the Source Code Area window, the Outline window, and the Messages windows, including a window that outputs programmer-specified debug messages that is called the LogCat window. There are other Messages windows available, but they are less important and won't be covered in this section.

Package Explorer

When you start on a new Android programming project, you will create a new package for it. In Eclipse, there is a window called Package Explorer, located by default on the left-hand side. This window lists all the Android packages located in the current work space. For example, Figure 1-5 lists packages such as "AndroidHelloWorld," "AndroidHelloWorldTest," and "ApiDemos."

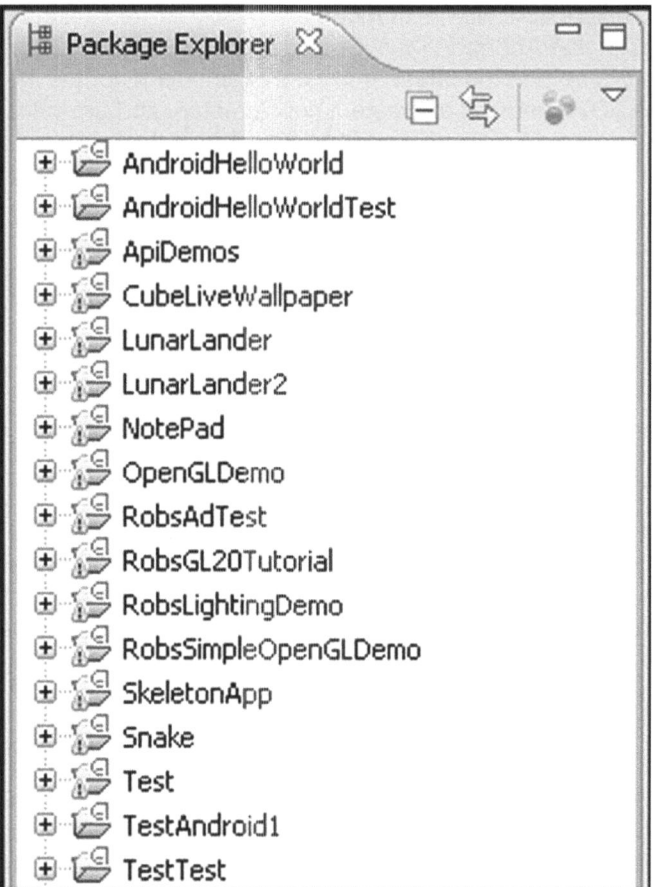

Figure 1-5. *Package Explorer*

You can also expand a package so that you can access all the files related to that package by clicking the "plus" symbol next to the package name. The Java source code files are located in the "src" directory, and the project-related resources, such as textures, 3D models, etc., are located in the "res" (short for *resources*) directory. Double-click a source code file or resource file to bring it up for viewing inside Eclipse. Source files can also be expanded so that you can get an overview of the class's variables and functions. You can double-click a variable or function to go to that variable or function within Eclipse's source view window. In Figure 1-6, there is only one function in the "AndroidHelloWorldActivity" class, which is "onCreate." Finally, every Android package has an AndroidManifest.xml file that defines such things as what permissions are needed to run the program, program-specific information such as version number, program icon, and program name, as well as what minimum Android operating system is needed to run the program.

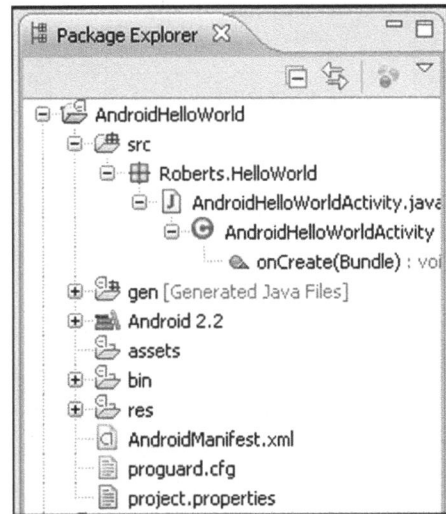

Figure 1-6. *A closer look into a package*

Source Code Area

By default, in the middle of Eclipse is the Java source code display window. Each different Java source code or .xml file is displayed here in its own tab (see Figure 1-7).

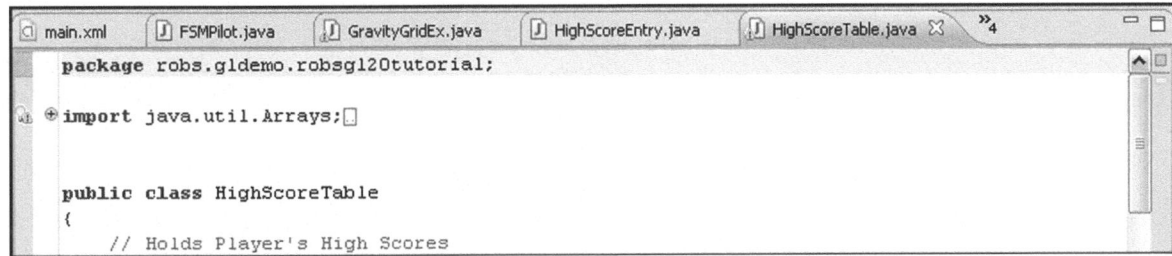

Figure 1-7. *Java source code area*

Notice that at the end of the last tab, there is a "➤" followed by "4." What this means is that there are four hidden files not shown. You can access these files by clicking the "➤4" region to bring up a complete list of files. Files listed in boldface type are not shown, and you can select these for viewing by highlighting them with your mouse pointer and left-clicking them (see Figure 1-8).

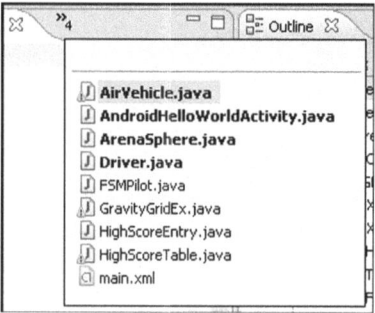

Figure 1-8. Accessing hidden Java source and .xml *files*

Outline

The Outline window in Eclipse is located by default on the right side, and it lists the variables and functions for the class that is selected in the source code window. You can easily jump to the corresponding class variable or class function in the source code window by clicking the variable or function in the Outline window (see Figure 1-9).

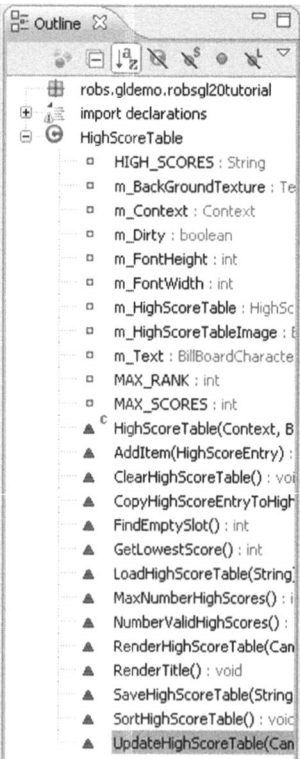

Figure 1-9. Outline window in Eclipse

In Figure 1-9, class variables or "fields" are listed first, followed by class functions. Some examples of class variables are HIGH_SCORES, m_BackGroundTexture, and m_Dirty. Some examples of class functions are FindEmptySlot(), RenderTitle(), and SortHighScoreTable().

Dalvik Debug Monitor Server (DDMS)

Eclipse with the ADT plug-in also provides a way to easily interface with actual Android hardware through the Dalvik Debug Monitor Server or DDMS. The button to access the DDMS is on the right upper corner of the Eclipse IDE. Click this button to switch views to the DDMS (see Figure 1-10).

Figure 1-10. *The DDMS button*

In the DDMS view, you can look at the actual directories and files on the Android device by using the File Explorer tab located on the right-hand side of the view. Figure 1-11 illustrates this.

Name	Size	Date	Time	Permissions	Ir
⊞ 📂 data		2012-05-15	04:02	drwxrwx--x	
⊟ 📂 mnt		2013-09-25	16:50	drwxrwxr-x	
⊞ 📂 asec		2013-09-25	16:50	drwxr-xr-x	
⊟ 📂 sdcard		1969-12-31	19:00	d---rwxr-x	
⊞ 📂 Android		2012-01-17	19:00	d---rwxr-x	
⊞ 📂 DCIM		2013-09-20	12:46	d---rwxr-x	
⊞ 📂 Eyeglasses		2013-08-18	11:06	d---rwxr-x	
⊞ 📂 LOST.DIR		2012-01-17	19:00	d---rwxr-x	
⊞ 📂 Music		2012-05-09	18:01	d---rwxr-x	
⊞ 📂 bluetooth		2012-05-09	11:48	d---rwxr-x	
⊞ 📂 data		2012-05-25	00:03	d---rwxr-x	
⊞ 📂 download		2013-07-28	19:30	d---rwxr-x	
📄 dronegrid.apk	591188	2013-08-16	21:46	----rwxr-x	
📄 dronegrid10.apk	591141	2013-08-18	01:06	----rwxr-x	
📄 startappdronegrid10.apk	687806	2013-09-20	12:44	----rwxr-x	
📄 startappexitbannerdroneg	687116	2013-09-22	15:25	----rwxr-x	
⊞ 📂 where		2013-02-12	07:37	d---rwxr-x	
⊞ 📂 secure		2013-09-25	16:50	drwx------	
⊟ 📂 system		2011-03-16	01:17	drwxr-xr-x	
⊞ 📂 app		2011-03-16	01:17	drwxr-xr-x	

Figure 1-11. *Exploring files on your Android device*

On the left side, if you have an actual physical Android device connected via the USB port, the device is displayed in the Devices tab, as shown in Figure 1-12.

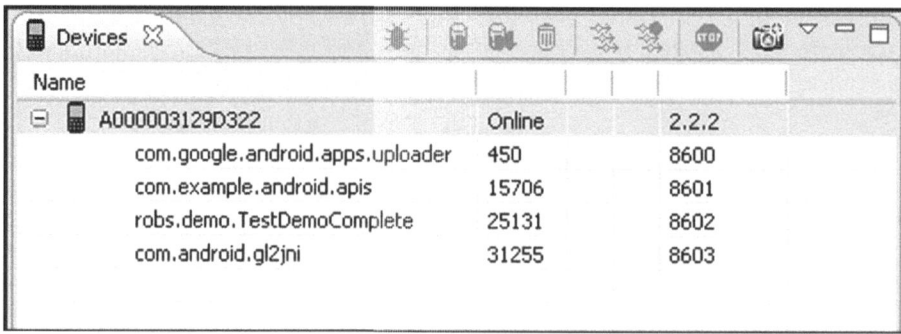

Figure 1-12. Devices tab on DDMS

Also notice the camera icon at the upper right-hand corner of the Devices window. If you click this button, then you capture a screenshot of what is currently on the Android device (see Figure 1-13).

Figure 1-13. Device Screen Capture in DDMS

From this pop-up window, you can rotate the image and save the image, if you so desire. This is a good way to get screenshots for promotional images when it is time to market your application to end users.

LogCat Window

At the bottom of the Eclipse IDE there are by default a few rectangular windows. One of the more important ones is called the LogCat window, and this window displays debug and error messages that are directly coming from a program that is running on the Android device that is attached to your computer via a USB cable (see Figure 1-14).

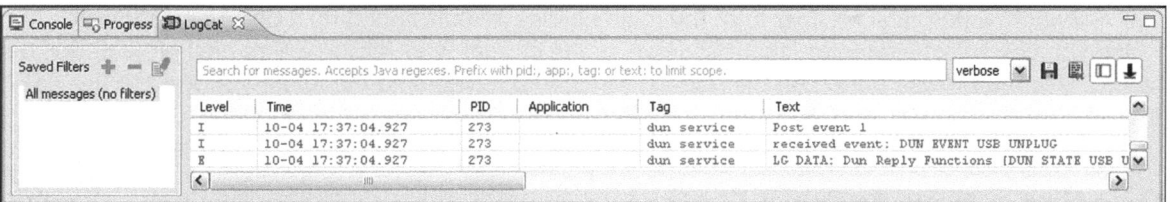

Figure 1-14. LogCat debugging tab

Launching the SDK Manager and AVD Manager from Eclipse

To launch the Android SDK Manager and the AVD Manager from within the Eclipse IDE, click "Window" in the top menu bar and use the menu items near the bottom of the list. The SDK Manager enables you to download new versions of the Android platform and other tools to develop with. The AVD Manager allows you to create and manage virtual Android devices for the Android device emulator (see Figure 1-15).

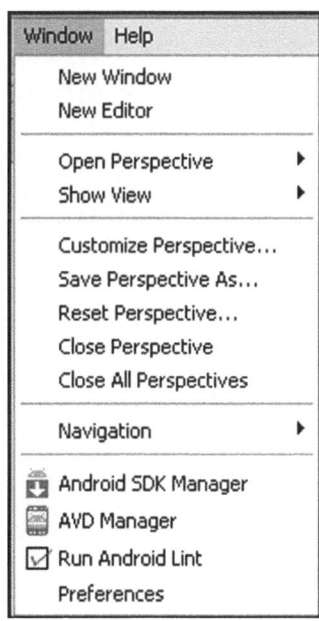

Figure 1-15. Launching SDK and AVD Managers from Eclipse

Hands-on Example: Non–OpenGL ES Text "Hello World" Program

In this hands-on example, we will create a new Android project that will output a simple "Hello World" text string. Start up the Eclipse IDE.

The first thing to do is to specify a work space where you will put this new project. Select File ➤ SwitchWorkSpace ➤ Other from the main Eclipse menu to bring up a pop-up window in which you can select a directory that will serve as the current work space where new projects will be stored. Use the Browse button on the pop-up window to navigate to the folder you want to use as the work space, then hit the OK button to set this folder as your current work space.

Creating a New Android Project

To create a new Android project, select "Android Application Project" under the File ➤ New menu (see Figure 1-16).

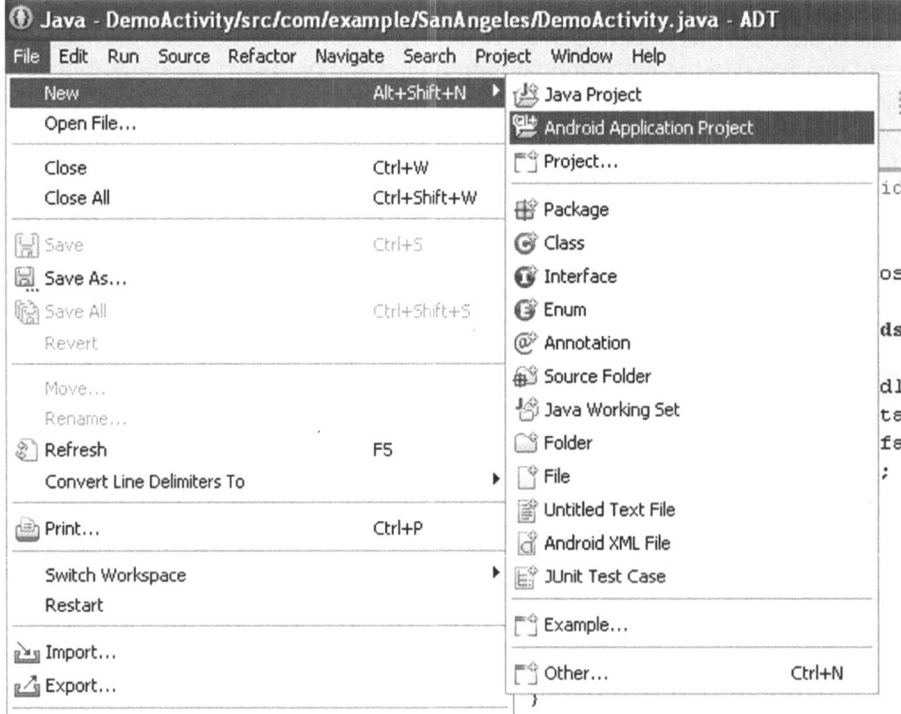

Figure 1-16. Creating a new Android project in Eclipse

This will bring up a pop-up window in which you can specify your application name, project name, package name, and SDK information (see Figure 1-17).

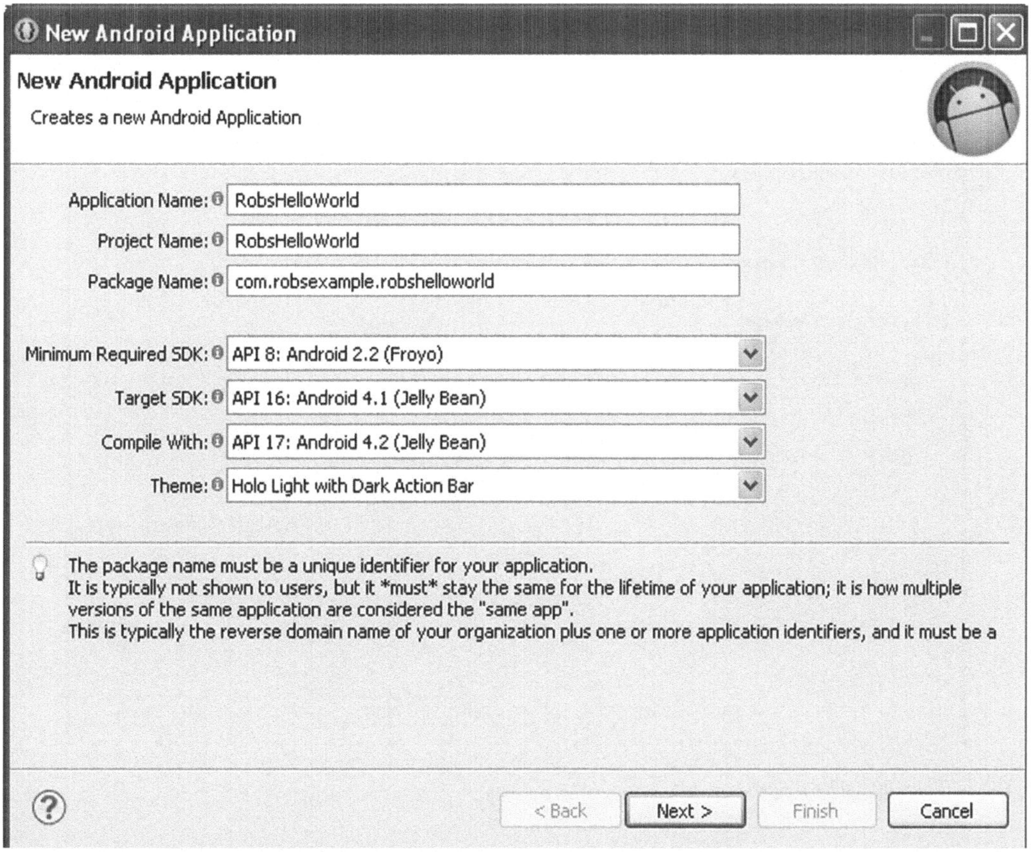

Figure 1-17. *Entering project and SDK info*

In the Application Name edit box, enter "RobsHelloWorld," which is the name of your application that will appear to users of your program. In the Project Name edit box, enter "RobsHelloWorld," which is the name of the project that is displayed in the Eclipse IDE. Enter "com.robsexample.robshelloworld" as the package name associated with this new Android project. This package name must be unique.

For the Minimum Required SDK select Android 2.2 (Froyo), because this is the lowest Android platform that supports OpenGL ES 2.0. For the Target SDK, select the highest Android platform API that you anticipate to have successfully tested your application against. For the Compile With list box, select the API version that you wish to compile your application for. You can leave the Theme list box at the default value. Click the Next button to move to the next screen.

The next thing to do is to configure the project. For this example, just accept the default values and click the Next button (see Figure 1-18).

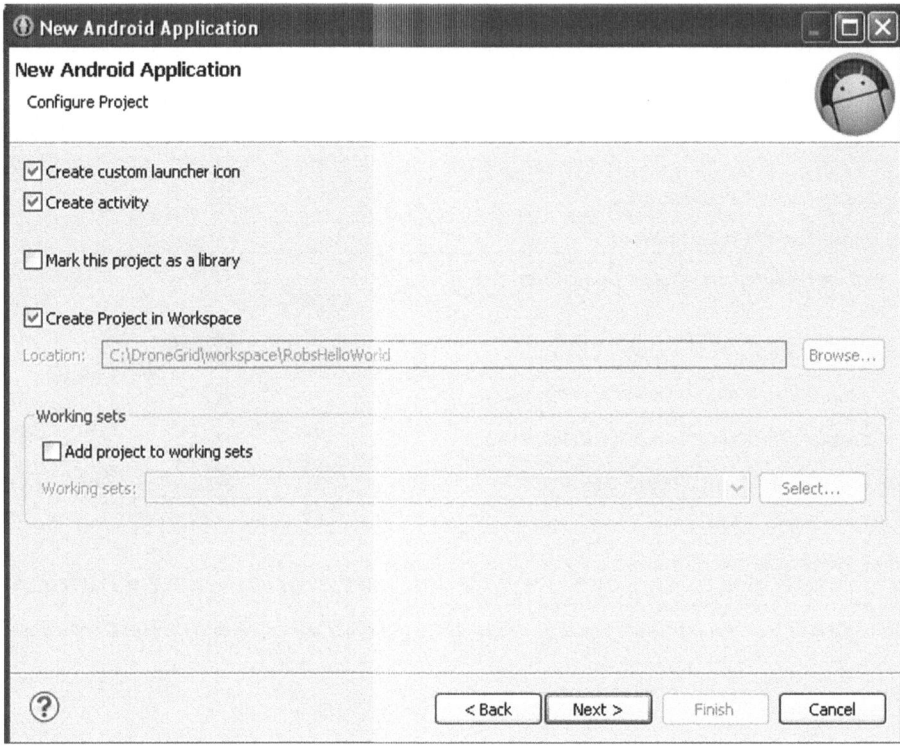

Figure 1-18. Configuring a new project

In the next screen, you can configure the launcher icon, if you wish. However, for this example, you can just accept the defaults (see Figure 1-19).

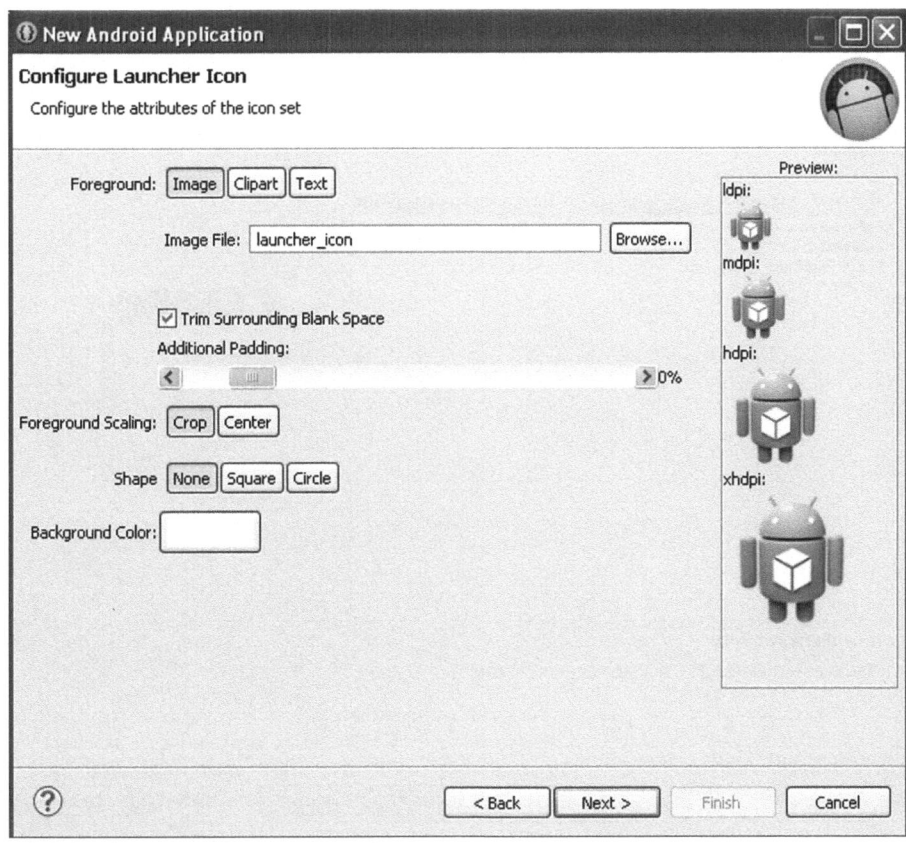

Figure 1-19. *Configure Launcher Icon*

Click the Next button. The next screen allows you to select the type of activity you want to create. Select the Blank Activity and click the Next button (see Figure 1-20).

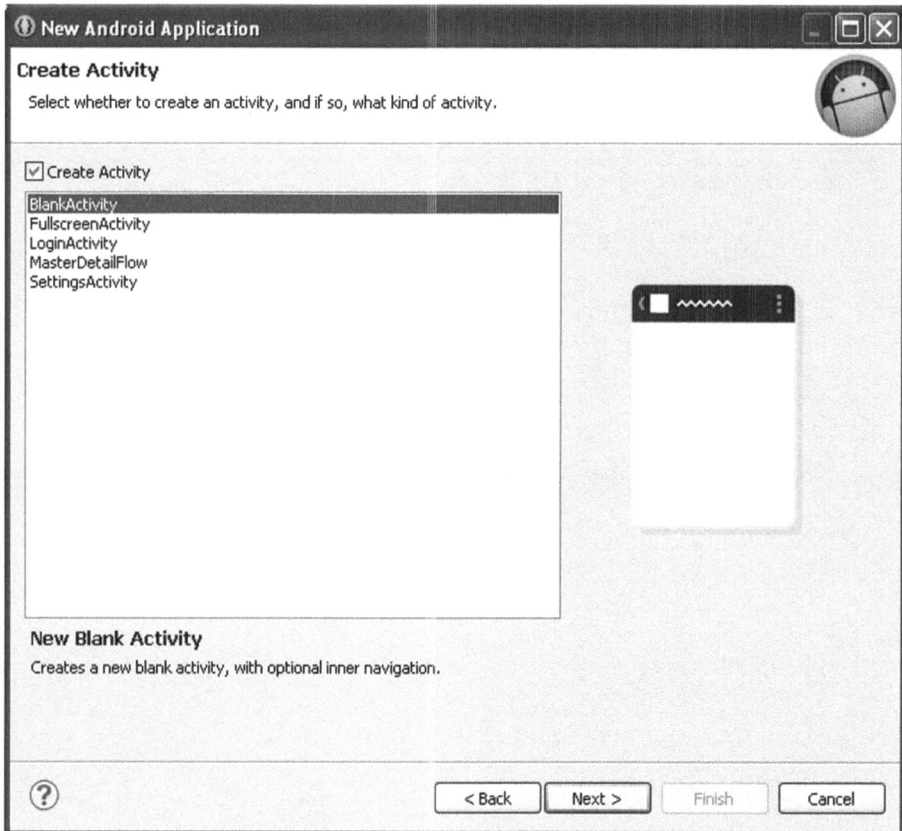

Figure 1-20. *Select activity type and Create Activity*

Accept the defaults for the blank activity. The default activity name is "MainActivity." The default Layout Name is "activity_main," and the default Navigation Type is "None." Click the Finish button to create the new Android application (see Figure 1-21).

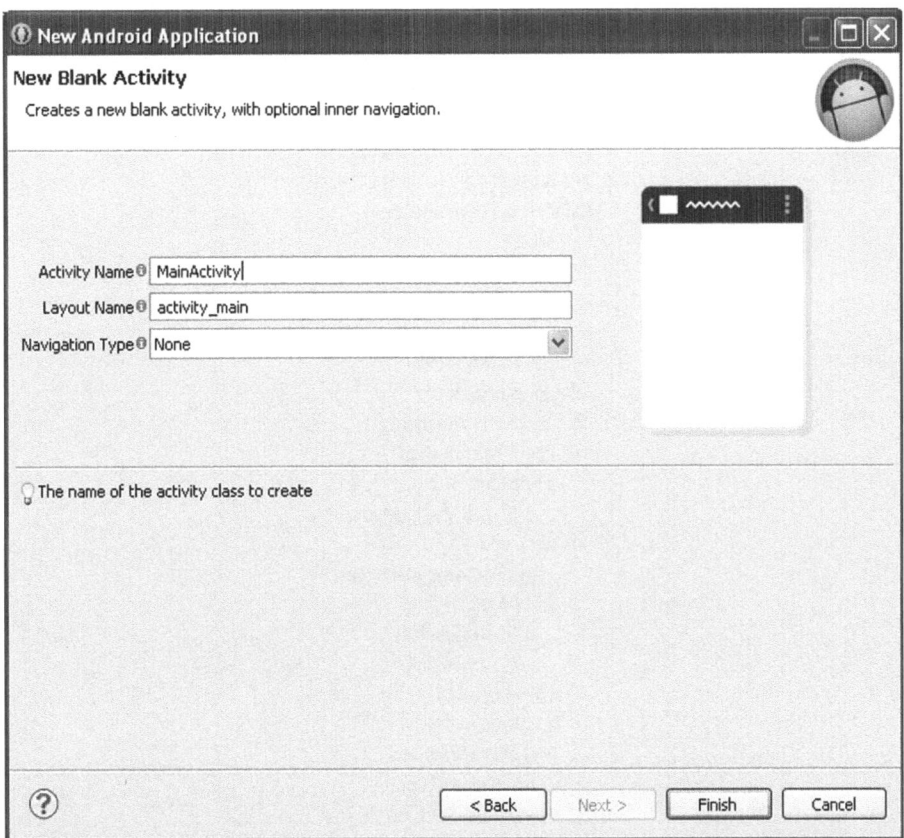

Figure 1-21. *Creating a New Blank Activity*

On the left-hand side of the Eclipse IDE in the Package Explorer window, you should see a new entry called "RobsHelloWorld," which is our new example program. The key directories are the "src" directory, where the Java source code is stored; the "libs" directory, where external libraries are stored; and the "res" directory, where resources such as graphics, 3D models, and layouts are stored. Under the "res" directory, the "layout" directory stores the application's graphical layout specifications; the "menu" directory stores the applications menu-related layout information; and the "values" directory stores the actual "Hello World" string that is displayed. Finally, a key file is the AndroidManifest.xml file, which contains information about permissions and other application specific information. (See Figure 1-22 for the layout of the "RobsHelloWorld" project.)

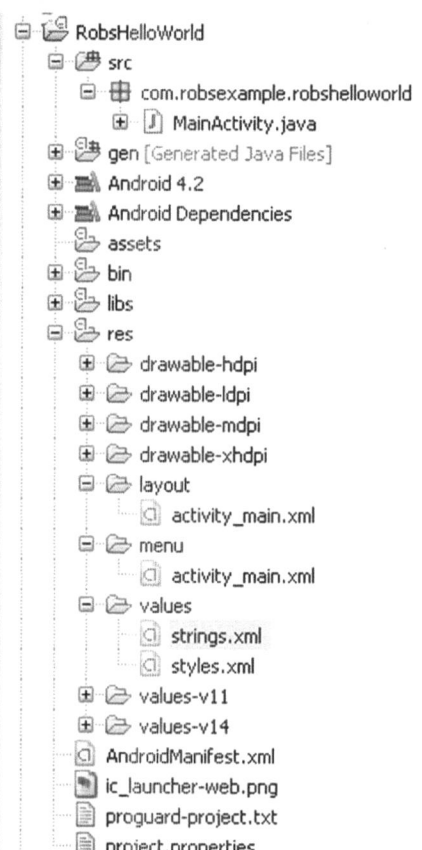

Figure 1-22. *"RobsHelloWorld" Android project*

Running on an Android Emulator

Before we run our example on the emulator, we must first set up an Android Virtual Device. Select Window ➤ Android Virtual Device Manager from the Eclipse menu to start up the Virtual Device Manager. Click the New button. Another window should pop up with the heading "Create new Android Virtual Device (AVD)." Enter a name for your virtual device in the "AVD Name:" field. Select a device to emulate and Target, as shown in Figure 1-23. Accept the default values for the rest of the inputs. Click the OK button.

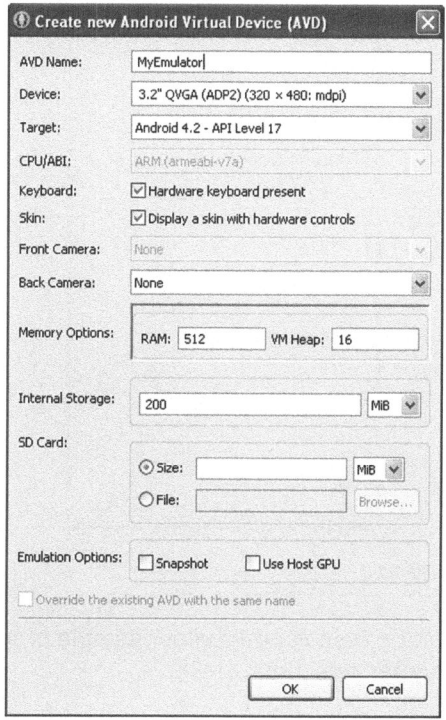

Figure 1-23. *Creating a new Android Virtual Device*

Next, we have to run our example. If you are running this for the first time, you will have to specify how to run this application. Make sure the "RobsHelloWorld" project is highlighted. Select Run ➤ Run from the Eclipse main menu.

When the pop-up window appears, select "Android Application" and click the OK button to run the example. If you do not have an actual Android device attached to your computer via USB cable, Eclipse will run the program on the Android emulator (see Figure 1-24).

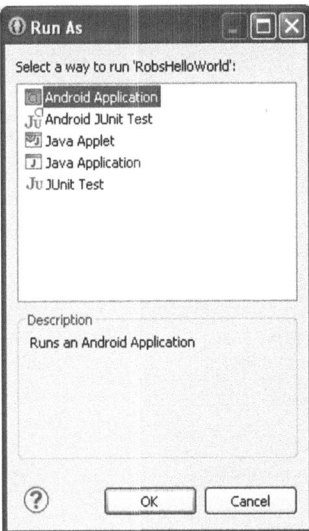

Figure 1-24. Running your "HelloWorld" example

The Android emulator should start by default and run our sample program. The actual code for this program is shown later in this chapter (see Figure 1-25).

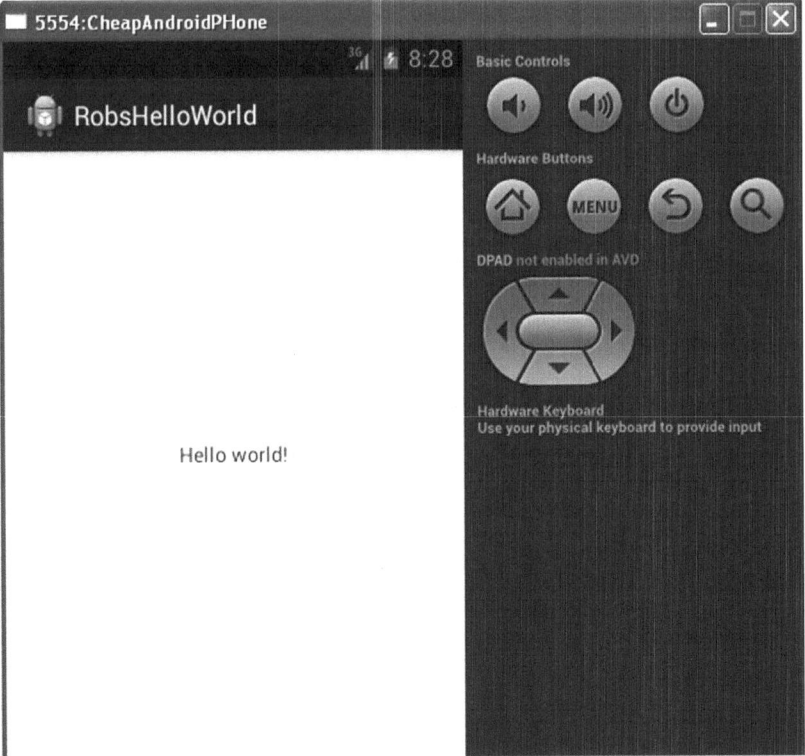

Figure 1-25. "RobsHelloWorld" example running on the Android emulator

Running on an Actual Android Device

In order to download and run the program on an actual Android device, the device must be put into USB Debugging mode. Press the Menu key, which is the left-most key on the bottom portion of an Android phone. Click the Settings button, then click the Applications button and then the Development button. Click the "USB Debugging" option. After doing so, the item should be checked, as in Figure 1-26.

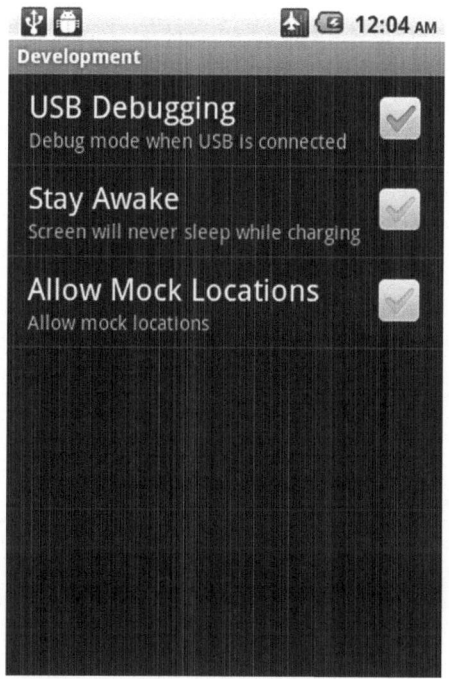

Figure 1-26. *Setting USB Debugging mode*

Note On Android 4.0 and newer models, the USB Debugging option is under the Settings ➤ Developer options. On Android 4.2 and newer models, the Developer options are hidden by default. To make it available, go to Settings ➤ About Phone and tap the Build number seven times. Return to the previous screen to find Developer options.

Next, you have to install the appropriate USB software driver for your model of Android phone onto your development system. Try to connect your Android device to your computer first, to see if it automatically installs the correct drivers. If you can't run the program on your device, then you will have to install a device driver from the manufacturer. Usually the manufacturer of your phone has a web site with downloadable drivers. After doing this, connect your phone to your development system, using a USB cable, which most likely was included with your mobile phone.

Now you are ready to start using the device. Select Run ➤ Run from the main Eclipse menu. A window should appear in which you can choose to run the program on an actual Android device or an Android Virtual Device (see Figure 1-27). Select the hardware device and click the OK button.

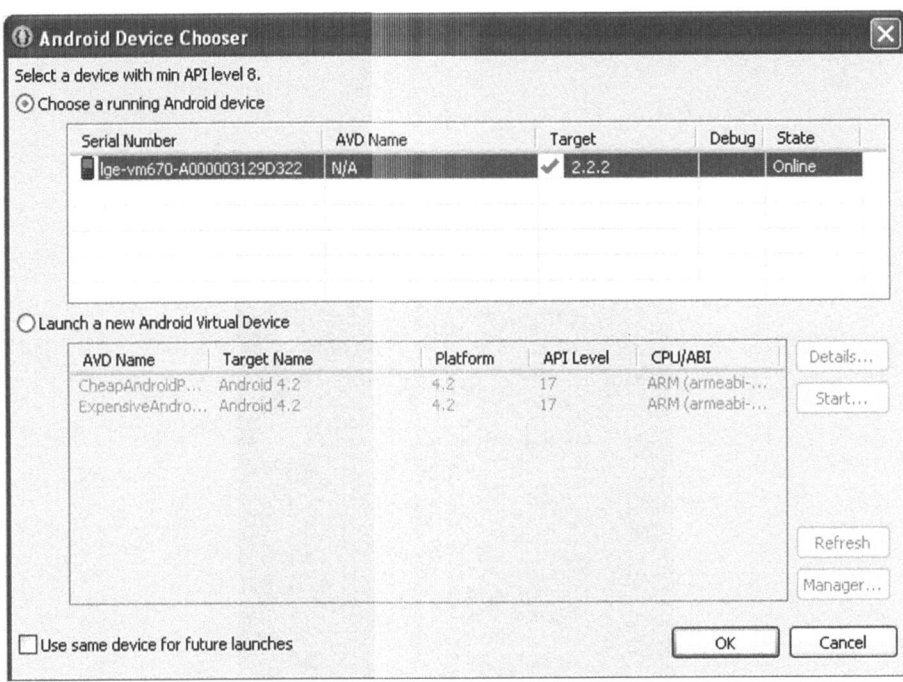

Figure 1-27. Choose a device on which to run your program

The program running on the device should be the same as depicted in Figure 1-25. Press the Back key to exit the program.

The Main Source Code

When you create a new program within the Android development framework, what you are actually doing in terms of coding is creating a new activity. What you need to do is derive a new class from the existing `Activity` class that is part of the standard Android code base (see Listing 1-1).

Listing 1-1. MainActivity.java Source Code for "RobsHelloWorld" Example

```
package com.robsexample.robshelloworld;

import android.os.Bundle;
import android.app.Activity;
import android.view.Menu;
public class MainActivity extends Activity {
        @Override
```

```
        protected void onCreate(Bundle savedInstanceState) {
                super.onCreate(savedInstanceState);
                setContentView(R.layout.activity_main);
        }

        @Override
        public boolean onCreateOptionsMenu(Menu menu) {
                // Inflate the menu; this adds items to the action bar if it is present.
                getMenuInflater().inflate(R.menu.activity_main, menu);
                return true;
        }
}
```

For example, our "HelloWorld" program consists of a new class MainActivity, which is derived from the Activity class.

The onCreate() function is called when this new activity is created. It first calls the onCreate() function in its parent class through the super.onCreate() statement. It then sets the activity's view to the layout specified in the activity_main.xml file located in the "res/layout" directory for the project. The R class is a generated class located in the "gen" directory and reflects the current files in the resources, or "res," directory (see Figure 1-2x2).

The OnCreateOptionsMenu() function creates the options menu for the program. The menu specifications are located in the activity_main.xml file located in the "res/menu" directory.

The Graphical Layout

The graphical layout .xml file for this example is referenced by the code R.layout.activity_main, which refers to the activity_main.xml file located in the "res/layout" directory for this project (see Listing 1-2.)

Listing 1-2. Graphical Layout for "RobsHelloWorld"

```
<RelativeLayout xmlns:android="http://schemas.android.com/apk/res/android"
    xmlns:tools="http://schemas.android.com/tools"
    android:layout_width="match_parent"
    android:layout_height="match_parent"
    tools:context=".MainActivity" >

    <TextView
        android:layout_width="wrap_content"
        android:layout_height="wrap_content"
        android:layout_centerHorizontal="true"
        android:layout_centerVertical="true"
        android:text="@string/hello_world" />

</RelativeLayout>
```

This graphical layout specification is a Relative Layout type with one TextView component, inside which static alphanumeric text can be displayed.

The code android:text sets the text to be displayed.

The text to be displayed is set to a string variable named "hello_world," located in the file strings.xml, which is located in the "res/values" directory.

You can hard-code a string value by removing the "@string/" portion and just have the text you want to display enclosed in quotes, such as

```
android:text="Hello World EveryBODY!!!"
```

However, this is not recommended.

You can also preview and edit the layout inside of Eclipse by selecting the layout file and clicking the Graphical Layout tab located at the bottom left of the file view (see Figure 1-28.)

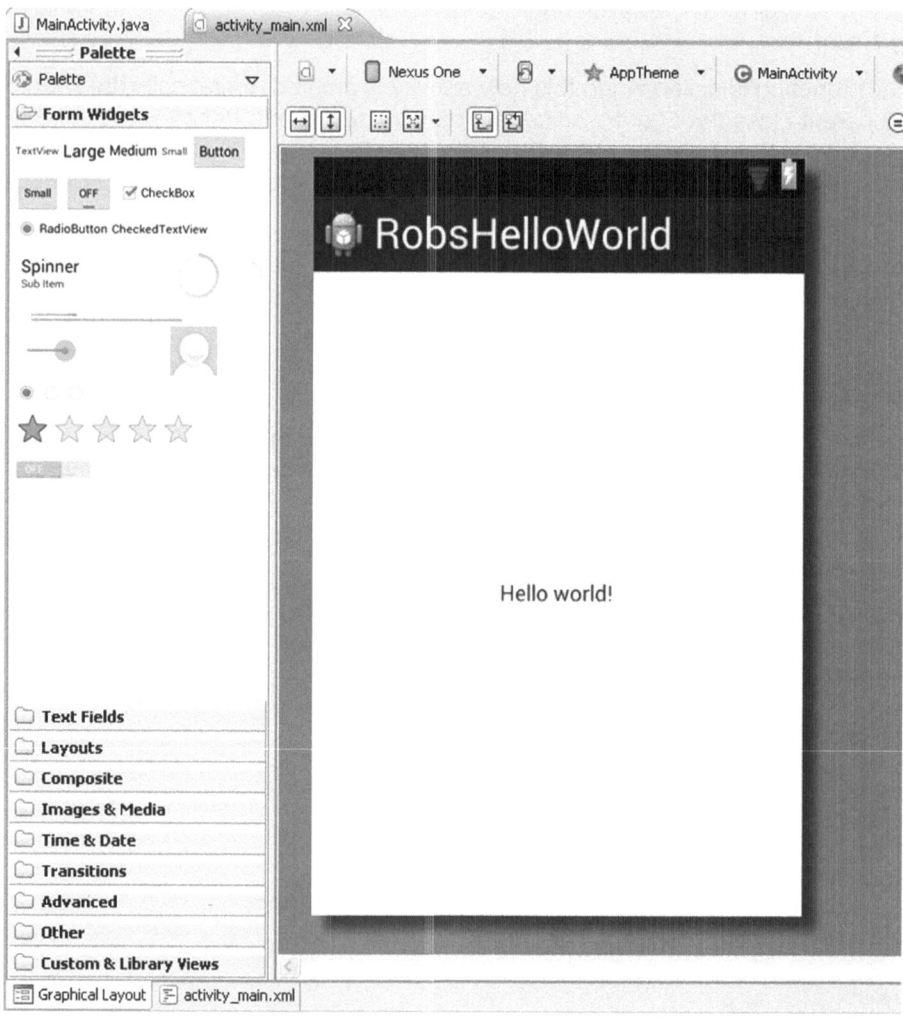

Figure 1-28. *Graphical layout preview in Eclipse*

The Actual "Hello World" Data

Finally, the file that has the actual "Hello World" data to display is the `strings.xml` file and is shown in Listing 1-3.

Listing 1-3. Data for "Hello World"

```xml
<?xml version="1.0" encoding="utf-8"?>
<resources>
    <string name="app_name">RobsHelloWorld</string>
    <string name="hello_world">Hello world!</string>
    <string name="menu_settings">Settings</string>
</resources>
```

The key variable that is used to display the text is "hello_world" and the associated text data is "Hello world!"

Summary

In this chapter, I gave overviews of the key components in Android game development. I first discussed the major components of the Android Software Development Kit, or SDK. I discussed the Eclipse IDE, Android SDK Manager, Android Virtual Device Manger, and the actual Android device emulator. Next, I explained how you would set up your development system to create and deploy Android programs. I discussed key components of the Eclipse IDE, such as the Project Explorer window, Source Code window, Outline window, and LogCat window. Next, I took you step by step through a hands-on example involving the creation of a "Hello World" program that was run on both the Android emulator and an actual Android device. Finally, I discussed exactly how this sample "Hello World" program is constructed.

Chapter 2

Java for Android

In this chapter, I will cover the Java language component in Android 3D games development. I will start with a brief overview and review of the basic Java language on Android. Then I cover the basic Java program framework on Android for all applications. Next, I cover the basic Java program framework on Android for applications that specifically utilize OpenGL ES graphics. Finally, I provide a hands-on example of a 3D Android OpenGL ES program.

Overview of the Java Language

This section on the Java language is intended as a quick-start guide for someone who has some knowledge of computer programming as well as some knowledge about object-oriented programming. This section is *not* intended to be a Java reference manual. It is also not intended to cover every feature of the Java programming language.

The Java language for Android is run on a Java virtual machine. This means that the same compiled Java Android program can run on many different Android phones with different central processing unit (CPU) types. This is a key feature in terms of future expandability to faster processing units, including those that will be specifically designed to enhance 3D games. The trade-off to this is speed. Java programs run slower than programs compiled for a CPU in its native machine language, because a Java virtual machine must interpret the code and then execute it on the native processor. A program that is already compiled for a specific native processor does not have to be interpreted and can save execution time by skipping this step.

However, you can compile C/C++ code for a specific Android processor type using the Android Native Development Kit or NDK. You can also call native C/C++ functions from within the Java programming framework. Thus, for key functions that require the speed of natively compiled code, you can put these functions into C/C++ functions that are compiled using the NDK and called from Java code in your main program.

Java Comments

Java comments can consist of single-line comments and multiline comments.

- Single-line comments start with two slash characters (//).

  ```
  // This is a single-line Java comment
  ```

- Multiline comments start with a slash followed by an asterisk (/*) and end with an asterisk followed by a slash (*/).

  ```
  /*
     This is
     a multiline
     comment
  */
  ```

Java Basic Data Types

Java data types can be numeric, character, or Boolean in nature.

byte: An 8-bit number with values from -128 to 127, inclusive

short: A 16-bit number with values from -32,768 to 32,767, inclusive

int: A 32-bit number with values from -2,147,483,648 to 2,147,483,647, inclusive

long: A 64-bit number with values from -9,223,372,036,854,775,808 to 9,223,372,036,854,775,807, inclusive

float: A single-precision 32-bit IEEE 754 floating-point number

double: A double-precision 64-bit IEEE 754 floating-point number

char: A single 16-bit Unicode character that has a range of '\u0000' (or 0) to '\uffff' (or 65,535, inclusive)

Boolean: Having a value of either true or false

Arrays

In Java, you can create arrays of elements from the basic Java data types listed in the preceding section. The following statement creates an array of 16 elements of type float called m_ProjectionMatrix.

```java
float[] m_ProjectionMatrix = new float[16];
```

Data Modifiers

Data modifiers allow the programmer to control how variables are accessed and stored. They include the following:

- **private:** Variables that are private are only accessible from within the class they are declared in. The following declares m_ProjectionMatrix to be private and only accessible from within its own class:

  ```
  private float[] m_ProjectionMatrix = new float[16];
  ```

- **public:** Variables that are public can be accessed from any class. The following variable is public:

  ```
  public float[] m_ProjectionMatrix = new float[16];
  ```

- **static:** Variables that are declared static have only one copy associated with the class they are declared in. The following static array is declared static and resides in the Cube class. This array defines the graphics data for a 3D cube. This 3D cube is the same for all instances of the Cube class, so it makes sense to make the CubeData array static.

  ```
  static float CubeData[] =
  {
  // x,      y,    z,    u,       v           nx,   ny, nz
  -0.5f,  0.5f, 0.5f, 0.0f,    0.0f,    -1,  1, 1,  // front top left
  -0.5f, -0.5f, 0.5f, 0.0f,    1.0f,    -1, -1, 1,  // front bottom left
   0.5f, -0.5f, 0.5f, 1.0f,    1.0f,     1, -1, 1,  // front bottom right
   0.5f,  0.5f, 0.5f, 1.0f,    0.0f,     1,  1, 1,  // front top right

  -0.5f,  0.5f, -0.5f, 0.0f,    0.0f,   -1,  1, -1, // back top left
  -0.5f, -0.5f, -0.5f, 0.0f,    1.0f,   -1, -1, -1, // back bottom left
   0.5f, -0.5f, -0.5f, 1.0f,    1.0f,    1, -1, -1, // back bottom right
   0.5f,  0.5f, -0.5f, 1.0f,    0.0f,    1,  1, -1  // back top right
  };
  ```

- **final:** The final modifier indicates that the variable will not change. For example, the following declares the variable TAG is of type String and is private, static, and cannot be changed.

  ```
  private static final String TAG = "MyActivity";
  ```

Java Operators

In this section, we cover arithmetic, unary, conditional, bitwise, and bit shift operators.

Arithmetic Operators

+   Additive operator (also used for `String` concatenation)

-       Subtraction operator

*   Multiplication operator

/   Division operator

%   Remainder operator

Unary Operators

+   Unary plus operator

-       Negates an expression

++   Increments a value by 1

-       Decrements a value by 1

!   Inverts the value of a Boolean

Conditional Operators

&&   Conditional-AND

|| Conditional-OR

=   Assignment operator

==   Equal to

!=   Not equal to

>   Greater than

>=   Greater than or equal to

<   Less than

<=   Less than or equal to

Bitwise and Bit Shift Operators

~   Unary bitwise complement

<<   Signed left shift

>>   Signed right shift

>>>   Unsigned right shift

&   Bitwise AND

^   Bitwise exclusive OR

| Bitwise inclusive OR

Java Flow Control Statements

- if then statement

```
if (expression)
{
    // execute statements here if expression evaluates to true
}
```

- if then else statement

```
if (expression)
{
    // execute statements here if expression evaluates to true
}
else
{
    // execute statements here if expression evaluates to false
}
```

- switch statement

```
switch(expression)
{
        case label1:
                // Statements to execute if expression evaluates to
                // label1:
        break;

        case label2:
                // Statements to execute if expression evaluates to
                // label2:
break;
}
```

- while statement

```
while (expression)
{
// Statements here execute as long as expression evaluates // to true;
}
```

- for statement

```
for (variable counter initialization;
          expression;
          variable counter increment/decrement)
{
// variable counter initialized when for loop is first
// executed

// Statements here execute as long as expression is true

      // counter variable is updated

}
```

Java Classes

Java is an object-oriented language. What this means is that you can derive or extend existing classes to form new customized classes of the existing classes. The derived class will have all the functionality of the parent class, in addition to any new functions that you may want to add in.

The following class is a customized version of the parent class from which it derives, which is the Activity class.

```
public class MainActivity extends Activity
{
      // Body of class
}
```

Packages and Classes

Packages are a way in Java to group together classes and interfaces that are related in some way. For example, a package can represent a game or other single application. The following is the package designation for the "Hello Droid" Android project that I cover at the end of this chapter.

```
package com.robsexample.glhelloworld;
```

Accessing Classes in Packages

In order to access classes that are located in other packages, you have to bring them into view using the "import" statement. For example, in order to use the GLSurfaceView class that is located inside the android.opengl.GLSurfaceView package, you have to import it with the following statement:

```
import android.opengl.GLSurfaceView;
```

Then, you can use the class definition without the full package name, such as

```
private GLSurfaceView m_GLView;
```

Refer to the main Android developer's web site to find out more information about Android's built-in classes, as well as the exact import you need to specify to use these classes in your own programs.

Java Interfaces

The purpose of a Java interface is to provide a standard way for programmers to implement the actual functions in an interface in code in a derived class. An interface does not contain any actual code, only the function definitions. The function bodies with the actual code must be defined by other classes that implement that interface. A good example of a class that implements an interface is the render class that is used for rendering graphics in OpenGL on the Android platform.

```
public class MyGLRenderer implements GLSurfaceView.Renderer
{
        // This class implements the functions defined in the
        // GLSurfaceView.Renderer interface

        // Custom code
private PointLight m_Light;
public PointLight m_PublicLight;
void SetupLights()
{
        // Function Body
}

// Other code that implements the interface

}
```

Accessing Class Variables and Functions

You can access a class's variables and functions through the ".\" operator, just as in C++. See the following examples:

```
MyGLRenderer m_Renderer;
m_Renderer.m_PublicLight = null; // ok
m_Renderer.SetupLights();        // ok
m_Renderer.m_Light = null;       // error private member
```

Java Functions

The general format for Java functions is the same as in other languages, such as C/C++. The function heading starts with optional modifiers, such as private, public, or static. Next is a return value that can be void, if there is no return value or a basic data type or class. This is followed by the function name and then the parameter list.

```
Modifiers Return_value FunctionName(ParameterType1 Parameter1, ...)
{
        // Code Body
}
```

An example of a function from our Vector3 class in our "Hello Droid" example at the end of this chapter is:

```
static Vector3 CrossProduct(Vector3 a, Vector3 b)
{
        Vector3 result = new Vector3(0,0,0);

        result.x= (a.y*b.z) - (a.z*b.y);
        result.y= (a.z*b.x) - (a.x*b.z);
        result.z= (a.x*b.y) - (a.y*b.x);

        return result;
}
```

Also in Java, all parameters that are objects are passed by reference.

Calling the Parent Function

A function in a derived class can override the function in the parent or superclass, using the @Override annotation. This is not required but helps to prevent programming errors. If the intention is to override a parent function but the function does not in fact do this, then a compiler error will be generated.

In order for the function in a derived class to actually call its corresponding function in the parent class, you use the super prefix as seen below.

```
@Override
public void onCreate(Bundle savedInstanceState)
{
        super.onCreate(savedInstanceState);

        // Create a MyGLSurfaceView instance and set it
        // as the ContentView for this Activity
        m_GLView = new MyGLSurfaceView(this);
        setContentView(m_GLView);
}
```

> **Note** Additional Java tutorials can be found on http://docs.oracle.com/javase/tutorial/.

The Basic Android Java Program Framework

In this section, I cover the basic Android Java program framework. This framework applies to all Android programs, not just Android 3D games or games in general. I start off with an overview of the activity life cycle. Then I cover key cases in the life cycle and follow up by code additions where you can see for yourself the changes in the Activity's life cycle through the use of debug statements.

Android Activity Life Cycle Overview

The Activity class is the main entry point within the Android framework where a programmer is able to create new Android applications and games. In order to effectively code within this framework, you must understand the Activity class life cycle. See Figure 2-1 for a graphical flowchart style overview.

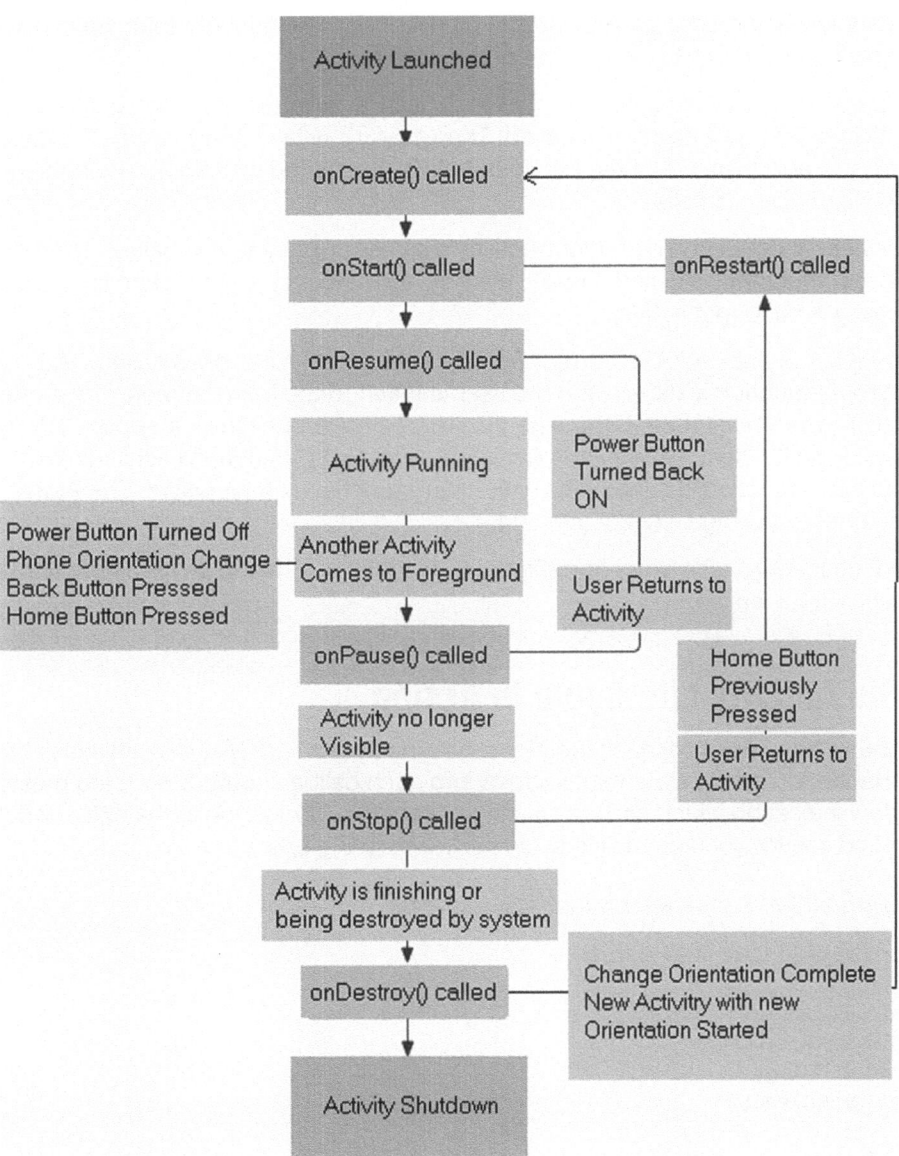

Figure 2-1. *Activity class callback life cycle*

Key Activity Life Cycle Cases

There are some key situations that you have to consider when programming your Activity class.

Another activity comes to the foreground: The current Activity is paused; that is, the Activity's onPause() function is called.

The Power key is turned off: The current Activity's onPause() function is called; the Power key is turned back on; and then the Activity's onResume() function is called, followed by a return to resumption of the activity.

Phone orientation changes: The current Activity's onPause() function is called. The Activity's onStop() function is called. The Activity's onDestroy() function is called. Finally, a new instance of the previous Activity is created with the new Orientation and onCreate() is called.

Back key is pressed: The current Activity's onPause() function is called. The Activity's onStop() function is called. Finally, the Activity's onDestroy() function is called. The Activity is no longer active.

Home key is pressed: The current Activity's onPause() function is called. The onStop() function is called, and the user is taken to the home screen where other activities can be started. If the user tries to begin the previously stopped Activity by clicking its icon, the previous Activity's onRestart() function is called. Next, the onStart() function is called. The onResume() function is then called. The Activity becomes active again and is running.

The important concept to take away from Figure 2-1 is that you should save the game state whenever onPause() is called.

Seeing the Activity Life Cycle in Action

Listing 2-1 shows how these callback functions look inside our new MainActivity class that we created in Chapter 1. The log statements added into each callback output error log messages to the LogCat window indicating which callback is being executed. Try to type in the extra code and run the program and see for yourself the life cycle callbacks being executed.

Listing 2-1. "RobsHelloWorld" Example with Life Cycle Callbacks Added

```
package com.robsexample.robshelloworld;

import android.os.Bundle;
import android.app.Activity;
import android.util.Log;
import android.view.Menu;

public class MainActivity extends Activity {
        private static final String TAG = "MyActivity";
        @Override
```

```java
    protected void onCreate(Bundle savedInstanceState) {
            super.onCreate(savedInstanceState);
            setContentView(R.layout.activity_main);
            Log.e(TAG,"onCreate() called!");
    }
    @Override
    public boolean onCreateOptionsMenu(Menu menu) {
            // Inflate the menu; this adds items to the action bar if it is present.
            getMenuInflater().inflate(R.menu.activity_main, menu);
            return true;
    }
    @Override
    protected void onStart() {
            super.onStart();
            Log.e(TAG, "onStart() called!");
    }
    @Override
    protected void onRestart() {
            super.onRestart();
            Log.e(TAG, "onRestart() called!");
    }
    @Override
    protected void onStop() {
            super.onStop();
            Log.e(TAG, "onStop() called!");
    }
    @Override
    protected void onResume() {
            // Ideally a game should implement onResume() and onPause()
            // to take appropriate action when the activity looses focus
            super.onResume();
            Log.e(TAG, "onResume() called!");
    }
    @Override
    protected void onPause() {
        // Ideally a game should implement onResume() and onPause()
        // to take appropriate action when the activity looses focus
        super.onPause();
        Log.e(TAG, "onPause() called!");
    }
    @Override
    protected void onDestroy()
    {
            // Implement onDestroy() to release objects and free up memory when
            // an Activity is terminated.
        super.onDestroy();
        Log.e(TAG , "onDestroy() called!");
    }
}
```

The Basic Android Java OpenGL Framework

In this section, I cover the basic Android Java OpenGL framework that is the basis for all OpenGL related applications, including games. I first cover the basic framework for a program with a single OpenGL view. Next, I cover a framework that contains multiple views that include an OpenGL view as part of the user interface.

Basic Android OpenGL ES Framework for a Single-View OpenGL ES Application

In this section, I will discuss how to create an OpenGL ES 2.0 Android application where there is only a single OpenGL ES 2.0 view. I first discuss a customized GLSurfaceView class. Then I discuss the custom renderer we need to do the actual drawing of the 3D OpenGL ES objects.

The Custom GLSurfaceView

In order to create your own custom OpenGL ES–based games, you have to create a custom GLSurfaceView, a custom Renderer that draws this custom GLSurfaceView, and then set this new custom GLSurfaceView as the main view through the setContentView() function in your custom Activity class.

The custom GLSurfaceView object must be notified when the Activity is paused or resumed. This means that the onPause() and onResume() functions in the GLSurfaceView object must be called when onPause() or onResume() is called in the Activity.

In the custom MyGLSurfaceView class below, which is derived from the GLSurfaceView class, you also have to set the OpenGL ES version to use to 2.0 by calling setEGLContextClientVersion(2) inside the constructor. You must also set your custom renderer, which is MyGLRenderer in the example below, using the setRenderer(new MyGLRenderer()) statement, also in the constructor. See Listing 2-2.

Listing 2-2. Activity Class for a Single OpenGL ES View Application

```
package robs.demo.robssimplegldemo;

import android.app.Activity;
import android.content.Context;
import android.opengl.GLSurfaceView;
import android.os.Bundle;

public class RobsSimpleOpenGLDemoActivity extends Activity
{
        private GLSurfaceView m_GLView;

        @Override
        public void onCreate(Bundle savedInstanceState)
        {
        super.onCreate(savedInstanceState);
```

```
                    // Create a MyGLSurfaceView instance and set it
                    // as the ContentView for this Activity
                    m_GLView = new MyGLSurfaceView(this);
                    setContentView(m_GLView);
        }

        @Override
        protected void onPause()
        {
        super.onPause();
                    m_GLView.onPause();
        }

        @Override
        protected void onResume()
        {
        super.onResume();
                    m_GLView.onResume();
        }
}
//////////////////////////////////////////////////////////////////////////
class MyGLSurfaceView extends GLSurfaceView {
    public MyGLSurfaceView(Context context) {
        super(context);

        // Create an OpenGL ES 2.0 context.
        setEGLContextClientVersion(2);

        // Set the Renderer for drawing on the GLSurfaceView
        setRenderer(new MyGLRenderer());
    }
}
```

The Custom Renderer

The custom MyGLRenderer class implements the interface for GLSurfaceView.Renderer. This means that this class needs to implement the functions onSurfaceCreated(), onSurfaceChanged(), and onDrawFrame().

The function onSurfaceCreated() is called when an OpenGL surface is created or when the EGL context that is used for OpenGL ES rendering is lost. Put the creation and initialization of any OpenGL objects and resources you need for your game here.

The onSurfaceChanged() function is called whenever the OpenGL surface changes size or a new surface is created.

The onDrawFrame() function is called when it's time to render the OpenGL surface to the Android screen. Put code to actually render your 3D objects here.

See Listing 2-3 for the full custom renderer class implementation.

Listing 2-3. MyGLRenderer Custom Renderer Class

```
package robs.demo.robssimplegldemo;

import java.nio.ByteBuffer;
import java.nio.ByteOrder;
import java.nio.FloatBuffer;
import javax.microedition.khronos.egl.EGLConfig;
import javax.microedition.khronos.opengles.GL10;
import android.opengl.GLES20;
import android.opengl.GLSurfaceView;

public class MyGLRenderer implements GLSurfaceView.Renderer
{
        @Override
        public void onSurfaceCreated(GL10 unused, EGLConfig config)
        {
                // Called when an new surface has been created
                // Create OpenGL resources here
        }

        @Override
        public void onSurfaceChanged(GL10 unused, int width, int height)
        {
                // Called when new GL Surface has been created or changes size
                // Set the OpenglES camera viewport here
        }

        @Override
        public void onDrawFrame(GL10 unused)
        {
                // Put code to draw 3d objects to screen here
        }
}
```

Basic Android OpenGL ES Framework for a Multiple View OpenGL ES Application

In this section, we will cover the basic framework for an OpenGL program that contains multiple View objects inside your user interface or layout, such as those including Text views, Edit Box views, as well as having an OpenGL view. For example, you can have a portion of your screen with an EditBox view where the user can enter his name using the standard virtual Android keyboard that is already built into software and another portion of the screen running an OpenGL animation.

The XML Layout File

The following XML layout file is a linear layout with three view components: a TextView component, an EditText component, and a custom GLSurfaceView component called MyGLSurfaceView.

The custom GLSurfaceView class to use in this view is specified by the following statement, which is the complete name of the class that includes the package that it is in:

```
robs.demo.TestDemoComplete.MyGLSurfaceView
```

The id for this view is specified by the following statement:

```
android:id="@+id/MyGLSurfaceView"
```

The "@" symbol tells the compiler to parse and expand the rest of the string as an identity resource. The "+" tells the compiler that this new id must be added to the resources file located in the gen/R. java file. "MyGLSurfaceView" is the actual id (see Listing 2-4).

Listing 2-4. XML Layout for Multiple View OpenGL ES Application

```
<?xml version="1.0" encoding="utf-8"?>
<LinearLayout xmlns:android="http://schemas.android.com/apk/res/android"
    android:id="@+id/layout"
    android:layout_width="fill_parent"
    android:layout_height="fill_parent"
    android:orientation="vertical">

    <TextView
        android:id="@+id/Text1"
        android:layout_width="fill_parent"
        android:layout_height="wrap_content"
        android:text="@string/hello"/>

    <EditText
        android:id="@+id/EditTextBox1"
        android:layout_width="fill_parent"
        android:layout_height="wrap_content"
        android:text="@string/hello"/>

    <robs.demo.TestDemoComplete.MyGLSurfaceView
        android:id="@+id/MyGLSurfaceView"
        android:layout_width="wrap_content"
        android:layout_height="wrap_content"/>

</LinearLayout>
```

The Activity Class and GLSurfaceView Class

The XML layout in the preceding section is set to be the user interface by the statement setContentView() located in our Activity class.

Within the Activity class, we use the findViewById() function to get a reference to the newly created MyGLSurfaceView object, so we can reference it in our Activity class.

A new constructor is added to the MyGLSurfaceView class. This is needed because of our addition of the MyGLSurfaceView class into the XML layout (see Listing 2-5).

Listing 2-5. Multiple View OpenGL ES Activity

```java
package robs.demo.TestDemoComplete;

import android.app.Activity;
import android.os.Bundle;
import android.content.Context;
import android.opengl.GLSurfaceView;
import android.view.MotionEvent;
import android.util.AttributeSet;

public class OpenGLDemoActivity extends Activity
{
        private GLSurfaceView m_GLView;
        @Override
        public void onCreate(Bundle savedInstanceState)
        {
                super.onCreate(savedInstanceState);
                setContentView(R.layout.main);
                MyGLSurfaceView V = (MyGLSurfaceView)this.findViewById (R.id.MyGLSurfaceView);
                m_GLView = V;
        }

        @Override
        protected void onPause()
        {
                super.onPause();
                m_GLView.onPause();
        }

        @Override
        protected void onResume()
        {
        super.onResume();
                m_GLView.onResume();
        }
}
//////////////////////////////////////////////////////////////////////////////
class MyGLSurfaceView extends GLSurfaceView
{
        private final MyGLRenderer m_Renderer;

        // Constructor that is called when MyGLSurfaceView is created
        // from within an Activity with the new statement.
        public MyGLSurfaceView(Context context)
        {
                super(context);
                // Create an OpenGL ES 2.0 context.
                setEGLContextClientVersion(2);
```

```
                    // Set the Renderer for drawing on the GLSurfaceView
                    m_Renderer = new MyGLRenderer();
                    setRenderer(m_Renderer);
            }
            // Constructor that is called when MyGLSurfaceView is created in the XML
            // layout file
            public MyGLSurfaceView(Context context,  AttributeSet attrs)
            {
                    super(context, attrs);

                    // Create an OpenGL ES 2.0 context.
                    setEGLContextClientVersion(2);

                    // Set the Renderer for drawing on the GLSurfaceView
                    m_Renderer = new MyGLRenderer();
                    setRenderer(m_Renderer);
            }
}
```

Hands-on Example: A 3D OpenGL "Hello Droid" Example

In this hands-on exercise, I cover a simple 3D OpenGL example that gives you a preview of the kind of things I will be covering later in this book.

Importing Project Examples into Eclipse

In order to run the project examples from this book, you will need to import them into the current Eclipse work space. Under the main Eclipse menu, select File ➤ Import. This should bring up another window. Select Android ➤ Existing Android Code Into Workspace to start the process of importing existing code into your current work space. Follow the directions in the next window to select a root directory. Select the projects you want to import and if you want to copy the code to the existing work space or not. Click the Finish button when done.

Start up the Eclipse IDE. Import the Chapter 2 projects into your current work space, if you haven't done so already. Select the GLHelloWorld project and bring up the source code listing into the Package Explorer window area of the Eclipse IDE.

The MainActivity and MyGLSurfaceView Classes

Double-click the MainActivity Java file in the Package Explorer window to bring it up in the source code area. This file defines the new program or Activity and follows the same format for the single OpenGL view layout discussed earlier (see Listing 2-6.)

Listing 2-6. MainActivity and MyGLSurfaceView Classes

```
package com.robsexample.glhelloworld;

import android.os.Bundle;
import android.app.Activity;
import android.view.Menu;
```

```java
import android.opengl.GLSurfaceView;
import android.content.Context;

public class MainActivity extends Activity {

        private GLSurfaceView m_GLView;

        @Override
        public void onCreate(Bundle savedInstanceState)
        {
                super.onCreate(savedInstanceState);
                // Create a MyGLSurfaceView instance and set it
                // as the ContentView for this Activity
                m_GLView = new MyGLSurfaceView(this);
                setContentView(m_GLView);
        }

        @Override
        protected void onPause()
        {
                super.onPause();
                m_GLView.onPause();
        }

        @Override
        protected void onResume()
        {
                super.onResume();
                m_GLView.onResume();
        }

        @Override
        public boolean onCreateOptionsMenu(Menu menu) {
                // Inflate the menu; this adds items to the action bar if it is present.
                getMenuInflater().inflate(R.menu.activity_main, menu);
                return true;
        }
}
/////////////////////////////////////////////////////////////////////////////
class MyGLSurfaceView extends GLSurfaceView
{
    public MyGLSurfaceView(Context context)
    {
        super(context);

        // Create an OpenGL ES 2.0 context.
        setEGLContextClientVersion(2);

        // Set the Renderer for drawing on the GLSurfaceView
        setRenderer(new MyGLRenderer(context));
    }
}
```

The MyGLRenderer Class

Double-click the MyGLRenderer source code file in the Package Explorer window to bring it up in the Eclipse IDE source code window area (see Listing 2-7).

Listing 2-7. MyGLRenderer

```
package com.robsexample.glhelloworld;

import javax.microedition.khronos.egl.EGLConfig;
import javax.microedition.khronos.opengles.GL10;

import android.opengl.GLES20;
import android.opengl.GLSurfaceView;
import android.content.Context;

public class MyGLRenderer implements GLSurfaceView.Renderer
{
        private Context m_Context;
        private PointLight m_PointLight;
        private Camera m_Camera;
        private int m_ViewPortWidth;
        private int m_ViewPortHeight;
        private Cube m_Cube;

        public MyGLRenderer(Context context)
        {
           m_Context = context;
        }

        void SetupLights()
        {
                // Set Light Characteristics
            Vector3 LightPosition = new Vector3(0,125,125);

            float[] AmbientColor = new float [3];
            AmbientColor[0] = 0.0f;
            AmbientColor[1] = 0.0f;
            AmbientColor[2] = 0.0f;

            float[] DiffuseColor = new float[3];
            DiffuseColor[0] = 1.0f;
            DiffuseColor[1] = 1.0f;
            DiffuseColor[2] = 1.0f;

            float[] SpecularColor = new float[3];
            SpecularColor[0] = 1.0f;
            SpecularColor[1] = 1.0f;
            SpecularColor[2] = 1.0f;
```

```java
        m_PointLight.SetPosition(LightPosition);
        m_PointLight.SetAmbientColor(AmbientColor);
        m_PointLight.SetDiffuseColor(DiffuseColor);
        m_PointLight.SetSpecularColor(SpecularColor);
}

void SetupCamera()
{
    // Set Camera View
    Vector3 Eye    = new Vector3(0,0,8);
    Vector3 Center = new Vector3(0,0,-1);
    Vector3 Up     = new Vector3(0,1,0);

    float ratio = (float) m_ViewPortWidth / m_ViewPortHeight;
    float Projleft   = -ratio;
    float Projright  = ratio;
    float Projbottom = -1;
    float Projtop    = 1;
    float Projnear   = 3;
    float Projfar    = 50; //100;

    m_Camera = new Camera(m_Context,
                          Eye,
                          Center,
                          Up,
                          Projleft, Projright,
                          Projbottom,Projtop,
                          Projnear, Projfar);
  }

void CreateCube(Context iContext)
{
        //Create Cube Shader
        Shader Shader = new Shader(iContext, R.raw.vsonelight, R.raw.fsonelight);    // ok

        //MeshEx(int CoordsPerVertex,
        //            int MeshVerticesDataPosOffset,
        //            int MeshVerticesUVOffset,
        //            int MeshVerticesNormalOffset,
        //            float[] Vertices,
        //            short[] DrawOrder)
        MeshEx CubeMesh = new MeshEx(8,0,3,5,Cube.CubeData, Cube.CubeDrawOrder);

        // Create Material for this object
        Material Material1 = new Material();
        //Material1.SetEmissive(0.0f, 0, 0.25f);

        // Create Texture
        Texture TexAndroid = new Texture(iContext,R.drawable.ic_launcher);
        Texture[] CubeTex  = new Texture[1];
        CubeTex[0]         = TexAndroid;
```

```
                         m_Cube = new Cube(iContext,
                                           CubeMesh,
                                           CubeTex,
                                           Material1,
                                           Shader);

                         // Set Intial Position and Orientation
                         Vector3 Axis = new Vector3(0,1,0);
                         Vector3 Position = new Vector3(0.0f, 0.0f, 0.0f);
                         Vector3 Scale = new Vector3(1.0f,1.0f,1.0f);

                         m_Cube.m_Orientation.SetPosition(Position);
                         m_Cube.m_Orientation.SetRotationAxis(Axis);
                         m_Cube.m_Orientation.SetScale(Scale);

                         //m_Cube.m_Orientation.AddRotation(45);
                     }

                 @Override
                 public void onSurfaceCreated(GL10 unused, EGLConfig config)
                 {
                         m_PointLight = new PointLight(m_Context);
                         SetupLights();
                         CreateCube(m_Context);
                 }

                 @Override
                 public void onSurfaceChanged(GL10 unused, int width, int height)
                 {
                     // Ignore the passed-in GL10 interface, and use the GLES20
                     // class's static methods instead.
                     GLES20.glViewport(0, 0, width, height);
                     m_ViewPortWidth = width;
                     m_ViewPortHeight = height;
                     SetupCamera();
                 }

                 @Override
                 public void onDrawFrame(GL10 unused)
                 {
                         GLES20.glClearColor(1.0f, 1.0f, 1.0f, 1.0f);
                         GLES20.glClear( GLES20.GL_DEPTH_BUFFER_BIT | GLES20.GL_COLOR_BUFFER_BIT);
                         m_Camera.UpdateCamera();
                         m_Cube.m_Orientation.AddRotation(1);
                         m_Cube.DrawObject(m_Camera, m_PointLight);
                 }
         }
```

First the onSurfaceCreated() function is called. In this function, a new light is created and initialized and the 3D cube object is also created.

Next, the onSurfaceChanged() function is called. In this function, the camera is created and initialized. The camera's properties, such as position, orientation, and camera lens qualities, are defined.

In the onDrawFrame() function, the background is cleared to the color white. The camera is then updated. Next, the cube is rotated by one degree, and finally, the cube object is drawn.

Class Overview

In this book, the base class for 3D objects is the Object3d class. Other 3D objects, such as the Cube class, derive or extend directly or indirectly from the Object3d class.

The Object3d class contains other key classes, such as the Orientation class, MeshEx class, Texture class, Material class, and Shader class.

The Orientation class holds a 3D object's position, rotation, and scaling data.

The MeshEx class defines one type of OpenGL 3D mesh that is used to represent a 3D object.

The Texture class defines a texture that consists of a bitmap image that can be applied across a 3D object.

The Material class defines an object's Material properties, which define the color and lighting properties of an object. The properties are Emissive, Ambient, Diffuse, Specular, Specular_Shininess, and Alpha.

> **Emissive** refers to the light emitted by the object itself.
>
> **Ambient** refers to the color the material reflects when hit with ambient light. Ambient light is constant all over the object and is not affected by the light's position or the viewer's position.
>
> The **Diffuse** property refers to the color the material reflects when hit with diffuse light. The intensity of diffuse light across an object depends on the angle the object's vertex normals make with the light direction.
>
> The **Specular** property refers to the specular color the material reflects. The specular color depends on the viewer's position, the light's position, as well as the object's vertex normals.
>
> The **Specular_Shininess** property refers to how intense specular light reflections on the object will be.
>
> The **Alpha** value is the object's transparency.

The Shader class defines how a 3D object will be drawn and lighted. It consists of vertex shaders and pixel or fragment shaders. Vertex shaders determine where the object's vertices are located in the 3D world. Fragment shaders determine the color of the object being shaded.

The Camera class represents the view into the OpenGL 3D world. Position, orientation, and the camera lens properties are all contained within this class.

The Cube class contains vertex position data, vertex texture data, and vertex normal data that are needed to render a 3D cube with textures and lighting.

The PointLight class represents our light source. This light is modeled after a point light source, such as the sun. This kind of light is located at a single point in space, with light radiating in all directions. Light characteristics include ambient color, diffuse color, and specular color.

The Vector3 class holds data for a 3D vector consisting of x, y, and z components, as well as 3D vector math functions.

I cover the classes mentioned above in more detail in later chapters, so don't worry if you don't completely understand all the concepts. The purpose of this chapter is to give you a brief overview of some of the key classes contained in this book and show you how they are used in an actual program.

Experimenting with "Hello Droid"

Let's do some hands-on experimentation and play around with the lighting. Run the "GLHelloWorld" program on your Android phone that has version 2.2 or more recent versions of the operating system. Figure 2-2 shows what you should see by default. You should be seeing a 3D rotating cube with a texture of the Android robot placed on two sides of it.

Figure 2-2. Default output

Stop the cube rotation by commenting out the statement in the onDrawFrame() function that rotates the cube, as follows:

```
//m_Cube.m_Orientation.AddRotation(1);
```

Change the color of the background to black by changing the statement

```
GLES20.glClearColor(1.0f, 1.0f, 1.0f, 1.0f);
```

to

```
GLES20.glClearColor(0.0f, 0.0f, 0.0f, 1.0f);
```

It is located in the onDrawFrame() function as well. Figure 2-3 shows what you should see now.

Figure 2-3. Light positioned in front of and above cube

Next, let's change the light position so that it is to the right side of the droid. We are looking down the negative z axis, and the positive x axis is pointing to the right, and the negative x axis is pointing left, and the positive y axis is up. The droid is located at the origin, which is location (0,0,0). Change the light position in SetupLights() to the following:

```
Vector3 LightPosition = new Vector3(125,0,0);
```

This will move the light to the right of the droid. Run the program (see Figure 2-4). You can clearly see that the left arm is darkened, since most of the light is falling on the right side of the cube.

Figure 2-4. *Light positioned on right side of cube*

Next, change the light position so that the light is on the left side of the cube.

```
Vector3 LightPosition = new Vector3(-125,0,0);
```

Run the program. You should see something like Figure 2-5.

Figure 2-5. Light positioned on left side of cube

Next, change the light position so that the light is high above the cube. Change the light position to the following:

```
Vector3 LightPosition = new Vector3(0,125,0);
```

You should see something like Figure 2-6. Note that the android's legs are darkened.

Figure 2-6. *Light positioned high above cube*

Next, position the light far below the cube (see Figure 2-7).

Figure 2-7. *Light positioned far below cube*

```
Vector3 LightPosition = new Vector3(0,-125,0);
```

Run the program. You should see something similar to Figure 2-7.

Feel free to experiment more with light properties in the SetupLights() function. For example, try to change the values of the diffuse, ambient, and specular values to see what effect they will have on the object.

Summary

In this chapter, I covered the Java programming language as it relates to Android programming. First I covered the basics of Java, such as data types, functions, classes, and operators. Then we looked at the basic Java program framework that applies to all Android applications. Next, I covered the specific Java program framework that applies specifically to OpenGL ES applications. Finally, I presented a "Hello Droid" project that gave you a preview of how the rest of the code in this book will be structured.

Chapter 3

3D Math Review

In this chapter, I cover vectors and matrices. Vectors and matrices are key to 3D game programming in terms of such things as determining the placement of 3D objects in the scene and how a 3D object is projected onto a 2D screen. Vectors can also be used for defining properties such as velocity and force. I start by discussing vectors and operations that can be performed with vectors. I then cover matrices and the essential operations related to 3D graphics that can be performed with matrices. Finally, I present a hands-on example that will demonstrate how vectors and matrices are actually used in a real 3D graphics program on an Android device.

Vectors and Vector Operations

Vectors are an essential topic related to 3D graphics. In this section, I will cover what vectors are and what they are used for. I also cover important vector functions, such as dot products and cross products.

What Is a Vector?

A vector is a quantity that has direction and magnitude. For the purposes of this book, vectors will be 3D vectors, with components in the x, y, and z direction in the 3D world. Vectors can represent things such as position, velocity, direction, an object's rotation axis, an object's local axes, and forces acting on an object. On Android in OpenGL ES, the coordinate system consists of the x and z axes forming the ground plane and the y axis indicating height (see Figure 3-1).

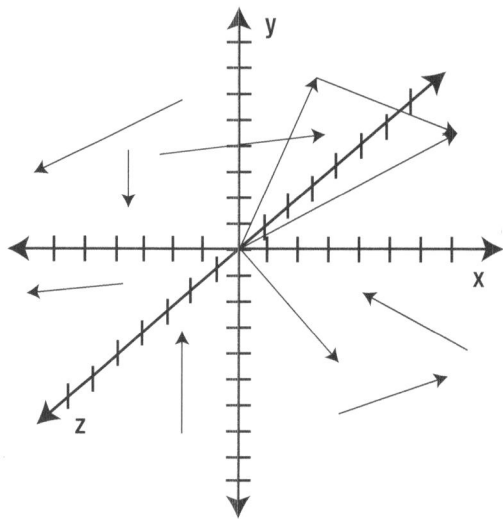

Figure 3-1. *3D vectors and the Android OpenGL ES coordinate system*

Vector Representing Position

A vector can represent an object's position in the Android 3D OpenGL ES world. In fact, in the Orientation class from the example in Chapter 2, the object's position is a 3D vector represented by the Vector3 class.

```
private Vector3 m_Position;
```

Graphically, you can see a vector representing an object's position in the 3D world in Figure 3-2.

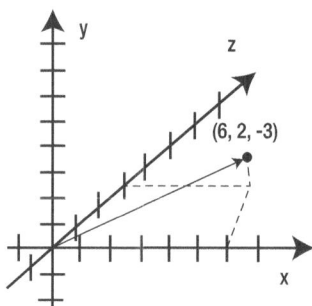

Figure 3-2. *Vector representing a position*

Vector Representing Direction

A vector can also represent a direction. A vector that is of length or magnitude 1 is called a unit vector (see Figure 3-3). Unit vectors are important because you can set properties such as an object's velocity or the force acting on an object by first finding the unit vector of the direction vector you want the object to move in, then multiplying this unit vector by a number. This number represents the magnitude of the speed of the object or the force you want to apply to the object. The final vector would contain both the direction of the object or direction of the force and also the speed of the object or amount of force to apply to the object.

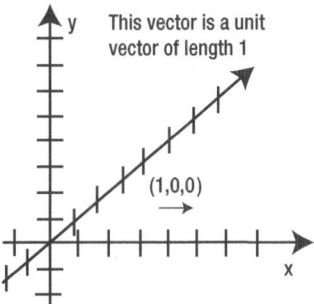

Figure 3-3. *Unit vector representing a direction*

Vector Representing Rotation Axis

A vector can also represent the axis of rotation of an object. The axis of rotation is the line about which an object rotates. In the Orientation class in Chapter 2, the variable m_RotationAxis is the local axis about which the object rotates.

```
private Vector3 m_RotationAxis;
```

See Figure 3-4 for a graphical representation of the local rotation axis.

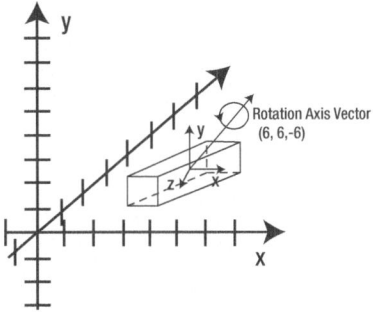

Figure 3-4. *Vector representing a rotation axis of an object*

Vector Representing Force

Vectors can also represent a force. A force has a direction and magnitude, so it is a good fit for representation by vectors. In Figure 3-5, you see a force vector acting on a ball. The direction of the force is in the negative x direction. I will discuss forces acting on 3D objects in more depth later in the book. More specifically, forces will be discussed in Chapter 5, "Motion and Collision."

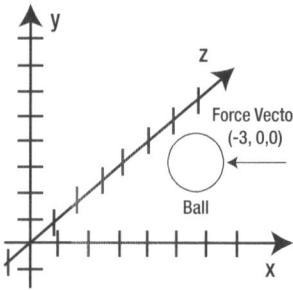

Figure 3-5. *Vector representing force*

Vectors Representing Local Axes

Vectors can also represent an object's local axes. Figure 3-6 shows the local x, y, and z axes of a 3D cube object. Local axes are important because they define an object's orientation. That is, they define which side of the object is considered up, which part of the object is the right side, and which part of the object is considered the front or forward part. For example, if a 3D object represents a vehicle such as a tank or car, then it would be good to know which part of the object is the front. If you want to move the tank or car forward, for example, you would need the front or forward vector in world coordinates as part of the calculation of the next position. The Orientation class defines an object's local axes as m_Right, m_Up, and m_Forward.

```
// Local Axes
private Vector3 m_Right;
private Vector3 m_Up;
private Vector3 m_Forward;
```

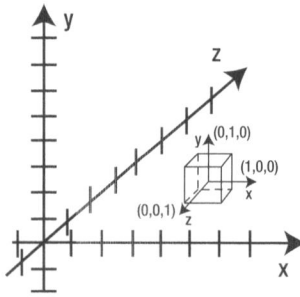

Figure 3-6. *Vectors representing local axes*

Our Vector Class

In terms of our code for this book, the vectors are represented in the Vector3 class (see Listing 3-1).

Listing 3-1. The Vector3 Class

```
class Vector3
{
    public float x;
    public float y;
    public float z;

    // Vector3 constructor
    public Vector3(float _x, float _y, float _z)
    {
        x = _x;
        y = _y;
        z = _z;
    }
}
```

In the Vector3 class, the x, y, and z components of the vector are represented by floats. The constructor accepts three float values that represent a 3D vector. For example:

```
Vector3 m_MyVector = new Vector3(1,2,3);
```

declares a new Vector3 class called `m_MyVector`, initialized with the values x = 1, y = 2, and z = 3.

The Vector Magnitude

The magnitude of the vector is the scalar value of a vector and is the length of the vector. Recall that a scalar value is a numerical value that has no direction associated with it. A velocity of an object can be represented by a vector and has the components of direction and speed. The speed is the scalar component and is calculated by finding the magnitude of the vector. The magnitude of a vector is found by squaring the x, y, and z components, adding them together, and then taking the square root (see Figure 3-7).

$$\text{Vector } V = (Vx, Vy, Vz)$$

$$\|v\| = \sqrt{v_x^2 + v_y^2 + v_z^2}$$

Figure 3-7. Vector magnitude calculation

In code, the magnitude of a vector is calculated by the Length() function in the Vector3 class, as shown in Listing 3-2.

Listing 3-2. Length or Magnitude Function

```
float Length()
{
    return FloatMath.sqrt(x*x + y*y + z*z);
}
```

Vector Normalization

Normalization in terms of vectors means that the length or magnitude of the vector is changed to 1, while maintaining the vector's direction. Normalization is a good way to set vector quantities, such as velocity and force. First, you would find a vector in the desired direction, then you would normalize it to change its length to 1. Finally, you would multiply the vector by the magnitude you want to assign to it, such as speed or the amount of force. In order to normalize a vector, you divide each of the vector's components by the length of the vector (see Figure 3-8).

$$\text{Normalized } V = \left(\frac{Vx}{||V||}, \frac{Vy}{||V||}, \frac{Vz}{||V||} \right)$$

Figure 3-8. Normalizing a vector

In code, the Normalize() function in the Vector3 class performs normalization (see Listing 3-3).

Listing 3-3. Normalize() Function

```
void Normalize()
{
    float l = Length();

    x = x/l;
    y = y/l;
    z = z/l;
}
```

Vector Addition

Vectors can be added together to produce a resultant vector that is the combination of the effects of all the individual vectors combined. You can add vectors graphically by putting them head to tail. The resultant vector, VR, is the one drawn from the tail of the starting vector to the head of the preceding vector (see Figure 3-9).

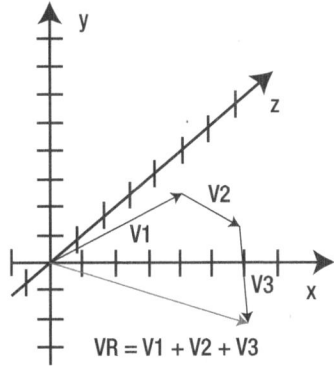

Figure 3-9. *Adding vectors together*

In terms of code, the Add() function in the Vector3 class adds two vectors and returns the resultant vector. Each component x, y, and z of the vectors is added together to form the new components of the resultant vector (see Listing 3-4).

Listing 3-4. The Add Function Adds Two Vectors

```
static Vector3 Add(Vector3 vec1, Vector3 vec2)
{
    Vector3 result = new Vector3(0,0,0);

    result.x = vec1.x + vec2.x;
    result.y = vec1.y + vec2.y;
    result.z = vec1.z + vec2.z;

    return result;
}
```

Vector Multiplication

You can also multiply a scalar value by a vector. For example, if you want to set an object's velocity, which is a combination of direction and speed, you would find a vector that points to the desired direction, normalize the vector so that the vector length would be 1, and then multiply that vector by the speed. The final resultant vector, VR, would be pointing in the desired direction and have a magnitude value of the speed (see Figure 3-10).

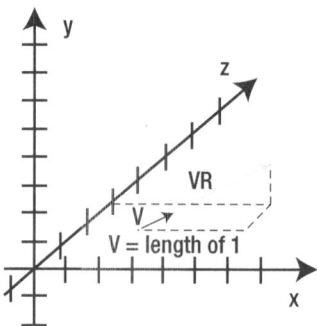

Figure 3-10. Multiplying a unit vector of length 1 with a scalar value

In terms of code, the Multiply() function in the Vector3 class multiplies a scalar value by the vector. Each component x, y, and z of the vector is multiplied by the scalar value (see Listing 3-5).

Listing 3-5. Multiply Function

```
void Multiply(float v)
{
    x *= v;
    y *= v;
    z *= v;
}
```

Vector Negation

Vector negation means that you multiply the vector by –1, which means you multiply every component of the vector by –1. Basically, the direction of the vector is reversed. Look at Figure 3-11, to see what this looks like graphically.

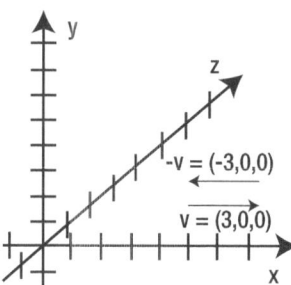

Figure 3-11. Vector negation

In terms of code, the Negate() function in the Vector3 class performs negation (see Listing 3-6).

Listing 3-6. The Negate Function

```
void Negate()
{
    x = -x;
    y = -y;
    z = -z;
}
```

The Right Triangle

The right triangle comes in handy when trying to break up vectors into components. For example, if you know the speed of a tank shell and the angle the shell's path makes with the ground, then you can get the horizontal and vertical speeds of the tank shell. The horizontal speed would be calculated using the formula for the adjacent side of the right triangle. The vertical speed would be calculated using the formula for the opposite side of the triangle. Let's review the basic trigonometric identities related to the right triangle, as depicted in Figure 3-12.

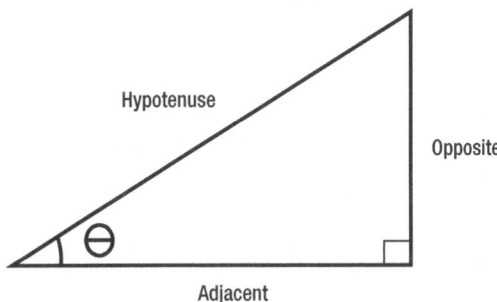

Figure 3-12. The right triangle

The following is a list of standard trigonometric identities that describe how the lengths of the sides of a right triangle relate to one another and the angle theta shown in Figure 3-12:

Sin(Theta) = Opposite/Hypotenuse

Cos(Theta) = Adjacent/Hypotenuse

Opposite = Hypotenuse * Sin(Theta)

Adjacent = Hypotenuse * Cos(Theta)

Vector Dot Product

The vector dot product of two vectors is the magnitude of vector A multiplied by the magnitude of vector B multiplied by the cosine of the angle between them. The dot product is commonly used to find the angle between two vectors. One application of the dot product is in billboarding, where a 2D rectangle with a complex image on it, such as a tree, is turned to face the camera. This is a way to achieve a 3D-like effect, by having an image face the camera at all times. If done with a complex background image, such as a tree, the viewer may not notice that it is the same image when viewed from different angles (see Figure 3-13).

$$A \text{ dot } B = \|A\|\|B\|\cos(\theta)$$

Figure 3-13. Dot product formula

You can also find the angle between two vectors using the dot product. The angle between two vectors is the angle whose cosine is given by the dot product of vector A and vector B, divided by the magnitude of vector A, multiplied by the magnitude of vector B (see Figure 3-14).

$$\theta = \arccos\left(\frac{A \text{ dot } B}{\|A\|\|B\|}\right)$$

Figure 3-14. Finding angle from dot product

You can simplify the above equation by normalizing both vectors, then taking the dot product. The denominator becomes 1, and the angle is the arc cosine of the dot product of vector A and vector B.

You can get the dot product directly from the vectors by multiplying each component in vector A by the corresponding component in vector B and adding the results together. Such as

```
Dot Product = (Ax * Bx) + (Ay * By) + (Az * Bz)
```

See Listing 3-7 for the Java code to do this.

Listing 3-7. DotProduct Function

```
float DotProduct(Vector3 vec)
{
     return (x * vec.x) + (y * vec.y) + (z * vec.z);
}
```

Vector Cross Product

The cross product of two vectors A and B is a third vector that is perpendicular to both A and B (see Figure 3-15). Cross products can be used in applications such as billboarding, where you need to find a rotation axis and know the vector that represents the front of the image and the vector that points toward the object you want to turn toward. The cross product is calculated in code in Listing 3-8.

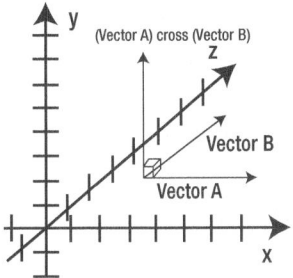

Figure 3-15. *Cross product*

Listing 3-8. *Cross Product Function*

```
void crossProduct(Vector3 b)
{
    Set((y*b.z) - (z*b.y),
        (z*b.x) - (x*b.z),
        (x*b.y) - (y*b.x));
}
```

Matrices and Matrix Operations

In this section, I cover matrices and matrix operations. I first discuss the definition of a matrix. I then cover various key topics related to matrix math and the key properties of matrices as they relate to 3D computer graphics needed to develop a 3D game on the Android mobile platform. This section is not meant to cover every aspect of matrices but is a quick-start guide to matrices and matrix math operations that are essential to 3D game programming.

What Is a Matrix?

Matrices are key in 3D graphics. They are used to determine properties such as the final position of 3D objects, the rotation of 3D objects, and the scaling of 3D objects. A matrix is defined in Figure 3-16. A matrix consists of columns and rows of numbers. The general notation we will use is that of Amn. The subscript *m* refers to the row number, and the *n* refers to the column number. For example, the notation A23 refers to the number at row 2, column 3.

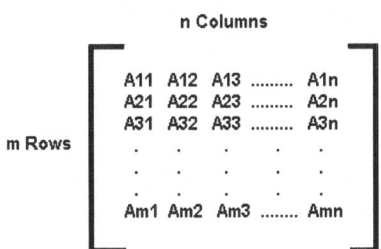

Figure 3-16. *Definition of a matrix*

In terms of code, we represent matrices as a float array of 16 elements. This translates to a 4-by-4 matrix; that is, a matrix that has four rows and four columns. The following declares a 4-by-4 matrix (total 16 elements) of type float that is private to the class that it is located in:

```
private float[] m_OrientationMatrix = new float[16];
```

Built-in Android Matrix Class

There is a Matrix class available in the standard Android class library that provides many matrix functions. You can access this class by using the following import statement:

```
import android.opengl.Matrix;
```

The Identity Matrix

The identity matrix is a square matrix that has an equal number of rows and columns that contain a 1 diagonally, with the rest of the values set to 0. The identity matrix can be used to initialize or reset the value of a matrix variable. A matrix multiplied by the identity matrix returns the original matrix. This is the equivalent of multiplying a number by 1. For example, let's say you have a matrix that keeps track of an object's rotation. In order to reset the object back to its original rotation, you would set the matrix to the identity matrix (see Figure 3-17).

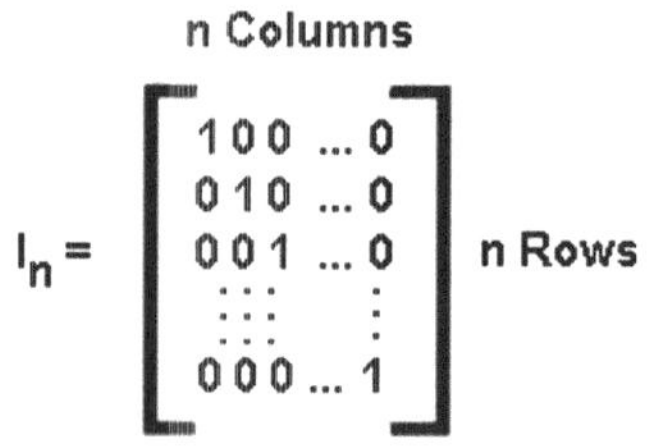

$$
\mathbf{I_n} = \begin{bmatrix} 1 & 0 & 0 & \dots & 0 \\ 0 & 1 & 0 & \dots & 0 \\ 0 & 0 & 1 & \dots & 0 \\ \vdots & \vdots & \vdots & & \vdots \\ 0 & 0 & 0 & \dots & 1 \end{bmatrix} \text{ n Rows}
$$

n Columns

Figure 3-17. The identity matrix

In terms of code, you can set a matrix to the identity matrix, using the following statement:

```
//static void setIdentityM(float[] sm, int smOffset)
Matrix.setIdentityM(m_OrientationMatrix, 0);
```

The matrix contained in the float array m_OrientationMatrix will be set to the identity matrix. There is 0 offset into the array to the start of the matrix data.

Matrix Transpose

The transpose of a matrix is created by rewriting the rows of the matrix as columns. You will have to use the matrix transpose to calculate the value of the normal matrix, which is used in lighting (see Figure 3-18).

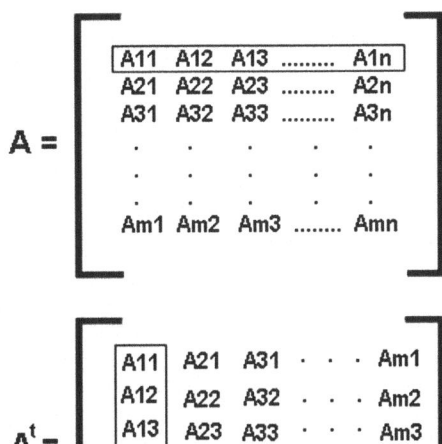

Figure 3-18. *Matrix transpose*

The following code statement transposes a 4-by-4 matrix `m_NormalMatrixInvert` and puts the result in `m_NormalMatrix`. The offsets into the data of both matrices are 0.

```
//static void transposeM(float[] mTrans, int mTransOffset,
//                       float[] m, int mOffset)
Matrix.transposeM(m_NormalMatrix, 0, m_NormalMatrixInvert, 0);
```

Matrix Multiplication

Matrix multiplication of matrix A and matrix B is done by multiplying elements in the rows of A into the corresponding elements in the columns of B and adding the products. Matrix multiplication is essential in translating objects, rotating objects, scaling objects, and displaying 3D objects on a 2D screen. For example, in Figure 3-19, matrix A is being multiplied by matrix B, and the result is put into matrix C.

```
C11 = (A11 * B11) + (A12 * B21) + (A13 * B31)
C12 = (A11 * B12) + (A12 * B22) + (A13 * B32)
C21 = (A21 * B11) + (A22 * B21) + (A23 * B31)
C22 = (A21 * B12) + (A22 * B22) + (A23 * B32)
```

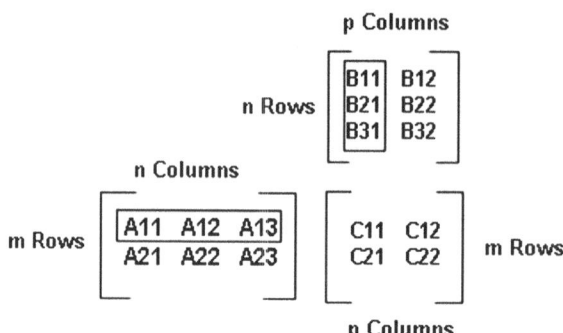

Figure 3-19. *Matrix multiplication*

In code you, use the `multiplyMM()` function located in the standard Android built-in Matrix class. This function multiplies two 4-by-4 matrices together and stores the result in a third 4-by-4 matrix. The following statement multiplies `m_PositionMatrix` by `m_RotationMatrix` and puts the result in `TempMatrix`. All of the array offsets to the matrix data are 0.

```
//static void multiplyMM(float[] result, int resultOffset,
//                       float[] lhs, int lhsOffset,
//                       float[] rhs, int rhsOffset)
Matrix.multiplyMM(TempMatrix, 0, m_PositionMatrix, 0, m_RotationMatrix, 0);
```

Matrix Inverse

If a matrix A is n rows by n columns and there exists another matrix B that is also n rows by n columns, such that AB = the identity matrix and BA = the identity matrix, then B is the inverse of A. This is the definition of the matrix inverse.

In code, you can use the function `invertM()` to find the inverse of a 4-by-4 matrix. You will have to use the matrix inverse to calculate the value of the normal matrix that is used for calculating the lighting of the 3D objects. The following code inverts the `m_NormalMatrix` and stores the result in `m_NormalMatrixInvert`.

```
//static boolean invertM(float[] mInv, int mInvOffset,
//                       float[] m, int mOffset)
Matrix.invertM(m_NormalMatrixInvert, 0, m_NormalMatrix, 0);
```

Homogeneous Coordinates

Homogeneous coordinates are a coordinate system used in projective geometry. They specify points in a 3D world. Homogeneous coordinates are important because they are used in building the matrices that are sent to the vertex shader to translate, rotate, and scale the 3D object's vertices. OpenGL represents all coordinates internally as 3D homogeneous coordinates. The coordinate system we have been using previously in this chapter to specify points had Cartesian coordinates in the Euclidean system.

The general form of homogeneous coordinates is (x,y,z,w). A point in homogeneous coordinates can be converted to normal 3D Euclidean space coordinates by dividing all the coordinates by the w coordinate. For example, given the point in homogeneous coordinates (x,y,z,w), the point in 3D Euclidean space is (x/w,y/w,z/w).

A point in 3D Euclidean space denoted by (x,y,z) can be represented in homogeneous space by (x,y,z,1).

We will discuss OpenGL ES 2.0 vertex and fragment shaders in more depth in Chapter 4, "3D Graphics Using OpenGL ES 2.0."

Using Matrices to Move Objects

Matrices can be used to do many things, such as translating objects, rotating objects, scaling objects, and projecting a 3D object onto a 2D screen. The matrix used to move objects in the 3D world is called the translation matrix. In Figure 3-20, the new position is calculated by converting the old position into homogeneous coordinates, creating a matrix from this homogeneous coordinate, and then multiplying it by the translation matrix. The values Tx, Ty, Tz indicate the amount to move the object in the x, y, and z direction on the plane. Using matrix multiplication to find the new x, y, and z coordinates results in the following:

```
x' = x + Tx
y' = y + Ty
z' = z + Tz
```

$$\begin{bmatrix} x' & y' & z' & 1 \end{bmatrix} = \begin{bmatrix} x & y & z & 1 \end{bmatrix} \begin{bmatrix} 1 & 0 & 0 & 0 \\ 0 & 1 & 0 & 0 \\ 0 & 0 & 1 & 0 \\ Tx & Ty & Tz & 1 \end{bmatrix}$$

Figure 3-20. *Translating an object*

In terms of code, we use the translateM() function to translate the input matrix in place by x, y, and z values.

For example, the following translates the m_PositionMatrix matrix in place:

```
//static void    translateM(float[] m, int mOffset, float x, float y, float z)
//Translates matrix m by x, y, and z in place.
Matrix.translateM(m_PositionMatrix, 0, position.x, position.y, position.z);
```

Using Matrices to Rotate Objects

Matrices are also used to rotate 3D objects. One example of how to build a rotation matrix that rotates around the x axis is shown in Figure 3-21.

$$Rx = \begin{bmatrix} 1 & 0 & 0 & 0 \\ 0 & \cos(\theta) & \sin(\theta) & 0 \\ 0 & -\sin(\theta) & \cos(\theta) & 0 \\ 0 & 0 & 0 & 1 \end{bmatrix}$$

Figure 3-21. Rotation matrix

In terms of code, there is a built-in function in the Matrix class in which you can rotate a matrix in place around an arbitrary rotation axis you specify by an angle that is specified in degrees.

The rotateM() function rotates the matrix m by angle a in degrees around the axis (x,y,z). You can also specify and offset to where the matrix data begins.

```
//rotateM(float[] m, int mOffset, float a, float x, float y, float z)
//Rotates matrix m in place by angle a (in degrees) around the axis (x, y, z)
Matrix.rotateM(m_RotationMatrix, 0,
                AngleIncrementDegrees,
                m_RotationAxis.x,
                m_RotationAxis.y,
                m_RotationAxis.z);
```

Using Matrices to Scale Objects

You can also use a matrix to scale objects. The diagonal of the square 4-by-4 matrix contains the scaling factors in the x, y, and z directions (see Figure 3-22).

$$S = \begin{bmatrix} Sx & 0 & 0 & 0 \\ 0 & Sy & 0 & 0 \\ 0 & 0 & Sz & 0 \\ 0 & 0 & 0 & 1 \end{bmatrix}$$

Figure 3-22. Scale matrix

In terms of code, the scaleM() function in the Matrix class scales a matrix in place in the x, y, and z directions.

```
//static void scaleM(float[] m, int mOffset, float x, float y, float z)
//Scales matrix m in place by sx, sy, and sz
Matrix.scaleM(m_ScaleMatrix, 0, Scale.x, Scale.y, Scale.z);
```

Combining Matrices

By multiplying matrices together, you can combine the effects of translation, rotation, and scaling on an object. One key combination matrix we will need in order to render 3D objects in our game will be the ModelMatrix. The ModelMatrix is a combination of the translation matrix, rotation matrix, and the scale matrix multiplied together to form a single final matrix (see Figure 3-23).

$$\text{ModelMatrix} = \overset{\substack{\text{Translation} \\ \text{Matrix}}}{\begin{bmatrix} 1 & 0 & 0 & 0 \\ 0 & 1 & 0 & 0 \\ 0 & 0 & 1 & 0 \\ Tx & Ty & Tz & 1 \end{bmatrix}} \overset{\substack{\text{Rotation} \\ \text{Matrix}}}{\begin{bmatrix} 1 & 0 & 0 & 0 \\ 0 & \cos(\theta) & \sin(\theta) & 0 \\ 0 & -\sin(\theta) & \cos(\theta) & 0 \\ 0 & 0 & 0 & 1 \end{bmatrix}} \overset{\substack{\text{Scale} \\ \text{Matrix}}}{\begin{bmatrix} Sx & 0 & 0 & 0 \\ 0 & Sy & 0 & 0 \\ 0 & 0 & Sz & 0 \\ 0 & 0 & 0 & 1 \end{bmatrix}}$$

Figure 3-23. *Model matrix*

The important thing to understand about matrix multiplication is that the order of multiplication matters. That is, matrix multiplication is not commutative. Thus, AB does not equal BA.

For example, if you want to rotate an object around an axis then translate it, you need to have the rotation matrix on the right-hand side and the translation matrix on the left-hand side. In code, this would look like the following:

```
// Rotates object around Axis then translates it
// public static void multiplyMM (float[] result, int resultOffset,
//                          float[] lhs, int lhsOffset,
//                          float[] rhs, int rhsOffset)

//                          Matrix A          Matrix B
Matrix.multiplyMM(TempMatrix, 0, m_PositionMatrix, 0, m_RotationMatrix, 0);
```

The multiplyMM() function multiplies two matrices A and B. In terms of effects, the matrix B is applied first then matrix A is applied. Thus, the above code first rotates the object around its rotation axis and then translates it to a new position.

Thus, the ModelMatrix in Figure 3-23 is set up to first scale an object, then rotate the object around its axis of rotation, and then translate the object.

Hands-on Example: Manipulating Objects in 3D Space

In this hands-on example, we will concentrate on manipulating a 3D object's position, rotation, and scaling to demonstrate the concepts of vectors and matrices covered in this chapter.

This example uses some vector functions like Negate(), presented earlier. Please make sure to add these functions, as well as other functions you wish to experiment with, into the Vector3 class from Chapter 2. You can also find the code for this example in the Source Code/Download area of apress.com.

Building a 3D Object's Model Matrix

The Orientation class holds data for a 3D object's position, rotation, and scaling. It also calculates the object's model matrix, which contains the object's position, rotation, and scaling information (refer to the preceding Figure 3-23).

The model matrix is called the m_OrientationMatrix in our code. The m_PositionMatrix is the translation matrix; the m_RotationMatrix is our rotation matrix; and the m_ScaleMatrix is our scale matrix. We also have a TempMatrix for use for the temporary storage of matrices.

The SetPositionMatrix() function creates the translation matrix by first initializing the matrix to the identity matrix, by calling setIdentity() and then creating the translation matrix by calling translateM() in the default Matrix class that is part of the standard Android library.

The SetScaleMatrix() function creates the scale matrix by first initializing the matrix to the identity matrix and then calling scaleM() from the Matrix class library to create the scale matrix.

The UpdateOrientation() function actually builds the model matrix.

1. It first creates the translation matrix by calling SetPositionMatrix().

2. Next, SetScaleMatrix() is called to create the Scale Matrix.

3. Then, the final model matrix starts to be built by calling Matrix.multiplyMM() to multiply the translation matrix by the rotation matrix.

4. Finally, the result matrix from step 3 is multiplied by the scale matrix and then returned to the caller of the function. The net result is that a matrix is created that first scales a 3D object, then rotates it around its axis of rotation, and then finally puts it into the 3D world at a location specified by m_Position (see Listing 3-9).

Listing 3-9. Building the Model Matrix in the Orientation Class

```
// Orientation Matrices
private float[] m_OrientationMatrix = new float[16];
private float[] m_PositionMatrix = new float[16];
private float[] m_RotationMatrix = new float[16];
private float[] m_ScaleMatrix = new float[16];
private float[] TempMatrix = new float[16];

// Set Orientation Matrices
void SetPositionMatrix(Vector3 position)
{
        // Build Translation Matrix
        Matrix.setIdentityM(m_PositionMatrix, 0);
        Matrix.translateM(m_PositionMatrix, 0, position.x, position.y, position.z);
}

void SetScaleMatrix(Vector3 Scale)
{
        // Build Scale Matrix
        Matrix.setIdentityM(m_ScaleMatrix, 0);
        Matrix.scaleM(m_ScaleMatrix, 0, Scale.x, Scale.y, Scale.z);
}
```

```
float[] UpdateOrientation()
{
        // Build Translation Matrix
        SetPositionMatrix(m_Position);

        // Build Scale Matrix
        SetScaleMatrix(m_Scale);

        // Then Rotate object around Axis then translate
        Matrix.multiplyMM(TempMatrix, 0, m_PositionMatrix, 0, m_RotationMatrix, 0);

        // Scale Object first
        Matrix.multiplyMM(m_OrientationMatrix, 0, TempMatrix, 0, m_ScaleMatrix, 0);

        return m_OrientationMatrix;
}
```

Adding a Rotation to an Object

Using the "Hello Droid" project from Chapter 2, let's play around with the code to demonstrate how vectors and matrices work on OpenGL ES 2.0 for Android.

In the onDrawFrame() function in the MyGLRenderer class, make sure the

```
m_Cube.m_Orientation.AddRotation(1)
```

statement is uncommented. This will add a rotation of 1 degree every time onDrawFrame() is executed, which will be continuously (see Listing 3-10).

Listing 3-10. onDrawFrame() Function in MyGLRenderer

```
@Override
public void onDrawFrame(GL10 unused)
{
    GLES20.glClearColor(0.0f, 0.0f, 0.0f, 1.0f);
    GLES20.glClear( GLES20.GL_DEPTH_BUFFER_BIT | GLES20.GL_COLOR_BUFFER_BIT);

    m_Camera.UpdateCamera();

    m_Cube.m_Orientation.AddRotation(1);
    m_Cube.DrawObject(m_Camera, m_PointLight);
}
```

The AddRotation() function is part of the Orientation class and is shown in Listing 3-11.

Listing 3-11. AddRotation() Function in the Orientation Class

```
void AddRotation(float AngleIncrementDegrees)
{
        m_RotationAngle += AngleIncrementDegrees;
```

```
//rotateM(float[] m, int mOffset, float a, float x, float y, float z)
//Rotates matrix m in place by angle a (in degrees) around the axis (x, y, z)
Matrix.rotateM(m_RotationMatrix, 0,
               AngleIncrementDegrees,
               m_RotationAxis.x,
               m_RotationAxis.y,
               m_RotationAxis.z);
}
```

What occurs here is the angle for the object to be rotated is added to the m_RotationAngle variable, which holds the current angle of rotation of the object. The rotation matrix, which is m_RotationMatrix, is then modified to reflect the addition of the new angle delta. Run the program and you see the cube rotating as in Figure 3-24.

Figure 3-24. Cube rotating

Moving an Object in 3D Space

Now, we will guide you through moving the cube back and forth along the z axis. Because the z axis is facing toward you, the cube will be getting larger and smaller.

First, we should stop the cube from rotating. Comment out the AddRotation() function, as shown in Listing 3-12.

Next, add in the m_CubePositionDelta variable to the MyGLRenderer class. This variable holds the direction and magnitude of the position change that will be applied to the cube each time onDrawFrame() is called.

The key part of the new code performs the actual position update, the testing of boundaries, and the change in direction of the m_CubePositionDelta variable.

That code does the following:

1. Gets the current position of the cube.

2. Tests the position to see if it is within the z position 4 to –4. If the cube is outside these boundaries, then the cube's direction is reversed. That is, the m_CubePositionDelta vector is negated.

3. The current position vector of the cube is added to the m_CubePositionDelta vector and then set as the new position of the cube.

Add in this new code that is highlighted in Listing 3-10 and run the program.

Listing 3-12. Adding Code to Move the Cube Along the Z Axis

```
private Vector3 m_CubePositionDelta = new Vector3(0.0f,0,0.1f);

@Override
public void onDrawFrame(GL10 unused)
{
    GLES20.glClearColor(0.0f, 0.0f, 0.0f, 1.0f);
    GLES20.glClear( GLES20.GL_DEPTH_BUFFER_BIT | GLES20.GL_COLOR_BUFFER_BIT);
    m_Camera.UpdateCamera();

    // Add Rotation to Cube
    //m_Cube.m_Orientation.AddRotation(1);

    // Add Translation to Cube
    Vector3 Position = m_Cube.m_Orientation.GetPosition();
    if ((Position.z > 4) || (Position.z < -4))
    {
        m_CubePositionDelta.Negate();
    }
    Vector3 NewPosition  = Vector3.Add(Position, m_CubePositionDelta);
    Position.Set(NewPosition.x, NewPosition.y, NewPosition.z);

    m_Cube.DrawObject(m_Camera, m_PointLight);
}
```

You should see the android image moving toward and away from you in a loop (see Figure 3-25).

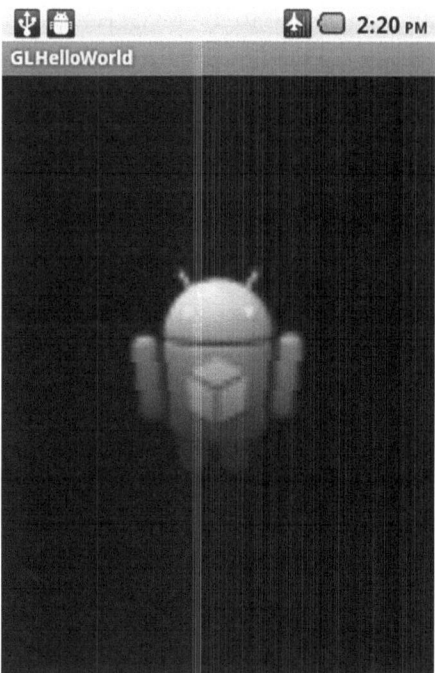

Figure 3-25. Translating an object on z axis

Scaling an Object

Here I will cover scaling objects. First add the following statement under the entry for the m_CubePositionDelta variable that you added previously.

```
private Vector3 m_CubeScale = new Vector3(4,1,1);
```

The m_CubeScale variable represents the amount to scale the object in the x, y, and z directions. In this example, the cube scales by four times the normal size in the local x axis direction and scales by the normal size, which is 1, in the y and z directions.

The following statement sets the scale for the cube. Type this into the onDrawFrame() function after the previous code you have added.

```
// Set Scale
m_Cube.m_Orientation.SetScale(m_CubeScale);
```

Run the program, and you should see what is in Figure 3-26.

Figure 3-26. *Scaling in the x direction*

Experiment with the code. I made some modifications that change the background color as well as the direction of the translation (to move diagonally back and forth) and scaling (see Figure 3-27).

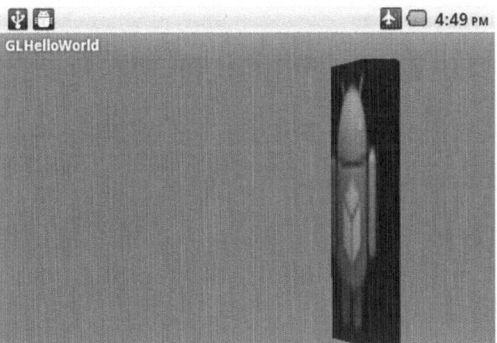

Figure 3-27. *Experimenting with the code*

See if you can replicate these changes.

Summary

In this chapter, I covered the basics of 3D math relating to vectors and matrices. I first covered vectors and the operations relating to vectors, such as addition, multiplication, dot products, and cross products. Next, I covered matrices and the operations involving matrices, such as matrix multiplication, that are essential to 3D game programming. Finally, I presented a hands-on example that demonstrated the practical use of vectors and matrices in translating, rotating, and scaling a 3D object.

Chapter 4

3D Graphics Using OpenGL ES 2.0

In this chapter, we will take a look at 3D graphics for OpenGL ES 2.0 for Android. I first give you a general overview of how OpenGL renders 3D objects. Next, I go into more detail on how this is done, using matrix math, matrix transformations, and vertex and fragment shaders. I then offer a look at the shader language used for the vertex and fragment shaders and give you a quick review of the language and some examples.

Next, I cover the custom classes that demonstrate OpenGL ES 2.0 concepts that are essential to rendering 3D graphics, which are the

- Shader class
- Camera class
- MeshEx class
- PointLight class
- Materials class
- Texture class

Overview of OpenGL ES 2.0 on Android

In this section, I first cover the general concepts behind rendering 3D objects in OpenGL. Then I get into more specifics, involving the exact steps required to transform a 3D object from our 3D virtual world to a 2D image on a flat screen.

General Overview of OpenGL Object Rendering

The general procedure to render a 3D object in OpenGL is discussed in this section. The following general steps are used to render a 3D object in OpenGL to the final viewing window.

1. *Put 3D objects in the world:* Place the 3D objects you are going to use in the 3D world (see Figure 4-1).

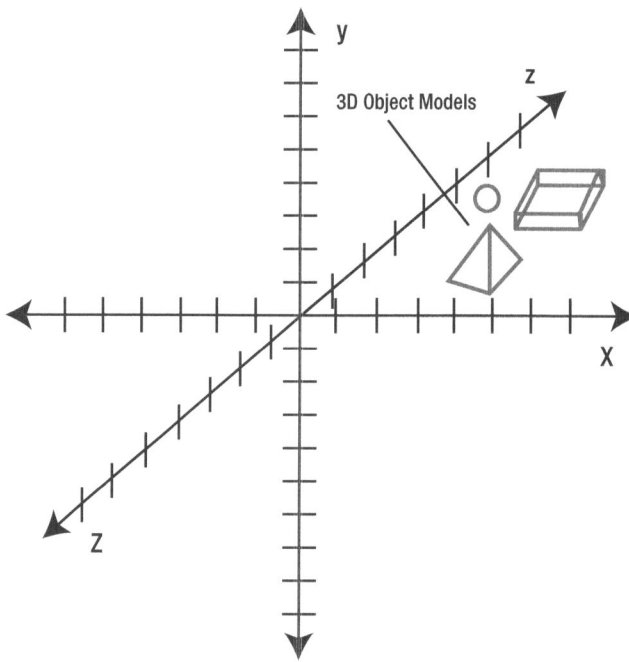

Figure 4-1. Putting 3D objects into the world

2. *Place camera into the world:* Place the viewer, which you can think of as a camera or as a human eye, into the 3D world (see Figure 4-2).

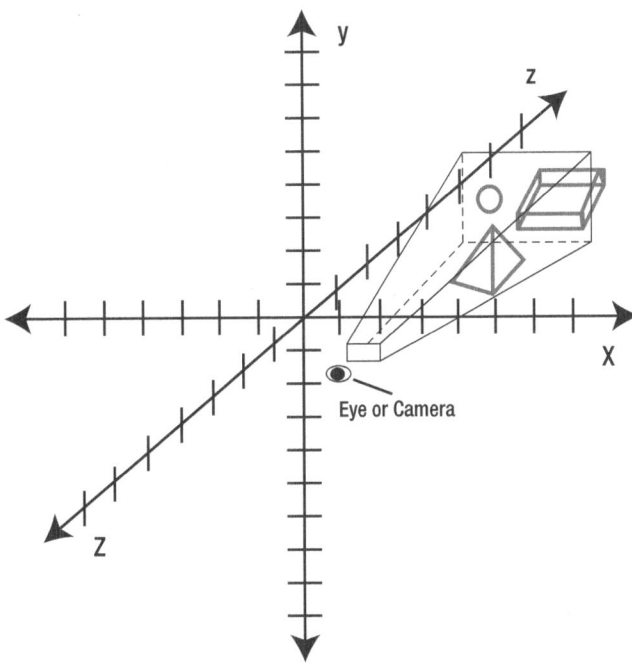

Figure 4-2. Placing the camera or viewer into the world

3. *Project the 3D view onto a 2D surface using a specific camera lens:* For the examples in this book, we use a frustrum, in which objects that are farther away are smaller than objects that are closer to the viewer. A frustrum is basically a pyramid with the top capstone cut off (see the section entitled "Building the Projection Matrix" later in this chapter for a more detailed description). The frustrum also defines which objects are included in the final view. In Figure 4-3, only the pyramid is included in the final view, because it is the only object included in the frustrum volume.

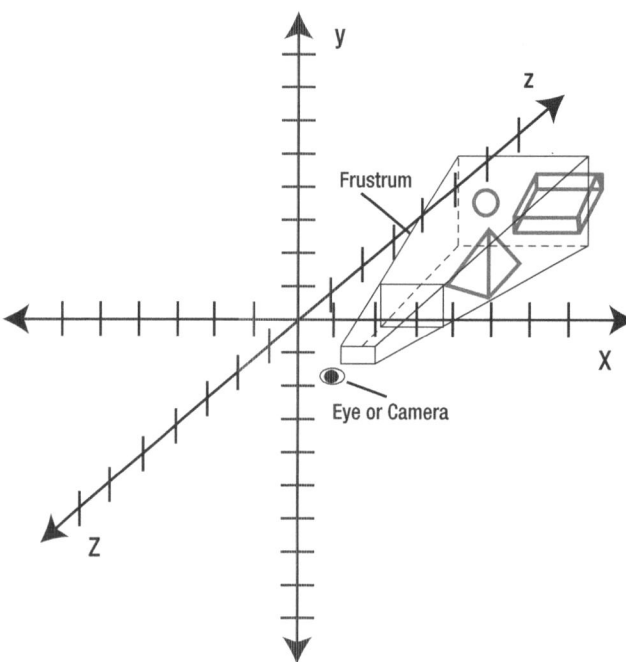

Figure 4-3. *Using a frustrum to project the 3D view onto a 2D surface*

4. *Transfer the projected view onto the final viewport for viewing:* Take the 2D image generated from the projection shown in Figure 4-3 and resize it to fit the final viewport (see Figure 4-4).

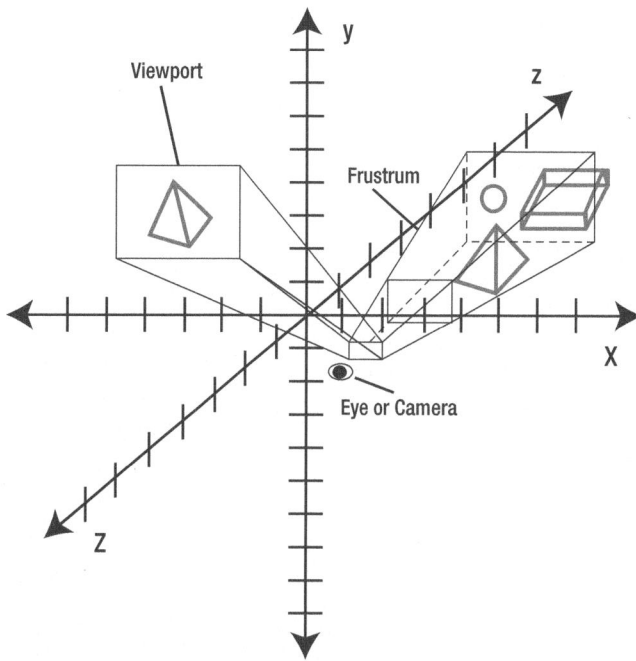

Figure 4-4. Viewport transformation

Specific Overview of the Rendering Procedure

Let's discuss in greater detail how a 3D object is rendered onscreen using OpenGL ES 2.0 specifically. Matrix transformations and vertex and fragment shaders are covered.

Transforming the 3D Object's Vertices

In this section, I will discuss in detail how to put a 3D object on the screen. In order to put a 3D object onscreen, we must transform each of the object's vertex coordinates, taking into account the object's position in the world, the camera position and orientation, the projection type, and the viewport specifications (see Figure 4-5). Remember that the coordinates are in (x,y,z) format but are internally represented in homogeneous vertex format, which is (x,y,z,1). This is why 4-by-4 matrices are needed to transform a vertex in OpenGL.

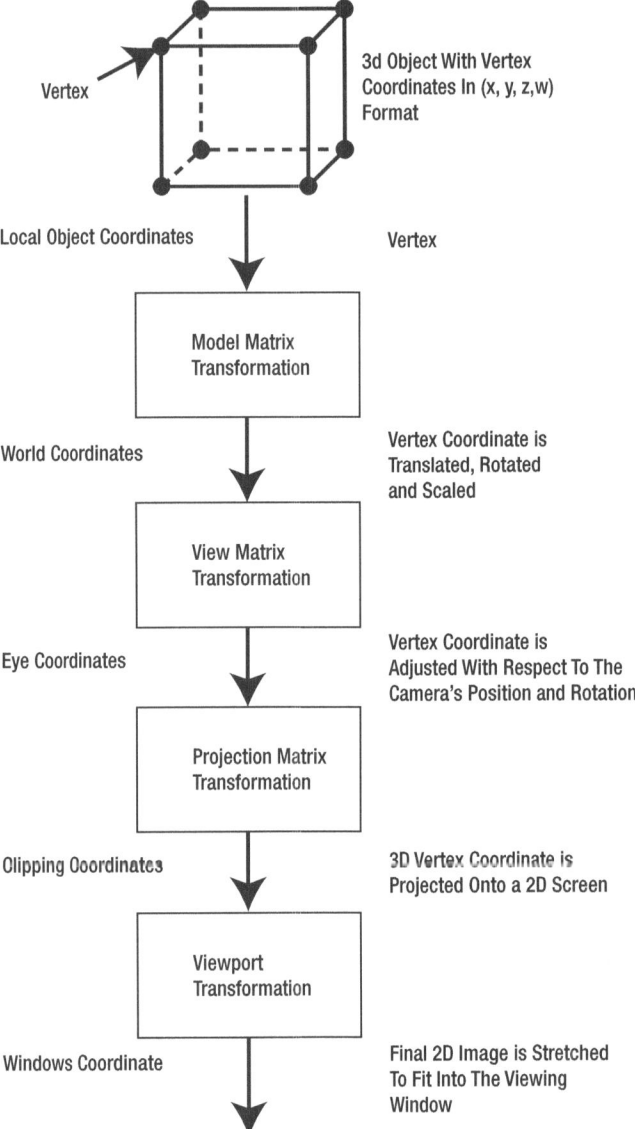

Figure 4-5. *Vertex transformation procedure*

Building the Model Matrix

The first thing that you need to build is the model matrix that puts the 3D object into the 3D world; that is, transform the model from local object coordinates that originally define the object into world coordinates. You do this by setting up translation, rotation, and scaling matrices and then multiplying them with the original vertex in object coordinates. This translates the object to where you want to put it in the world. It also rotates and scales the matrix if you so desire.

As mentioned previously in Chapter 3, the order of multiplication of matrices is key. Matrix multiplication is *not* commutative, and the order of multiplication does matter. For example, let's say we want to rotate the object first, then translate it along the x axis, as shown in Figure 4-6.

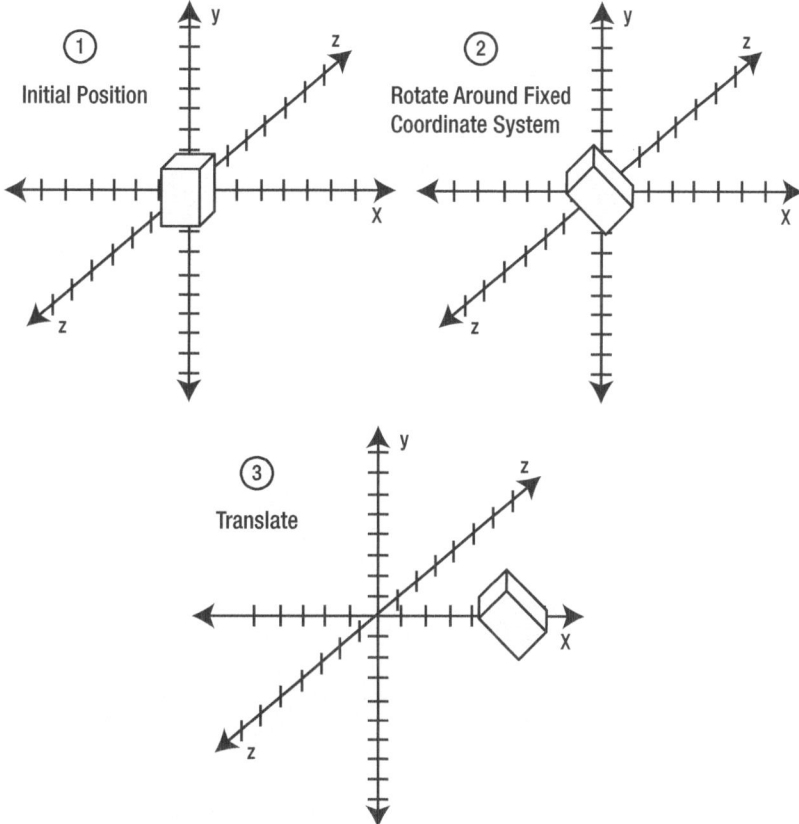

Figure 4-6. Rotate then translate object

The model matrix for this would be of the form as in Figure 4-7.

$$ModelMatrix = \begin{bmatrix} \text{Translation} \\ \text{Matrix} \end{bmatrix} \begin{bmatrix} \text{Rotation} \\ \text{Matrix} \end{bmatrix}$$

Figure 4-7. Model matrix form for rotating then translating an object

Note that the order of multiplication in which the rotation matrix is applied first to the object's vertices then the translation matrix is applied.

Take another example where you want to translate then rotate the object around an axis (see Figure 4-8).

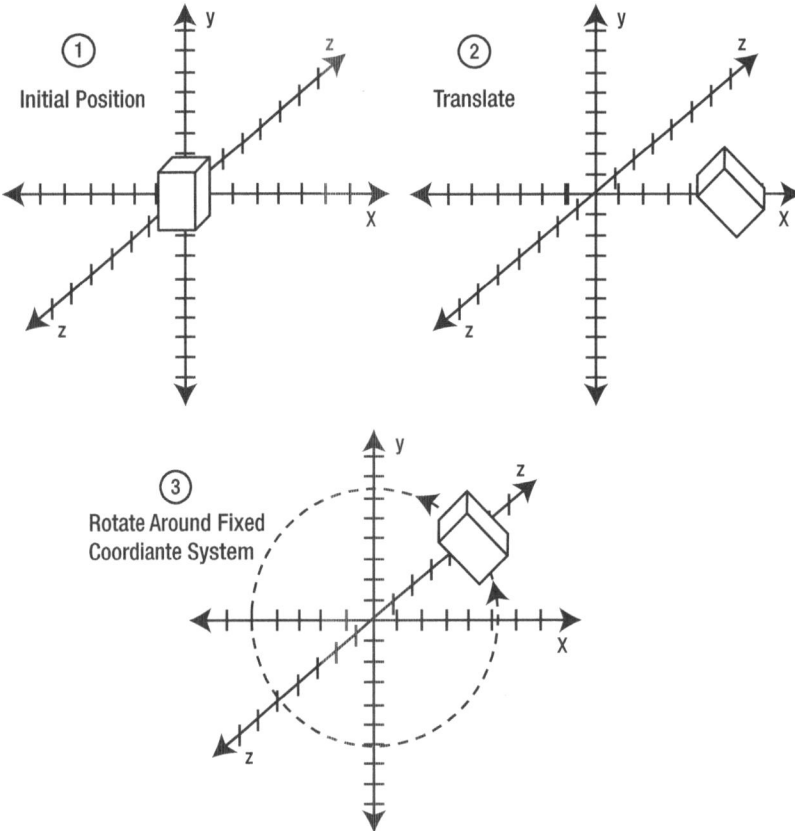

Figure 4-8. *Object is translated and then rotated around the rotation axis*

The general format for the model matrix for this situation is shown in Figure 4-9.

$$ModelMatrix = \begin{bmatrix} Rotation \\ Matrix \end{bmatrix} \begin{bmatrix} Translation \\ Matrix \end{bmatrix}$$

Figure 4-9. *Model matrix for object that is translated then rotated*

Building the View Matrix

By default, the camera view for OpenGL is located at the origin, with the lens pointed down the negative z axis. The camera can be moved and rotated anywhere in the 3D world. What this actually means is that the camera is stationary, but the object vertices are translated and rotated to simulate the movement of the camera. The matrix that does this is called the view matrix. The resulting coordinates are called the eye coordinates.

There is an easy way to do this in Android, and that is by using the `Matrix.setLookAtM()` function, which generates a view matrix based on the camera or eye position; center of the camera, where the view is focused; and the up direction of the camera. The Camera class represents our camera, and Figure 4-10 shows the up, center, and right vectors that specify the camera's orientation.

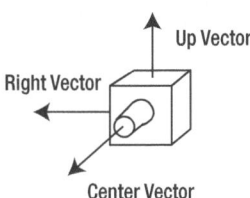

Figure 4-10. Camera or eye

Listing 4-1 shows in code how the view matrix is created in our Camera class in the `SetCameraView()` function.

Listing 4-1. Setting the View Matrix in the Camera Class

```
void SetCameraView(Vector3 Eye,
                   Vector3 Center,
                   Vector3 Up)
{
        // Create Matrix
        Matrix.setLookAtM(m_ViewMatrix,0,
                          Eye.x, Eye.y, Eye.z,
                          Center.x, Center.y, Center.z,
                          Up.x, Up.y, Up.z);
}
```

Building the Projection Matrix

The projection matrix transforms the 3D object vertices onto a 2D viewing surface. For 3D game programming, we need to use the frustrum to project a 3D image onto a 2D surface. As noted previously, a frustrum is a viewing area similar in shape to a pyramid with the top capstone cut off. It is defined by six clip planes, which are named top, bottom, right, left, near, and far, as indicated in Figure 4-11. By using a frustrum, objects that are closer to the viewer are larger than objects that are farther away. A frustrum also limits the viewing area to the area within the frustrum and excludes vertices outside of the frustrum. The vertices are transformed from world coordinates into clip coordinates.

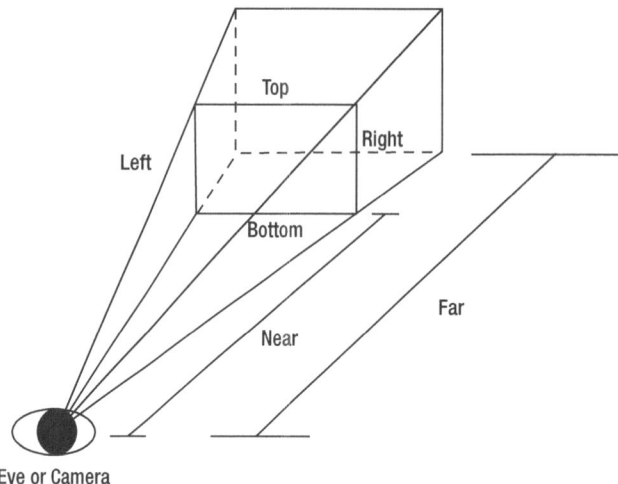

Figure 4-11. The projection frustrum

In terms of code, the function SetCameraProjection() in the Camera class creates the projection matrix and puts it in the m_ProjectionMatrix variable. See Listing 4-2.

Listing 4-2. Building the Projection Matrix in the Camera Class

```
void SetCameraProjection(float Projleft,
                         float Projright,
                         float Projbottom,
                         float Projtop,
                         float Projnear,
                         float Projfar)
{
        m_Projleft   = Projleft;
        m_Projright  = Projright;
        m_Projbottom = Projbottom;
        m_Projtop    = Projtop;
        m_Projnear   = Projnear;
        m_Projfar    = Projfar;
        Matrix.frustumM(m_ProjectionMatrix, 0,
                        m_Projleft, m_Projright,
                        m_Projbottom, m_Projtop,
                        m_Projnear, m_Projfar);
}
```

Setting the Viewport

The viewport transformation maps the 2D image generated from the projection matrix to the actual viewing window. It stretches the image horizontally or vertically to make it fit into the viewport. The glViewport() function sets the viewport specifications. It is located in the onSurfacedChanged()

function in the Renderer class. In our example from Chapter 3, that would be the MyGLRenderer class. On Android, the exact statement to set the final viewport specifications is as follows:

```
GLES20.glViewport(0, 0, width, height);
```

The first two parameters specify the lower left coordinates of the viewport. The next two parameters specify the width and the height of the viewport.

Sending the Matrices and Lighting Information to the Vertex and Fragment Shaders

Next, you have to actually send the matrices to the vertex and fragment shaders that OpenGL ES 2.0 uses to render 3D objects.

Vertex shaders place the object's vertices into the 3D world using the model, view, and projection matrix transforms that were mentioned in the preceding few sections. They also can determine the diffuse and specular lighting at a vertex and pass this information along to the fragment shader.

Fragment or pixel shaders actually determine the final color at the vertex. In the fragment shader, the texture, lighting, and material properties of the object all can provide input as to the final color. The color of the entire object can then be interpolated from these vertex colors.

In our code framework, both vertex and fragment shaders are represented in the Shader class.

Within the Shader class, the GLES20.glUniformXXXX() series of functions actually are responsible for sending the matrix, lighting, and material property data to the vertex and fragment shaders.

I discuss lighting and materials later in this chapter.

Rendering the Scene

Next, we have to actually render the scene. We render the scene in the onDrawFrame() function in our Renderer class MyGLRenderer. The code statement that actually does the rendering is

```
m_Cube.DrawObject(m_Camera, m_PointLight);
```

DrawObject() is located in the Object3d class and takes two parameters as input: the camera and the point light that is used to illuminate the scene.

Later in this chapter, I will go into more detail as to the OpenGL specific functions that draw an object.

Overview of the OpenGL ES 2.0 Shading Language

This section lists the basics of the shader language used for vertex and pixel shaders in OpenGL ES 2.0. It is not meant to be a full reference guide to every aspect of the language but a quick overview. The purpose of this section is to familiarize you with the fundamentals of the language. If you wish to learn more details about the shader language, see the web site at www.khronos.org/registry/gles/.

Basic Data Types

The basic data types for the OpenGL ES 2.0 shader language are:

void: No function return value

bool: Boolean

int: Signed integer

float: Floating scalar

vec2, vec3, vec4: 2, 3, and 4 component floating-point vector

bvec2, bvec3, bvec4: 2, 3, and 4 component Boolean vector

ivec2, ivec3, ivec4: 2, 3, and 4 component signed integer vector

mat2, mat3, mat4: 2-by-2, 3-by-3, and 4-by-4 float matrices

sampler2D: Used to represent and access a 2D texture

samplerCube: Used to represent and access a cube mapped texture

float floatarray[3]: One-dimensional arrays; can be of types such as floats, vectors, and integers

Vector Components

In the vertex and fragment shader language, you can refer to vector components in different ways. For example, you can address components in a vec4 type in the following ways:

- **{x, y, z, w}:** You can use this representation when accessing vectors that represent points or normals.

  ```
  vec3 position;
  position.x = 1.0f;
  ```

- **{r, g, b, a}:** You can use this representation when accessing vectors that represent colors.

  ```
  vec4 color;
  color.r = 1.0f;
  ```

- **{s, t, p, q}:** You can use this representation when accessing vectors that represent texture coordinates.

  ```
  vec2 texcoord;
  texcoord.s = 1.0f;
  ```

Operators and Expressions

Some of the important operators involved in statements and expressions in the vertex and fragment shader language are specified in the following list. The operators are basically similar to corresponding operators in the Java and C++ languages.

++	Increment operator
--	Decrement operator
+	Addition operator
-	Subtraction operator
!	Not operator
*	Multiply operator
/	Divide by operator
<	Less than relational operator
>	Greater than relational operator
<=	Less than or equal to relational operator
>=	Greater than or equal to relational operator
==	Conditional equals operator
!=	Not equal to conditional operator
&&	Logical and operator
^^	Logical exclusive or operator
||	Logical inclusive or operator
=	Assignment operator
+=	Addition and assignment operator
-=	Subtraction and assignment operator
*=	Multiplication and assignment operator
/=	Division and assignment operator

Program Flow Control Statements

The important program flow control statements for the OpenGL ES 2.0 shader language follow.

for loop: To use a for loop, initialize the counter value before starting the loop; execute the loop, if the expression evaluates to true; at the end of the loop, update the counter value; and repeat the loop, if the expression in the for loop evaluates to true.

```
for(Initial counter value;
    Expression to be evaluated;
    Counter increment/decrement value)
{
        // Statements to be executed.
}
```

while loop: Execute the statements in the while loop, as long as the expression to be evaluated is true.

```
while( Expression to evaluate )
{
        // Statement to be executed
}
```

if statement: If the expression evaluates to true, then the statements within the if block are executed.

```
if (Expression to evaluate )
{
        // Statements to execute
}
```

if else statement: Execute the statements in the if block; if expression evaluates to true else, execute the statements in the else block.

```
if (Expression to evaluate)
{
        // Statement to execute if expression is true
}
else
{
        // Statement to execute if expression is false
}
```

Storage Qualifiers

Storage qualifiers indicate how variables will be used in the shader program. From this information, the compiler can more efficiently process and store the shader variables.

Const: The const qualifier specifies a compile time constant or a read-only function parameter.

```
Const int NumberLights = 3;
```

Attribute: The attribute qualifier specifies a linkage between a vertex shader and the main OpenGL ES 2.0 program for per-vertex data. Some examples of the types of variables where the attribute qualifier can be used are vertex position, vertex textures, and vertex normal.

```
attribute vec3 aPosition;
attribute vec2 aTextureCoord;
attribute vec3 aNormal;
```

Uniform: The uniform qualifier specifies that the value does not change across the primitive being processed. Uniform variables form the linkage between a vertex or fragment shader and the main OpenGL ES 2.0 application. Some examples of variables where the uniform qualifier would be appropriate are lighting values, material values, and matrices.

```
uniform vec3 uLightAmbient;
uniform vec3 uLightDiffuse;
uniform vec3 uLightSpecular;
```

Varying: The varying qualifier specifies a variable that occurs both in the vertex shader and fragment shader. This creates the linkage between a vertex shader and fragment shader for interpolated data. These are usually values for the diffuse and the specular lighting that will be passed from the vertex shader to the fragment shader. Texture coordinates, if present, are also varying. The values of the diffuse and specular lighting, as well as textures, are interpolated or "varying" across the object the shader is rendering. Some examples of varying variables are the vertex texture coordinate, the diffuse color of the vertex, and the specular color of the vertex.

```
varying vec2 vTextureCoord;
varying float vDiffuse;
varying float vSpecular;
```

Reserved Variables

The main important reserved variable names within the OpenGL ES 2.0 shader language that we will be using in this book are specified in the following list.

vec4 gl_Position: Reserved variable within the vertex shader that holds the final transformed vertex to be displayed on the screen

vec4 gl_FragColor: Reserved variable within the fragment shader that holds the color of the vertex that has just been processed by the vertex shader

Built-in Functions

This list provides some of the important built-in functions in the shading language.

float radians(float degrees): Converts degrees to radians and returns radians

float degrees(float radians): Converts radians to degrees and returns degrees

float sin(float angle): Returns the sine of an angle measured in radians

float cos(float angle): Returns the cosine of an angle measured in radians

float tan(float angle): Returns the tangent of an angle measured in radians

float asin(float x): Returns the angle whose sine is x

float acos(float x): Returns the angle whose cosine is x

float atan(float y, float x): Returns the angle whose tangent is specified by the slope y/x

float atan(float slope): Returns the angle whose tangent is slope

float abs(float x): Returns the absolute value of x

float length(vec3 x): Returns the length of vector x

float distance(vec3 point0, vec3 point1): Returns the distance between point0 and point1

float dot(vec3 x, vec3 y): Returns the dot product between two vectors x and y

vec3 cross(vec3 x, vec3 y): Returns the cross product of two vectors x and y

vec3 normalize(vec3 x): Normalizes the vector to a length of 1 and then returns it

float pow(float x, float y): Calculates x to the power of y and return it

float min(float x, float y): Returns the minimum value between x and y

float max(float x, float y): Returns the maximum value between x and y

Overview of Vertex Shaders

In OpenGL ES 2.0, vertex shaders and fragment shaders are required to render 3D objects. The general purpose of the vertex shader is to position vertices in the 3D world and to determine vertex properties such as diffuse and specular lighting at that vertex. A simple vertex shader example is given in Listing 4-3.

Listing 4-3. Simple Vertex Shader

```
// Vertex Shader
uniform mat4 uMVPMatrix;
attribute vec3 aPosition;

uniform vec3 vColor;
varying vec3 Color;

void main()
{
    gl_Position = uMVPMatrix * vec4(aPosition,1);
    Color = vColor;
}
```

The variable uMVPMatrix holds the 4-by-4 ModelViewProjection matrix that will be used to transform the vertex. The ModelViewProjection matrix is simply the multiplication of the model matrix, the view matrix, and the projection matrix into one single matrix. The uniform qualifier is used, because this matrix does not change when the object is rendered. That is, that same matrix is used for all the vertices on the object being rendered.

The aPosition variable is a vector that holds the (x,y,z) position of the vertex of the object in its initial local object coordinates. The attribute qualifier indicates that this variable will receive input from the main OpenGL ES 2.0 application. This is how a vertex shader program is sent vertex data.

The vColor variable is a vector that holds the (r,g,b) input color values of the object to be rendered.

The vColor is copied into the Color variable and is sent to the fragment shader for processing. Variables that are linked to the fragment shader must be declared varying.

The main code itself is located within a main() block. The gl_Position variable is a reserved variable that transforms the vertex by multiplying its position by the uMVPMatrix (ModelViewProjection Matrix). This produces the projection of the 3D vertex on a 2D surface and puts the vertex in clip coordinates. Also, note how a vec3 is converted to a vec4 by using the vec4(aPosition,1) constructor.

A Complex Vertex Shader

The preceding vertex example shader is very simple. A more complex vertex shader is shown in Figure 4-12.

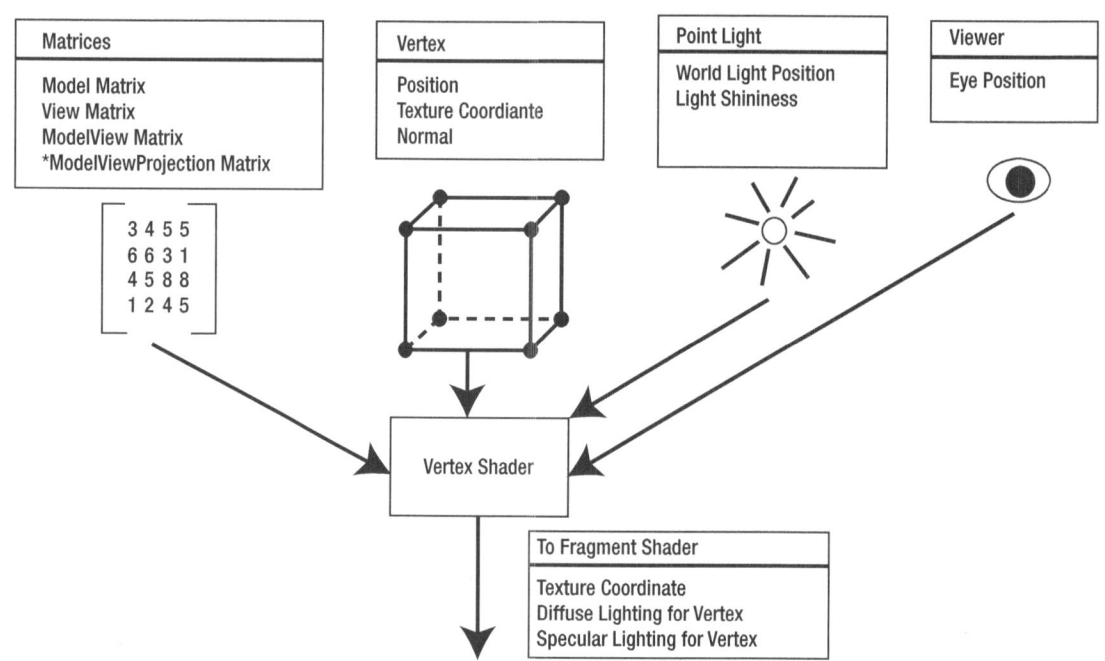

Figure 4-12. *Complex vertex shader*

The vertex data being sent to the vertex shader now includes the texture coordinates and the vertex normals, as well as the vertex position.

There is also new data in the form of lighting. The world position of the light and the light's shininess are input to the vertex shader. The location of the viewer or eye position is also input to the vertex shader.

The lighting and viewer information is used to determine the diffuse and specular colors at each vertex.

The output to the fragment shader includes the texture coordinate, the diffuse and specular lighting for the vertex. I will discuss lighting, including the vertex and fragment shaders that are needed to perform this lighting, later in this chapter.

Finally, in this book, we store the actual individual vertex and fragment shaders files in the res/raw directory.

Overview of Fragment or Pixel Shaders

Fragment shaders are used to determine the color of the pixels on the screen of a rendered 3D object. Listing 4-4 outputs the color passed in from the vertex shader in the previous section through the varying vector variable Color.

Listing 4-4. Simple Fragment Shader

```
// Fragment Shader
varying vec3 Color;

void main()
{
    gl_FragColor = vec4(Color,1);
}
```

The gl_FragColor variable is a reserved variable that outputs the actual color of the fragment.

An example of a more complex fragment shader is shown graphically in Figure 4-13. Here a texture, texture coordinate, diffuse lighting, specular lighting, a light source, and an object material all contribute to the final color of the object being rendered. I discuss these subjects in more detail later in this chapter.

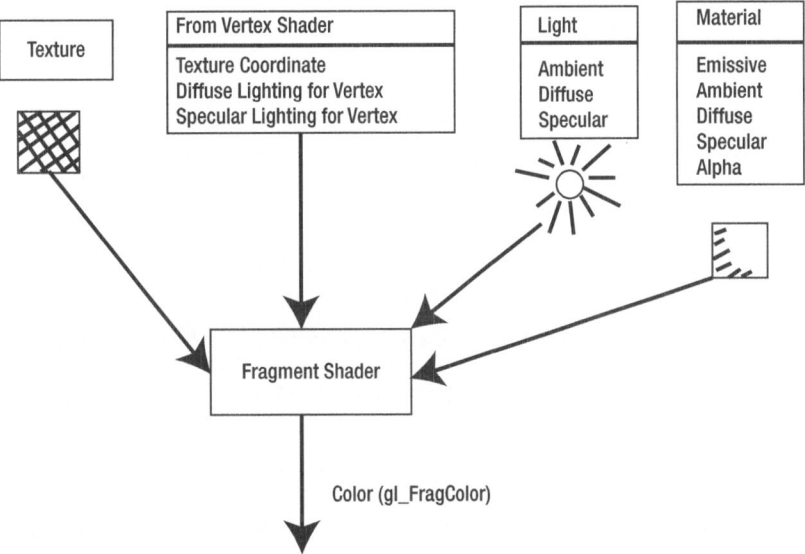

Figure 4-13. Complex fragment shader

Overview of the Shader Class

The Shader class holds the vertex and fragment shaders, as well as functions that are used to send data from the main application to the shader programs (see Figure 4-14).

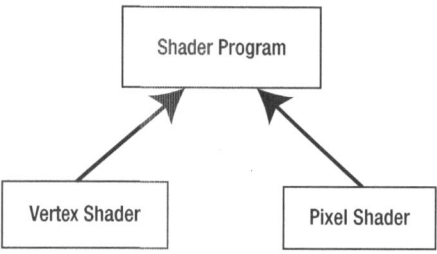

Figure 4-14. Shader program overview

The Shader class holds a handle to the fragment shader in m_FragmentShader. The handle to the vertex shader is in m_VertexShader. The handle to the main shader program that the vertex shader and fragment or pixel shader is attached to is m_ShaderProgram. Finally, the m_Context holds a reference to the activity that this shader object belongs to (see Listing 4-5). These variables are all private and cannot be accessed outside the class. Only class member functions may access these variables.

Listing 4-5. Shader Class Data

```
private Context m_Context;
private int     m_FragmentShader;
private int     m_VertexShader;
private int     m_ShaderProgram;
```

The Shader class's constructor initializes the vertex, fragment, and main shader program handles to 0. It also takes as input the vertex and fragment shader resource ids that represent the actual shader source text files. It then calls the InitShaderProgram() function with these shader resource ids (see Listing 4-6).

Listing 4-6. Shader Class Constructor

```
public Shader(Context context, int VSResourceId, int FSResourceId)
{
        // Shader Variables
        m_FragmentShader = 0;
        m_VertexShader   = 0;
        m_ShaderProgram  = 0;

        m_Context = context;
        InitShaderProgram(VSResourceId, FSResourceId);
}
```

The InitShaderProgram() function receives the vertex and fragment shader resource ids and creates the actual main shader program.

First, the main shader program is created through the GLES20.glCreateProgram() function.

Next, InitVertexShader() is called to create the vertex shader, and InitFragmentShader() is called to create the fragment shader.

After both shaders are created, they are linked using the GLES20.glLinkProgram(m_ShaderProgram) statement (see Listing 4-7).

Listing 4-7. InitShaderProgram Function

```
void InitShaderProgram(int VSResourceId, int FSResourceId)
{
        m_ShaderProgram = GLES20.glCreateProgram();

        InitVertexShader(VSResourceId);
        InitFragmentShader(FSResourceId);
        GLES20.glLinkProgram(m_ShaderProgram);

        String DebugInfo = GLES20.glGetProgramInfoLog(m_ShaderProgram);
        Log.d("DEBUG - SHADER LINK INFO ", DebugInfo);
}
```

Finally, the log file for the link operation is retrieved and displayed by calling glGetProgramInfoLog() to get the log information and then displaying it to the Android Log window through the Log() statement.

> **Note**   The GLES20 prefix indicates that the function is from the standard **implementation** of the OpenGL ES 2.0 specification for Android.

The InitVertexShader() function is called from the InitShaderProgram() function in Listing 4-7. The vertex shader source code is read into a temporary string buffer called tempBuffer.

An empty vertex shader is then created using the glCreateShader() function.

Next, the tempBuffer that holds the source code is associated with the vertex shader using the glShaderSource() function.

Then, the vertex shader itself is compiled using the glCompileShader() function.

Next, the compile error status is checked by calling the glGetShaderiv() function. If there is no error, the vertex shader is attached to the main shader program (see Listing 4-8).

Listing 4-8. InitVertexShader() Function

```
void InitVertexShader(int ResourceId)
{
        StringBuffer tempBuffer = ReadInShader(ResourceId);

      m_VertexShader= GLES20.glCreateShader(GLES20.GL_VERTEX_SHADER);
       GLES20.glShaderSource(m_VertexShader,tempBuffer.toString());
       GLES20.glCompileShader(m_VertexShader);
```

```
        IntBuffer CompileErrorStatus = IntBuffer.allocate(1);
        GLES20.glGetShaderiv(m_VertexShader,
                                GLES20.GL_COMPILE_STATUS,
                                CompileErrorStatus);
        if (CompileErrorStatus.get(0) == 0)
        {
                Log.e("ERROR - VERTEX SHADER ",
                        "Could not compile Vertex shader!! " +
                        String.valueOf(ResourceId));
                Log.e("ERROR - VERTEX SHADER ",
                        GLES20.glGetShaderInfoLog(m_VertexShader));
                GLES20.glDeleteShader(m_VertexShader);
                m_VertexShader = 0;
        }
        else
        {
            GLES20.glAttachShader(m_ShaderProgram,m_VertexShader);
            Log.d("DEBUG - VERTEX SHADER ATTACHED ", "In InitVertexShader()");
        }
}
```

The InitFragmentShader() function is called from the InitShaderProgram(), shown in Listing 4-7.

First the fragment shader source code is read in and stored in the tempBuffer string buffer.

Then, an empty fragment shader is created. The source code in the tempBuffer variable is then linked to the fragment shader that was just created. Next, the shader is compiled then attached to the main shader program, if there are no errors (see Listing 4-9).

Listing 4-9. InitFragmentShader() Function

```
void InitFragmentShader(int ResourceId)
{
        StringBuffer tempBuffer = ReadInShader(ResourceId);

        m_FragmentShader= GLES20.glCreateShader(GLES20.GL_FRAGMENT_SHADER);
        GLES20.glShaderSource(m_FragmentShader,tempBuffer.toString());
        GLES20.glCompileShader(m_FragmentShader);

        IntBuffer CompileErrorStatus = IntBuffer.allocate(1);
        GLES20.glGetShaderiv(m_FragmentShader,
                                GLES20.GL_COMPILE_STATUS,
                                CompileErrorStatus);
        if (CompileErrorStatus.get(0) == 0)
        {
                Log.e("ERROR - FRAGMENT SHADER ",
                        "Could not compile Fragment shader file =  " +
                        String.valueOf(ResourceId));
                Log.e("ERROR - FRAGMENT SHADER ",
                        GLES20.glGetShaderInfoLog(m_FragmentShader));
                GLES20.glDeleteShader(m_FragmentShader);
                m_FragmentShader = 0;
        }
```

```
            else
            {
                    GLES20.glAttachShader(m_ShaderProgram,m_FragmentShader);
                    Log.d("DEBUG - FRAGMENT SHADER ATTACHED ",
                            "In InitFragmentShader()");
            }
    }
```

The ReadInShader() function is called by the InitFragmentShader() and InitVertexShader() functions (see Listing 4-10).

Listing 4-10. ReadInShader() Function

```
StringBuffer ReadInShader(int ResourceId)
{
        StringBuffer TempBuffer = new StringBuffer();
        InputStream inputStream = m_Context.getResources().openRawResource(ResourceId);
        BufferedReader in = new BufferedReader(new InputStreamReader(inputStream));
        try
        {
                String read = in.readLine();
                while (read != null)
                {
                        TempBuffer.append(read + "\n");
                        read = in.readLine();
                }
        }
        catch (Exception e)
        {
                //Send a ERROR log message and log the exception.
                Log.e("ERROR - SHADER READ ERROR",
                        "Error in ReadInShader(): " +
                        e.getLocalizedMessage());
        }
        return TempBuffer;
}
```

First, a new string buffer object is created, called TempBuffer.

Next, a new InputStream object is created from the shader source file, using the file's resource id. This input stream is then used to create an InputStreamReader, which is then used to create a BufferedReader object called "in." The in object is then used to read in the shader source code line by line, and each line is added to the TempBuffer string buffer that was created first. If any error occurs, an error message is printed out in the Android LogCat window within the Eclipse IDE.

Finally, the shader source code is returned in a StringBuffer object.

The imports that bring BufferedReader, InputStreamReader, and InputStream into the current namespace are as follows:

```
import java.io.BufferedReader;
import java.io.InputStreamReader;
import java.io.InputStream;
```

> **Note**   These classes are standard in the Android development library, and you can learn more about them on
> the official Android web site at `http://developer.android.com`.

Once the shader has been successfully created, it is activated just before use by calling the
ActivateShader() function (see Listing 4-11). This is actually a wrapper function that just calls the
glUseProgram() function, which is a standard OpenGL 2.0 call. Again, note that the GLES20 prefix
denotes standard OpenGL 2.0 calls located in the standard Android library.

Listing 4-11. Activating the Shader

```
void ActivateShader()
{
        GLES20.glUseProgram(m_ShaderProgram);
}
```

You can also call DeActivateShader() to deactivate a shader from functioning. This function calls
the glUseProgram() function with an input of 0 to indicate that no shader program should be used in
rendering an object. This will set the rendering pipeline to the fixed rendering pipeline of OpenGL
ES 1.0. (see Listing 4-12).

Listing 4-12. Deactivating the Shader

```
void DeActivateShader()
{
    GLES20.glUseProgram(0);
}
```

The function GetShaderVertexAttributeVariableLocation() shown in Listing 4-13 is used to retrieve
the location of user-defined vertex shader variables, such as those used for the vertex position,
texture coordinates, and normals. These locations will be tied to the actual vertex data streams
through the glVertexAttribPointer() function. I will discuss this function that is used to draw 3D
meshes later. This function calls the standard OpenGL ES 2.0 function glGetAttribLocation().

Listing 4-13. Get Vertex Attribute Variable Location

```
int GetShaderVertexAttributeVariableLocation(String variable)
{
    return (GLES20.glGetAttribLocation(m_ShaderProgram, variable));
}
```

To set a uniform variable in a vertex or fragment shader, you can use the glUniformXXX series of
functions included in the standard GLES20 library. First, you would get the uniform variable location
index with the glGetUniformLocation() function, and then set it using the glUniformXXX() function
specific to the type of variable you want to set.

For example, in Listing 4-14 a uniform variable that has a value of type float is set using the glUniform1f() function.

Listing 4-14. Setting a Float Uniform Shader Variable

```
void SetShaderUniformVariableValue(String variable, float value)
{
        int loc = GLES20.glGetUniformLocation(m_ShaderProgram,variable);
        GLES20.glUniform1f(loc, value);
}
```

The SetShaderUniformVariableValue() function in Listing 4-15 that follows sets a uniform shader vec3 variable, taking as input a Vector3 object and using a glUniform3f() function to set the actual shader variable. As before, the glGetUniformLocation() function gets the index of the desired variable from the shader program.

Listing 4-15. Setting a Vector3 Uniform Shader Variable Using a Vector3 Object

```
void SetShaderUniformVariableValue(String variable, Vector3 value)
{
    int loc = GLES20.glGetUniformLocation(m_ShaderProgram,variable);
    GLES20.glUniform3f(loc, value.x, value.y, value.z);
}
```

The following SetShaderUniformVariableValue() function in Listing 4-16 sets a vec3 shader variable, taking as input a float array. The function sets the vec3 shader variable using the first three values of the float array.

Listing 4-16. Setting a vec3 Uniform Shader Variable Using a Float Array

```
void SetShaderUniformVariableValue(String variable, float[] value)
{
        int loc = GLES20.glGetUniformLocation(m_ShaderProgram,variable);
        GLES20.glUniform3f(loc, value[0], value[1], value[2]);
}
```

Function SetShaderVariableValueFloatMatrix4Array() in Listing 4-17 sets a uniform 4-by-4 matrix shader variable or array. The following code sets a mat4 shader variable called uModelViewMatrix to the data from the float array ModelViewMatrix. The count variable is set to 1, because there is only one 4-by-4 matrix. Transpose is set to false, to indicate that the default OpenGL matrix format will be used. Offset is the offset into ModelViewMatrix, where the matrix data starts.

```
m_Shader.SetShaderVariableValueFloatMatrix4Array("uModelViewMatrix", 1, false, ModelViewMatrix, 0);
```

Listing 4-17. Setting a Uniform Mat4 Shader Variable

```
void SetShaderVariableValueFloatMatrix4Array(String variable,
                                             int count,
                                             boolean transpose,
                                   float[] value,
                                             int offset)
```

```
{
        int loc = GLES20.glGetUniformLocation(m_ShaderProgram,variable);
        GLES20.glUniformMatrix4fv (loc, count, transpose, value, offset);
}
```

The Camera

The camera that provides our view into the 3D world is represented by the Camera class and was previously illustrated in Figure 4-10.

The camera's position and raw orientation are held in the variable m_Orientation, which is an object of the Orientation class. The final camera position is held in m_Eye. The final look at the point where the camera is pointing at is held in m_Center. The final up vector that denotes the top of the camera is held in m_Up (see Listing 4-18).

Listing 4-18. Camera Orientation

```
// Camera Location and Orientation
private Vector3 m_Eye = new Vector3(0,0,0);
private Vector3 m_Center= new Vector3(0,0,0);
private Vector3 m_Up = new Vector3(0,0,0);
private Orientation m_Orientation = null;
```

The viewing frustrum data for the camera is defined by six clipping planes, indicated in Listing 4-19.

Listing 4-19. Viewing Frustrum Variables

```
// Viewing Frustrum
private float m_Projleft    = 0;
private float m_Projright   = 0;
private float m_Projbottom = 0;
private float m_Projtop     = 0;
private float m_Projnear    = 0;
private float m_Projfar     = 0;
```

The key variables for this Camera class are the matrices that hold the camera's projection matrix and the camera's view matrix. These are 4-by-4 matrices of type float, allocated as a one-dimensional array of 16 elements (4 times 4 elements in the matrix;. see Listing 4-20). Remember: These are two of the key matrices that you need to transform an object's vertex, so that it can be displayed on the screen.

Listing 4-20. The Camera's Matrices

```
private float[] m_ProjectionMatrix = new float[16];
private float[] m_ViewMatrix       = new float[16];
```

The Camera class's constructor is shown in Listing 4-21. The constructor initializes the camera with the input parameters and performs the following functions:

■ Creates a new Orientation class object called m_Orientation.

■ Sets the camera projection frustrum using the user specified input parameters Projleft, Projright, Projbottom, Projtop, Projnear, and Projfar.

■ Sets the camera's local coordinate axis and position. Sets the forward local axis (z axis), up local axis (y axis), and the right local axis (x axis). The right local axis is calculated from the cross product of the Center and Up vectors.

Listing 4-21. The Camera Class's Constructor

```
Camera(Context context,
        Vector3 Eye,  Vector3 Center, Vector3 Up,
        float Projleft, float Projright,
        float Projbottom, float Projtop,
        float Projnear, float Projfar)
{
        m_Orientation = new Orientation(context);

        // Set Camera Projection
        SetCameraProjection(Projleft, Projright, Projbottom, Projtop, Projnear, Projfar);

        // Set Orientation
        m_Orientation.GetForward().Set(Center.x, Center.y, Center.z);
        m_Orientation.GetUp().Set(Up.x, Up.y, Up.z);
        m_Orientation.GetPosition().Set(Eye.x, Eye.y, Eye.z);

        // Calculate Right Local Vector
        Vector3 CameraRight = Vector3.CrossProduct(Center, Up);
        CameraRight.Normalize();
        m_Orientation.SetRight(CameraRight);
}
```

The SetCameraProjection() function actually sets the projection matrix, using the Matrix. frustumM() function to generate the matrix, and puts it in m_ProjectionMatrix (see Listing 4-22).

Listing 4-22. Setting the Camera Frustrum

```
void SetCameraProjection(float Projleft,
                          float Projright,
                          float Projbottom,
                          float Projtop,
                          float Projnear,
                          float Projfar)
{
        m_Projleft   = Projleft;
        m_Projright  = Projright;
        m_Projbottom = Projbottom;
        m_Projtop    = Projtop;
        m_Projnear   = Projnear;
```

```
        m_Projfar        =         Projfar;
        Matrix.frustumM(m_ProjectionMatrix, 0,
                        m_Projleft, m_Projright,
                        m_Projbottom, m_Projtop,
                        m_Projnear, m_Projfar);
}
```

The function to actually create and set the camera's view matrix is the SetCameraView() function (see Listing 4-23). The function calls the setLookAtM() function, which is part of the standard Android Matrix class, to create the actual matrix that is put into m_ViewMatrix. The function takes as parameters the location of the camera or eye, the center or focus point of the camera, and the vector that points in the up direction with respect to the camera.

Listing 4-23. Setting the Camera View Matrix

```
void SetCameraView(Vector3 Eye,
                   Vector3 Center,
                   Vector3 Up)
{
        // Create Matrix
        Matrix.setLookAtM(m_ViewMatrix,0,
                          Eye.x, Eye.y, Eye.z,
                          Center.x, Center.y, Center.z,
                          Up.x, Up.y, Up.z);
}
```

In order to produce accurate camera views, you must have accurate camera Center, or LookAt, and Up, and Eye vectors in world coordinates, not just local coordinates. In the function CalculateLookAtVector()(see Listing 4-24), you find the Center, or LookAt, vector by

1. Finding the Forward camera vector that represents the direction the camera lens is pointing in world coordinates, which is how the Forward vector is pointing with respect to the world coordinate system. The Forward vector will be returned normalized with length 1.

> **Note** In order to convert an object's local axis orientation from local object coordinates to world coordinates, you must multiply the local axis by the rotation matrix, using the Matrix.multiplyMV() function. This is done in the GetForwardWorldCoords() function in the Orientation class.

2. Lengthening the Forward vector by how far you wish to look into the scene. In this case, we chose 5.

3. Adding your camera's current position to the lengthened Forward vector you calculated in the previous step to determine the final Center, or LookAt, vector. This result is stored in m_Center.

Listing 4-24. Calculate Camera LookAt Vector

```
void CalculateLookAtVector()
{
        m_Center.Set(m_Orientation.GetForwardWorldCoords().x,
                        m_Orientation.GetForwardWorldCoords().y,
                        m_Orientation.GetForwardWorldCoords().z);
        m_Center.Multiply(5);
        m_Center = Vector3.Add(m_Orientation.GetPosition(), m_Center);
}
```

The CalculateUpVector() function sets the Camera's Up vector in world coordinates (see Listing 4-25).

Listing 4-25. Calculating the Camera Up Vector

```
void CalculateUpVector()
{
    m_Up.Set(m_Orientation.GetUpWorldCoords().x,
                m_Orientation.GetUpWorldCoords().y,
                m_Orientation.GetUpWorldCoords().z);
}
```

The way you convert the camera's Up vector local coordinates into world coordinates is the same as with the Forward vector in Listing 4-24 that we covered in the previous example. To see this visually, let's take a look at Figure 4-15 and Figure 4-16.

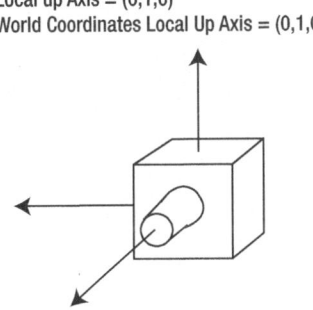

Figure 4-15. Camera local y axis before rotation

Local up Axis = (0,1,0)

World Coordinates Local Up Axis =
Local Up Axis *Rotation Matrix

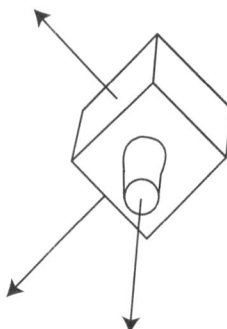

Figure 4-16. *Camera local y axis after rotation*

Figure 4-15 shows the camera before rotation, where the local up axis and the world coordinates of the local up axis are the same, which is pointing 1 unit along the positive y axis.

Figure 4-16 shows the camera after rotation. The local axis still points 1 unit along the y axis; however, the world coordinates of the local axis has now changed. You can clearly see from the figure that the Up camera vector is no longer pointing straight up along the positive y axis. The new world coordinates for the local axis is found by multiplying the local up axis vector (0,1,0) by the rotation matrix for the object which in this case is the camera.

Next, the CalculatePosition() function sets the current camera position held in the m_Orientation variables into the m_Eye variable (see Listing 4-26).

Listing 4-26. Calculate Position() function

```
void CalculatePosition()
{
        m_Eye.Set(m_Orientation.GetPosition().x,
                m_Orientation.GetPosition().y,
                m_Orientation.GetPosition().z);
}
```

The UpdateCamera() function is called continuously to update the player's viewpoint. It calculates the Center (LookAt) vector, the Up vector, and the Eye (Position) vector and then sets the camera view by setting the m_ViewMatrix (see Listing 4-27).

Listing 4-27. UpdateCamera() Function

```
void UpdateCamera()
{
        CalculateLookAtVector();
        CalculateUpVector();
        CalculatePosition();

        SetCameraView(m_Eye, m_Center, m_Up);
}
```

The Camera class also provides functions to retrieve the frustrum width, height, and depth (see Listing 4-28).

Listing 4-28. Camera Frustrum Parameters

```
// Camera Dimensions
float GetCameraViewportWidth() {return (Math.abs(m_Projleft-m_Projright));}
float GetCameraViewportHeight(){return (Math.abs(m_Projtop-m_Projbottom));}
float GetCameraViewportDepth(){return (Math.abs(m_Projfar-m_Projnear));}
```

There are also many functions to access private variables within the Camera class, such as variables related to orientation, camera vectors, camera frustrum information, and the camera's projection and view matrices (see Listing 4-29).

Listing 4-29. Functions to Access Private Variables

```
// Get Orientation
Orientation GetOrientation() {return m_Orientation;}

// Camera Vectors
Vector3 GetCameraEye() {return m_Eye;}
Vector3 GetCameraLookAtCenter() {return m_Center;}
Vector3 GetCameraUp(){return m_Up;}

// Camera Frustrum
float GetProjLeft(){return m_Projleft;}
float GetProjRight(){return m_Projright;}
float GetProjBottom() {return m_Projbottom;}
float GetProjTop(){return m_Projtop;}
float GetProjNear(){return m_Projnear;}
float GetProjFar(){return m_Projfar;}
 // Camera Matrices
 float[] GetProjectionMatrix(){return m_ProjectionMatrix;}
 float[] GetViewMatrix(){return m_ViewMatrix;}
```

The 3D Object Mesh

In this section, I discuss in detail the individual components of a 3D object, such as an object's vertices, and how exactly these vertices are rendered using OpenGL ES for Android.

Mesh Vertex Data

A 3D object in OpenGL is composed of vertices. Each vertex can have various attributes, such as vertex position, vertex texture coordinates, and a vertex normal (see Figure 4-17).

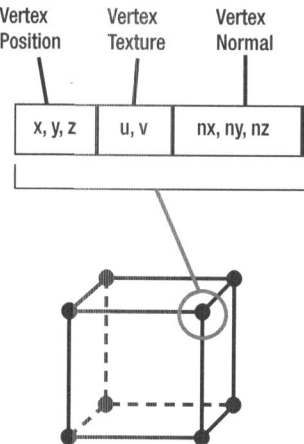

Figure 4-17. Vertex data format

The cube 3D model with the Android Texture on it from our program in Chapter 3 has the following mesh data in Listing 4-30:

Listing 4-30. Cube Mesh Data

```
static float CubeData[] =
{
        // x,      y,     z,   u,      v    nx,  ny, nz
        -0.5f,  0.5f, 0.5f, 0.0f,   0.0f,   -1,  1, 1,  // front top left
        -0.5f, -0.5f, 0.5f, 0.0f,   1.0f,   -1, -1, 1,  // front bottom left
        0.5f, -0.5f, 0.5f, 1.0f,   1.0f,    1, -1, 1,   // front bottom right
        0.5f,  0.5f, 0.5f, 1.0f,   0.0f,    1,  1, 1,   // front top right

        -0.5f,  0.5f, -0.5f, 0.0f,   0.0f,  -1,  1, -1, // back top left
        -0.5f, -0.5f, -0.5f, 0.0f,   1.0f,  -1, -1, -1, // back bottom left
         0.5f, -0.5f, -0.5f, 1.0f,   1.0f,   1, -1, -1, // back bottom right
         0.5f,  0.5f, -0.5f, 1.0f,   0.0f,   1,  1, -1  // back top right
};
```

In terms of vertex data, it has three coordinates for position, two coordinates for vertex texture coordinates, as well as three coordinates for vertex normals. Position coordinates are the location of the object's vertices in local model or object space coordinates. Texture coordinates allow you to place an image on a 3D object and range from 0 to 1. Normal coordinates allow you to use diffuse lighting effects to render the object and simulate the effects of a light and shadows.

The MeshEx Class

The MeshEx class holds the graphics data and related functions for a 3D object. Here, I give you an overview of this class, as well as details on key class functions.

MeshEx Class Overview

The MeshEx class uses the glDrawElements() function to actually render the mesh. This function uses the index method of rendering vertices, where a list of vertices is stored in the m_VertexBuffer FloatBuffer variable along with an index list into these vertices of the triangles that need to be drawn that are stored in the m_DrawListBuffer ShortBuffer variable.

The m_VertexBuffer holds a list of vertex data. Each vertex can have values for the position coordinates, texture coordinates, and the vertex normal coordinates. Also, there are offsets into the vertex data that indicate where the data actually starts for position, texture coordinates, and vertex normals (see Figure 4-18). The figure also shows the vertex stride, which is the length in bytes of the data of a single vertex.

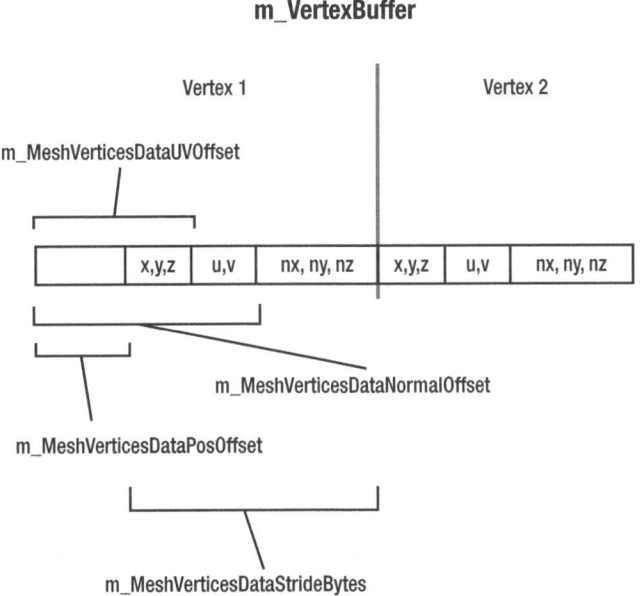

Figure 4-18. *The vertex buffer*

The m_DrawListBuffer holds an array of numbers that map into a vertex in the m_VertexBuffer array (see Figure 4-19).

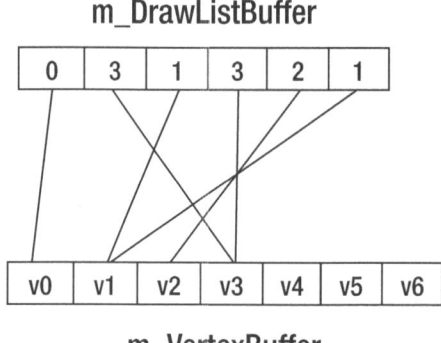

Figure 4-19. *The* m_DrawListBuffer

An example of a vertex index list is provided in Listing 4-31.

Listing 4-31. Vertex Draw Order Index List

```
static final short CubeDrawOrder[] =
{
        0, 3, 1, 3, 2, 1,          // Front panel
        4, 7, 5, 7, 6, 5,          // Back panel
        4, 0, 5, 0, 1, 5,          // Side
        7, 3, 6, 3, 2, 6,          // Side
        4, 7, 0, 7, 3, 0,          // Top
        5, 6, 1, 6, 2, 1           // Bottom
}; // order to draw vertices
```

MeshEx Class Constructor

The MeshEx constructor is shown in Listing 4-32. An example of this constructor in use is from the hands-on example from Chapter 3. The following code creates a new MeshEx object with

- Eight (8) coordinates per vertex: 3 position coords, 2 texture coords, and 3 vertex normal coords

- Zero (0) offset to the position coordinates

- Three (3) coordinate offset to the texture uv coordinates

- Five (5) coordinate offset to the vertex normal

- Cube.CubeData for the vertex data

- Cube.CubeDrawOrder for the index list of vertices to use for drawing the mesh triangles

    ```
    MeshEx CubeMesh = new MeshEx(8,0,3,5,Cube.CubeData, Cube.CubeDrawOrder);
    ```

The constructor takes the input vertices in the form of a float array and creates a FloatBuffer. The FloatBuffer is originally created as a ByteBuffer. A ShortBuffer is created from the input short array vertex index list.

If the UV texture offset is negative, there are no texture coordinates. If it is 0 or greater, there are texture coordinates. If the Normal offset is negative, there are no vertex normals. If it is greater or equal to 0, there are vertex normals. Thus, you can have a vertex with or without texture or lighting information.

> **Note**   FloatBuffer, ShortBuffer, and ByteBuffer are standard Android classes. You can find out more about them on the official Android developer's web site.

Listing 4-32. The MeshEx Constructor

```
public MeshEx(int CoordsPerVertex,
              int MeshVerticesDataPosOffset,
              int MeshVerticesUVOffset,
              int MeshVerticesNormalOffset,
              float[] Vertices,
              short[] DrawOrder)
{
      m_CoordsPerVertex            = CoordsPerVertex;
      m_MeshVerticesDataStrideBytes = m_CoordsPerVertex * FLOAT_SIZE_BYTES;
      m_MeshVerticesDataPosOffset   = MeshVerticesDataPosOffset;
      m_MeshVerticesDataUVOffset    = MeshVerticesUVOffset ;
      m_MeshVerticesDataNormalOffset = MeshVerticesNormalOffset;

      if (m_MeshVerticesDataUVOffset >= 0)
      {
            m_MeshHasUV = true;
      }

      if (m_MeshVerticesDataNormalOffset >=0)
      {
            m_MeshHasNormals = true;
      }

      // Allocate Vertex Buffer
      ByteBuffer bb = ByteBuffer.allocateDirect(
                  // (Number of coordinate values * 4 bytes per float)
                  Vertices.length * FLOAT_SIZE_BYTES);
      bb.order(ByteOrder.nativeOrder());
      m_VertexBuffer = bb.asFloatBuffer();
```

```
        if (Vertices != null)
        {
                m_VertexBuffer.put(Vertices);
                m_VertexBuffer.position(0);
                m_VertexCount = Vertices.length / m_CoordsPerVertex;
        }
        // Initialize DrawList Buffer
        m_DrawListBuffer = ShortBuffer.wrap(DrawOrder);
}
```

MeshEx Class Error Debug Function

The function CheckGLError() checks for errors that occur after an OpenGL operation by calling the GLES20.glGetError() function and by throwing an exception that stops the program and displays an error (see Listing 4-33).

Listing 4-33. Class Error Debug Function

```
public static void CheckGLError(String glOperation)
{
        int error;
        while ((error = GLES20.glGetError()) != GLES20.GL_NO_ERROR)
        {
                Log.e("ERROR IN MESHEX", glOperation + " IN CHECKGLERROR() : glError - " + error);
                throw new RuntimeException(glOperation + ": glError " + error);
        }
}
```

MeshEx Class Mesh Draw Function

The MeshEx class's function to actually draw the 3D object is called DrawMesh(). The function to perform the required set up is done in SetUpMeshArrays(). What actually happens is that triangles are drawn based on the vertex indices in the m_DrawListBuffer variable.

For example, the m_VertexBuffer contains eight vertices for the cube in Figure 4-20. The m_DrawListBuffer contains the values 0, 3, 1, 3, 2, 1. Then the two triangles drawn would contain v0, v3, v1 for the first triangle and v3, v2, v1 for the second. Together, these would form the front solid face of the cube.

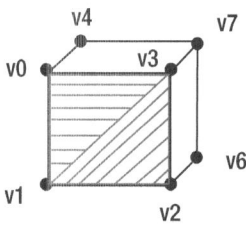

Figure 4-20. Drawing triangles

> **Note**   Quad primitives are not supported in OpenGL ES as they are in regular OpenGL. Thus, you will need two triangles to replace the 1 quad that would have covered all four vertices.

In the `SetUpMeshArrays()` function there are basically three preparation steps for setting up the vertex position data, vertex texture coordinates (if any), and the vertex normal data (if any). They are as follows:

1. Setting the starting position in the `m_VertexBuffer` for the vertex property being activated using the `position()` function.

2. Linking the vertex attribute data to a variable in the vertex shader that handles that vertex attribute by using the `glVertexAttribPointer()` function. The parameters to this function are

    a. the attribute handle, which is the link to the shader variable;

    b. the size of the attribute in number of coordinates;

    c. the type of the coordinate;

    d. whether the data is normalized, which is false;

    e. the stride in bytes, that is, the length of a single vertex;

    f. the vertex buffer that holds the vertex data.

3. Enabling the sending of that vertex attribute to the shader by calling `glEnableVertexAttribArray()` with the handle to that shader variable (see Listing 4-34).

Listing 4-34. Setting Up the Mesh for Drawing

```
void SetUpMeshArrays(int PosHandle, int TexHandle, int NormalHandle)
{
        // Set up stream to position variable in shader
        m_VertexBuffer.position(m_MeshVerticesDataPosOffset);
        GLES20.glVertexAttribPointer(PosHandle, 3, GLES20.GL_FLOAT, false,
m_MeshVerticesDataStrideBytes, m_VertexBuffer);
        GLES20.glEnableVertexAttribArray(PosHandle);

        if (m_MeshHasUV)
        {
                // Set up Vertex Texture Data stream to shader
                m_VertexBuffer.position(m_MeshVerticesDataUVOffset);
                GLES20.glVertexAttribPointer(TexHandle, 2, GLES20.GL_FLOAT, false,
m_MeshVerticesDataStrideBytes, m_VertexBuffer);
                GLES20.glEnableVertexAttribArray(TexHandle);
        }
```

```
            if (m_MeshHasNormals)
            {
                    // Set up Vertex Texture Data stream to shader
                    m_VertexBuffer.position(m_MeshVerticesDataNormalOffset);
                    GLES20.glVertexAttribPointer(NormalHandle, 3, GLES20.GL_FLOAT, false,
m_MeshVerticesDataStrideBytes, m_VertexBuffer);
                    GLES20.glEnableVertexAttribArray(NormalHandle);
            }
}
```

The DrawMesh() function in Listing 4-35 first calls the SetUpMeshArrays() function to prepare the vertex attributes for sending to the shader.

Listing 4-35. Main Drawing Function DrawMesh

```
void DrawMesh(int PosHandle, int TexHandle, int NormalHandle)
{
        SetUpMeshArrays(PosHandle, TexHandle, NormalHandle);
        GLES20.glDrawElements(GLES20.GL_TRIANGLES, m_DrawListBuffer.capacity(),
GLES20.GL_UNSIGNED_SHORT, m_DrawListBuffer);

        // Disable vertex array
        GLES20.glDisableVertexAttribArray(PosHandle);
        CheckGLError("glDisableVertexAttribArray ERROR - PosHandle");

        if (m_MeshHasUV)
        {
                GLES20.glDisableVertexAttribArray(TexHandle);
                CheckGLError("glDisableVertexAttribArray ERROR - TexHandle");
        }
        if (m_MeshHasNormals)
        {
                GLES20.glDisableVertexAttribArray(NormalHandle);
                CheckGLError("glDisableVertexAttribArray ERROR - NormalHandle");
        }
}
```

Next, the glDrawElements() function is called with the following parameters:

1. Primitive drawing type, which is GL_TRIANGLES

2. The number of vertex indices to be processed

3. The type of the vertex index, which is GL_UNSIGNED_SHORT

4. The buffer that contains the vertex indices to process, which is
    m_DrawListBuffer

Finally, each vertex attribute that was activated is now disabled with the glDisableVertexAttribArray() function. The CheckGLError() function is also called to check to see if any OpenGL errors have occurred.

DrawMesh() is called from the DrawObject() function in the Object3d class. The GetVertexAttribInfo() function gets the vertex attribute handles that are needed for the DrawMesh() call.

The DrawObject() function in the Object3d class will be the main entry point where 3D objects are rendered throughout this book.

Lighting

In this section, I cover OpenGL ES 2.0 lighting in depth. I start off with an overview of lighting. Then I discuss our custom PointLight class and how lighting is performed in the vertex and fragment shaders.

Overview of Lighting

You can model lighting by dividing lighting into three separate types:

> **Ambient Lighting:** This type of lighting models the lighting on an object that is constant at all object vertices and does not depend on the position of the light.

> **Diffuse Lighting:** This type of lighting models the lighting on an object that depends on the angle that the object's vertex normals make with the light source.

> **Specular Lighting:** This type of lighting models the lighting on an object that depends on the angle that the object's vertex normals make with the light source and also the position of the viewer or camera.

The key components in determining lighting are shown in Figure 4-21, which are

> **Light Vector:** The vector that represents the direction from the object's vertex to the light source.

> **Normal Vector:** The vector that is assigned to an object's vertex that is used to determine that vertex's lighting from a diffuse light source. Normal vectors are also vectors perpendicular to surfaces.

> **Eye Vector:** The vector that represents the direction from the object's vertex to the location of the viewer or camera. This is used in specular lighting calculations.

> **Point Light:** A light source that radiates light in all directions and has a position in 3D space. A point light contains ambient, diffuse, and specular lighting components.

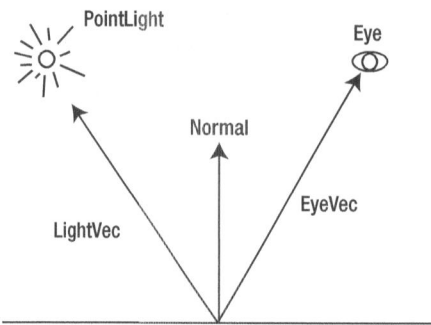

Figure 4-21. Important lighting vectors

In OpenGL ES 2.0, the lighting calculations are carried out in the programmable vertex and fragment shaders. I discuss these shaders later in this section; however, first let's go over the class that represents our point light.

The PointLight Class

The PointLight class will provide the lighting for our 3D scenes. The data contained in this class is shown in Listing 4-36.

Listing 4-36. Point Light Data

```
private float[] m_light_ambient = new float[3];
private float[] m_light_diffuse = new float[3];
private float[] m_light_specular = new float[3];
private float m_specular_shininess = 5;

private Vector3 m_Position;
```

The array variable m_light_ambient holds the color value of the ambient light emitted from this light.

The array variable m_light_diffuse holds the color value of the diffuse light emitted from this light.

The array variable m_light_specular holds the color value of the specular light emitted from this light.

For all of the preceding variable arrays, the first array element is the red value, the second element is the green value, and the third element is the blue value. These values range from 0 to 1. A value of all 1's for the r, g, b color would represent white, and a value of all 0's would represent black.

The m_specular_shininess variable sets the level of specular shininess caused by m_light_specular.

The m_Position variable holds the position of the light in 3D space.

The PointLight constructor is shown in Listing 4-37. By default, the PointLight object is initialized by the constructor to emit a light that has a maximum intensity of white light (red, green, blue values are all 1.0) for ambient, diffuse, and specular light. The default position of the light is at the origin.

Listing 4-37. PointLight Constructor

```
public PointLight(Context context)
{
        m_light_ambient[0] = 1.0f;
        m_light_ambient[1] = 1.0f;
        m_light_ambient[2] = 1.0f;

        m_light_diffuse[0] = 1.0f;
        m_light_diffuse[1] = 1.0f;
        m_light_diffuse[2] = 1.0f;

        m_light_specular[0] = 1.0f;
        m_light_specular[1] = 1.0f;
        m_light_specular[2] = 1.0f;

        m_Position = new Vector3(0,0,0);
}
```

The SetAmbientColor() function sets the ambient color of the light taking a float array as input.

The SetDiffuseColor() function sets the diffuse color of the light taking a float array as input.

The SetSpecularColor() function sets the specular color of the light taking a float array as input (see Listing 4-38).

Listing 4-38. Set Light Functions

```
void SetAmbientColor(float[] ambient)
{
        m_light_ambient[0] = ambient[0];
        m_light_ambient[1] = ambient[1];
        m_light_ambient[2] = ambient[2];
}

void SetDiffuseColor(float[] diffuse)
{
        m_light_diffuse[0] = diffuse[0];
        m_light_diffuse[1] = diffuse[1];
        m_light_diffuse[2] = diffuse[2];
}

void SetSpecularColor(float[] spec)
{
        m_light_specular[0] = spec[0];
        m_light_specular[1] = spec[1];
        m_light_specular[2] = spec[2];
}
```

The input float array holds the red, green, and blue values in the array position 0, 1, and 2, respectively.

There are two SetPosition() functions that set the light's position in the 3D world. One function takes x, y, and z float values for the light position, and the other function takes a Vector3 object (see Listing 4-39).

Listing 4-39. SetPosition Functions

```
void SetPosition(float x, float y, float z)
{
        m_Position.x = x;
        m_Position.y = y;
        m_Position.z = z;
}

void SetPosition(Vector3 Pos)
{
        m_Position.x = Pos.x;
        m_Position.y = Pos.y;
        m_Position.z = Pos.z;
}
```

Finally, the PointLight class has functions that allow you to retrieve private data contained within the class (see Listing 4-40).

Listing 4-40. Accessor Functions

```
Vector3 GetPosition(){return m_Position;}
float[] GetAmbientColor(){return m_light_ambient;}
float[] GetDiffuseColor(){return m_light_diffuse;}
float[] GetSpecularColor(){return m_light_specular;}
float GetSpecularShininess(){return m_specular_shininess;}
```

Building the Normal Matrix

In order to determine diffuse and specular lighting for an object, you need to transform the vertex normals that are sent into the vertex shader into eye space coordinates. In order to do so, you must use the Normal Matrix.

The Normal Matrix is created by

1. Finding the matrix inverse of the ModelView Matrix.

2. Finding the matrix transpose of the matrix inverse calculated in step 1.

Listing 4-41 shows how building a normal matrix is done in code in the GenerateMatrices() function in the Object3d class. Here are the steps:

1. The Model Matrix is multiplied by the View Matrix to generate the ModelView Matrix, which is placed in the m_NormalMatrix variable.

2. This ModelView Matrix is inverted using the Matrix.invertM() function.

3. The transpose of this inverted ModelView Matrix is taken and put in m_NormalMatrix.

Listing 4-41. Building the Normal Matrix in the `GenerateMatrices()` *Function*

```
// Create Normal Matrix for lighting
Matrix.multiplyMM(m_NormalMatrix, 0, Cam.GetViewMatrix(), 0, m_ModelMatrix, 0);
Matrix.invertM(m_NormalMatrixInvert, 0, m_NormalMatrix, 0);
Matrix.transposeM(m_NormalMatrix, 0, m_NormalMatrixInvert, 0);
```

This normal matrix is then sent to the vertex shader and is placed in the `NormalMatrix` shader variable, which is a 4-by-4 matrix (see Listing 4-42).

Listing 4-42. Using the Normal Matrix in the Vertex Shader

```
uniform mat4 NormalMatrix; // Normal Matrix
attribute vec3 aNormal;

// Put Vertex Normal Into Eye Coords
vec3 EcNormal = normalize(vec3(NormalMatrix * vec4(aNormal,1)));
```

The eye coordinates of the vertex normal is then calculated by multiplying the input vertex normal aNormal by NormalMatrix, then converting it to a vec3 vector and then normalizing the resulting vector to length 1. The variable EcNormal, which now contains the vertex normal in eye coordinates, can now be used to calculate diffuse and specular lighting.

Lighting in the Vertex Shader

I will now turn to how lighting is done in the vertex shader. For the purposes of our simulation, light will be composed of ambient, diffuse, and specular components.

Ambient

Ambient lighting is the same across the entire object and, thus, the same at all vertices. Because the lighting is the same for all of the object's vertices, ambient lighting can be handled in the fragment shader. There is no unique value that has to be calculated for a specific vertex for ambient lighting in the vertex shader.

Diffuse

The value of the diffuse lighting for the vertex is the maximum value of either 0 or the Vertex Normal Vector dot product with the Light Vector.

Recall that the value of the dot product of two vectors is the product of their magnitudes times the cosine of the angle between them. If the vectors have been normalized to length 1, the dot product is just the cosine of the angle between the vectors.

If the Normal Vector and the Light Vector are perpendicular, the cosine between the vectors is 0, and thus the dot product and the diffuse lighting is 0 (see Figure 4-22).

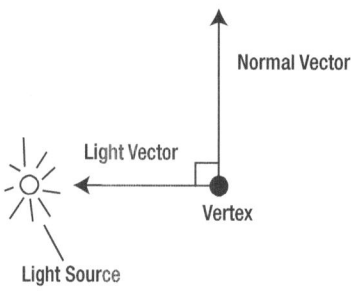

(Normal Vector) Dot (Light Vector) = 0

Figure 4-22. *Diffuse light value is 0*

If the angle between the Normal Vector and the Light Vector is 0, the cosine of the angle between the vectors is 1, and the dot product and diffuse lighting are at the maximum, which is 1 (see Figure 4-23).

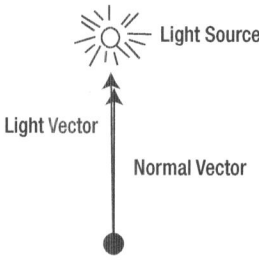

(Normal Vector) Dot (Light Vector) = 1

Figure 4-23. *Diffuse light value is 1*

In code, we find the diffuse lighting by

1. Finding the world coordinates of the Vertex Position, by multiplying the incoming vertex position by the ModelMatrix

    ```
    vec3 WcVertexPos = vec3(uModelMatrix * vec4(aPosition,1));
    ```

2. Finding the world coordinates of the Light Vector which is from the vertex to the light source through standard vector math by subtracting the vertex world position from the light source world position

    ```
    vec3 WcLightDir = uWorldLightPos - WcVertexPos;
    ```

3. Finding the Light Vector in eye coordinates by multiplying the Light Vector in world coordinates by the ViewMatrix

    ```
    vec3 EcLightDir = normalize(vec3(uViewMatrix * vec4(WcLightDir,1)));
    ```

4. Finding the diffuse light amount by taking the dot product between the Vertex Normal in eye coordinates and the Light Vector in eye coordinates or 0, whichever is greater

```
vDiffuse = max(dot(EcLightDir, EcNormal), 0.0);
```
Listing 4-43 is the full shader code.

Listing 4-43. Diffuse Calculations in the Vertex Shader

```
// Calculate Diffuse Lighting for vertex
// maximum of ( N dot L, 0)
vec3 WcVertexPos = vec3(uModelMatrix * vec4(aPosition,1));
vec3 WcLightDir = uWorldLightPos - WcVertexPos;
vec3 EcLightDir = normalize(vec3(uViewMatrix * vec4(WcLightDir,1)));
vDiffuse = max(dot(EcLightDir, EcNormal), 0.0);
```

The vDiffuse shader variable is then passed into the fragment shader as a varying float variable:

```
varying float vDiffuse;
```

Specular

The specular value of light is meant to simulate the shine an object gives off when viewed from a certain angle when being illuminated by a light source—for example, the shine on a chrome bumper on a car when being hit by the sun's rays and viewed at a certain angle.

The specular light amount for a vertex is calculated by

1. Calculating the vector S, which is the sum of the Light Vector and the Eye Vector (see Figure 4-24).

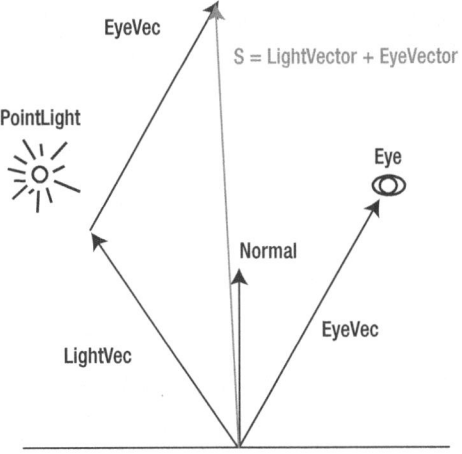

Figure 4-24. Specular lighting

2. Calculating the dot product of the S value and the Vertex Normal value. If this value is less than or equal to zero, the value of the specular color amount is 0. Otherwise, raise this value to the power of the light shininess level. The result is the final specular light value for that vertex.

The code in Listing 4-44 implements the following steps in order to find the specular value for the vertex.

1. The EyeVec or EyeDir from the vertex to the viewer or eye position is determined by subtracting the vertex position in world coordinates from the Eye or Viewer position in world coordinates.

    ```
    vec3 EyeDir = uEyePosition - WcVertexPos;
    ```

2. The S vector is determined by adding the light vector or WcLightDir and the eye vector or EyeDir using vector addition.

    ```
    vec3 S  = WcLightDir + EyeDir;
    ```

3. The S vector is converted from world coordinates to eye coordinates by multiplying it by the ViewMatrix.

    ```
    vec3 EcS = normalize(vec3(uViewMatrix * vec4(S,1)));
    ```

4. The specular lighting for the vertex is calculated by taking the dot product between S and the Vertex Normal, taking the greater value between 0 and the dot product to filter out negative results and raising the result to the power of uLightShininess. uLightShininess is input to the vertex shader as a uniform float variable that is defined by the programmer.

    ```
    vSpecular = pow(max(dot(EcS, EcNormal), 0.0), uLightShininess);
    ```

Listing 4-44. Calculating the Specular Term in the Vertex Shader

```
// S = LightVector + EyeVector
// N = Vertex Normal
// max (S dot N, 0) ^ Shininess
vec3 EyeDir = uEyePosition - WcVertexPos;
vec3 S  = WcLightDir + EyeDir;
vec3 EcS = normalize(vec3(uViewMatrix * vec4(S,1)));
vSpecular = pow(max(dot(EcS, EcNormal), 0.0), uLightShininess);
```

The vSpecular variable is then passed into the fragment shader as a varying variable:

```
varying float vSpecular;
```

Lighting in the Fragment Shader

I will now discuss the lighting in the fragment shader. Lighting will be divided into ambient, diffuse, and specular components.

Ambient Lighting

Ambient lighting is sent to the fragment shader through the uniform vector variable uLightAmbient.

```
uniform vec3 uLightAmbient;
```

This can be directly sent to the AmbientTerm, which is one of the components that determine the final color of the fragment produced from the fragment shader.

```
vec3 AmbientTerm = uLightAmbient;
```

Diffuse Lighting

The color of the diffuse portion of the light source is sent directly to the uLightDiffuse uniform vector variable.

```
uniform vec3 uLightDiffuse;
```

The Diffuse vertex value vDiffuse is received from the vertex shader, as indicated by the varying qualifier. Remember: Varying variables provide the link from the vertex shader to the fragment shader.

```
varying float vDiffuse;
```

The Diffuse component of the final fragment color is determined by multiplying the Diffuse value calculated from the vertex shader by the color of the diffuse light source. Remember that the vDiffuse term ranges from 0 to 1. If it is 1, the diffuse light color will be at full strength. If it is 0, the diffuse light color will be black and, thus, not contributing to the final color of the fragment.

```
vec3 DiffuseTerm  = vDiffuse * uLightDiffuse;
```

Specular Lighting

The specular color of the light source is input as a uniform vector variable called uLightSpecular.

```
uniform vec3 uLightSpecular;
```

The vSpecular variable holds the value of the specular light from the vertex shader.

```
varying float vSpecular;
```

The Specular term of the final fragment color is calculated by multiplying the amount of specular light at the current vertex calculated from the vertex shader times the specular color of the light source.

```
vec3 SpecularTerm = vSpecular * uLightSpecular;
```

Final Fragment Color

The final fragment color is the sum of the AmbientTerm, DiffuseTerm, and SpecularTerm. The gl_FragColor is a reserved variable within the fragment shader that returns the fragment color. The final color vector values are in (r, g, b, alpha) format. Alpha is the transparency level that is active if blending is enabled.

```
vec4 tempColor = (vec4(DiffuseTerm,1) + vec4(SpecularTerm,1) + vec4(AmbientTerm,1));
gl_FragColor = vec4(tempColor.r,tempColor.g, tempColor.b, 1);
```

Materials

The materials that an object is made of affect the color that it reflects and any color that it emits. Here, I cover the Material class, along with how materials are used in the fragment shader.

The Material Class

An object can also have a material, which is represented by the Material class. This material has the color components of

> **Emissive Color:** Color of the light that is emitted from the object
>
> **Ambient Color:** Color of the ambient light that is reflected by the material
>
> **Diffuse Color:** Color of the diffuse light that is reflected by the material
>
> **Specular Color:** Color of the specular light that is reflected by the material
>
> **Alpha:** The degree of opaqueness of the material, with 1 being fully opaque and 0 being completely transparent

See Listing 4-45 for the implementation of the preceding list of color components.

Listing 4-45. Material Class's Data

```
private float[] m_Emissive = new float[3];
private float[] m_Ambient  = new float[3];
private float[] m_Diffuse  = new float[3];
private float[] m_Specular = new float[3];
private float m_Alpha      = 1.0f;
```

The Material class also has functions to set and retrieve these private data items. Please refer to the actual code from the Chapter 3 example for more details.

Materials in the Fragment Shader

Listing 4-46 shows the additions you have to make to the fragment shader code to add in an object's material. The key additions are in bold. A new term, the EmissiveTerm, is used to derive the final output color. The material's ambient, diffuse, and specular properties are used to calculate the ambient, diffuse, and specular components of the final color. The material alpha value is used for the alpha value of the final color.

Listing 4-46. Fragment Shader with Object Materials Added

```
uniform vec3 uMatEmissive;
uniform vec3 uMatAmbient;
uniform vec3 uMatDiffuse;
uniform vec3 uMatSpecular;
uniform float uMatAlpha;

vec3 EmissiveTerm = uMatEmissive;
vec3 AmbientTerm  = uMatAmbient * uLightAmbient;
vec3 DiffuseTerm  = vDiffuse * uLightDiffuse * uMatDiffuse;
vec3 SpecularTerm = vSpecular * uLightSpecular * uMatSpecular;
vec4 tempColor = vec4(DiffuseTerm,1) + vec4(SpecularTerm,1) + vec4(AmbientTerm,1) +
vec4(EmissiveTerm,1);
gl_FragColor = vec4(tempColor.r,tempColor.g, tempColor.b, uMatAlpha);
```

Textures

A 3D object can have an image or texture mapped onto it. A texture has texture coordinates along the U or S horizontal direction and the V or T vertical direction. A texture can be mapped using these coordinates to an object whose vertices match these coordinates. Basically, the texture is wrapped around the 3D object according to the vertex UV texture coordinates (see Figure 4-25).

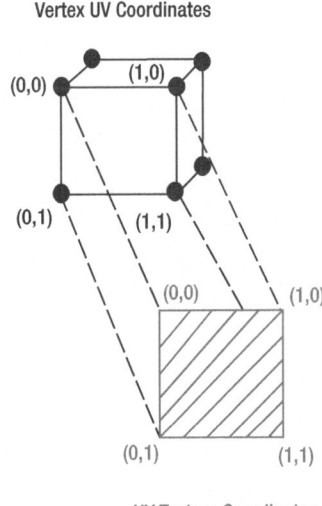

Figure 4-25. Texture UV coordinate mapping

Texture Magnification and Minification

Texture mapping of fixed size textures to 3D objects that can be scaled, rotated, and translated out to variable distances from the viewer require that individual texels (pixels of the texture) be mapped to individual pixels on the final viewing screen. If the texels are enlarged when they are displayed, texture magnification has occurred. If texels are shrunk when they are displayed on the screen, texture minification has occurred (see Figure 4-26).

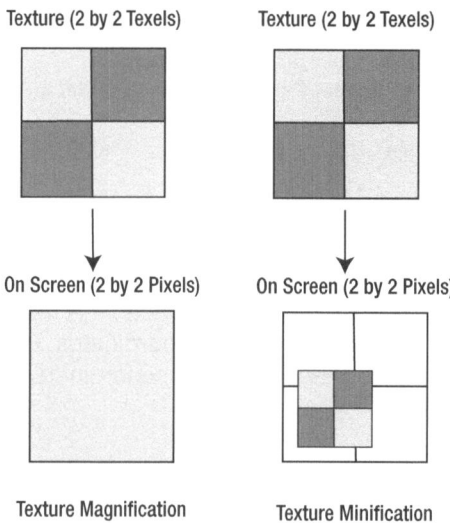

Figure 4-26. *Texture magnification and minification*

In both cases, the textures will have to be filtered to map the color from the texture to the final color displayed on the screen. Two of the ways to do this are the following:

- GLES20.GL_NEAREST: A filtering method that finds the closest texel to the pixel of that texture that is being displayed on screen.

- GLES20.GL_LINEAR: A filtering method that uses a weighted average of the closest 2-by-2 group of texels to the pixel of that texture that is being displayed on the screen.

Texture Clamping and Repeating

Texture coordinates are generally 0 through 1, inclusive. However, you can assign texture coordinates outside this range to vertices and have the textures clamped or repeated (see Figure 4-27). Texture clamping uses 1.0 for texture coordinates that are greater than 1 and 0 for texture coordinates that are less than 0. Texture repeating tries to fit multiple copies of the texture into texture coordinates that are greater than 1 or less than 0.

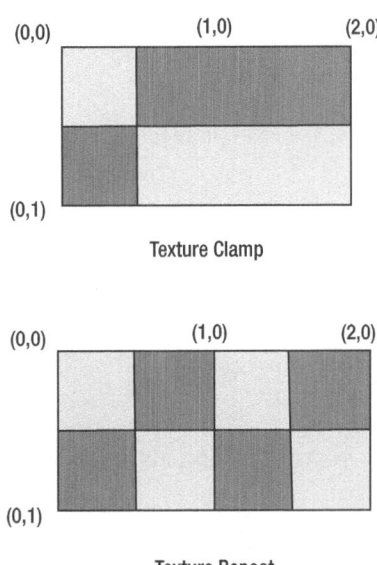

Figure 4-27. *Texture clamping and repeating*

The Texture Class

In our Texture class, we use a bitmap variable to hold the actual texture image that is read in (see Listing 4-47). In addition, the m_TextureId holds a handle to the created texture, and m_Context holds a reference to the activity that this Texture class object belongs to.

Listing 4-47. Texture Class's Data

```
private Context m_Context;
private int m_TextureId;
Bitmap m_Bitmap;
```

The Texture class constructor is shown in Listing 4-48. This constructor creates and initializes a Texture object by

1. First calling the InitTexture() function to load in the Texture from a resource using the ResourceId that is supplied an input parameter to the constructor

2. Setting the filter type for texture minification, which is set to GL_NEAREST

3. Setting the filter type for texture magnification, which is set to GL_LINEAR

4. Setting the way textures are mapped in the U or S (horizontal) and V or T (vertical) to clamping

Listing 4-48. Texture Class Constructor

```
public Texture(Context context, int ResourceId)
{
        // Create new Texture resource from ResourceId
        m_Context = context;
        InitTexture(ResourceId);

        // Setup Default Texture Parameters
        SetTextureWRAP_MIN_FILTER(GLES20.GL_NEAREST);
        SetTextureWRAP_MAG_FILTER(GLES20.GL_LINEAR);
        SetTextureWRAP_S(GLES20.GL_CLAMP_TO_EDGE);
        SetTextureWRAP_T(GLES20.GL_CLAMP_TO_EDGE);
}
```

The InitTexture() function loads in the texture and initializes it to be a 2D texture object (see Listing 4-49).

Listing 4-49. Initializing the Texture

```
boolean InitTexture(int ResourceId)
{
        int[] textures = new int[1];
        GLES20.glGenTextures(1, textures, 0);

        m_TextureId = textures[0];
        GLES20.glBindTexture(GLES20.GL_TEXTURE_2D, m_TextureId);

        // Loads in Texture from Resource File
        LoadTexture(ResourceId);

        GLUtils.texImage2D(GLES20.GL_TEXTURE_2D, 0, m_Bitmap, 0);
        return true;
}
```

The function initializes the texture by

1. Calling glGenTextures() to get an unused texture name from OpenGL

2. Calling nexgt glBindTexture() to create a new texture object of 2 dimensions that has length and width

3. Calling next LoadTexture() to read in the texture from a resource file and store it in our bitmap variable m_Bitmap

4. Calling finally the GLUtils.texImage2D() function to define the 2D texture as that data in our m_Bitmap variable that holds our texture data that we loaded in from the LoadTexture() function

The LoadTexture() function does the actual work of loading in the texture image from a file. The resource image file is opened for reading and is attached to an InputStream. BitmapFactory.decodeStream() is then used to read in the file and convert the data into bitmap form. See Listing 4-50.

Listing 4-50. LoadTexture Function

```
void LoadTexture(int ResourceId)
{
        InputStream is = m_Context.getResources()
                                .openRawResource(ResourceId);
        try
        {
                m_Bitmap = BitmapFactory.decodeStream(is);
        }
        finally
        {
                try
                {
                        is.close();
                }
                catch(IOException e)
                {
                        Log.e("ERROR - Texture ERROR", "Error in LoadTexture()! ");
                }
        }
}
```

You must also set the active texture unit where all the texture-related functions operate. This is done with the glActiveTexture() function in the SetActiveTextureUnit() function. For our purposes in this book, the active texture unit will always be 0 (see Listing 4-51).

Listing 4-51. Setting the Active Texture Unit

```
static void SetActiveTextureUnit(int UnitNumber)
{
        GLES20.glActiveTexture(UnitNumber);
}
```

In order to select this texture object as the current one, you have to active it by calling ActivateTexture(). The glBindTexture() is called with the m_TextureId to activate it (see Listing 4-52).

Listing 4-52. Activating the Texture

```
void ActivateTexture()
{
        // Activate Texture
        if (m_TextureId != 0)
        {
                GLES20.glBindTexture (GLES20.GL_TEXTURE_2D, m_TextureId);
        }
        else
        {
                Log.e("ERROR - Texture ERROR- m_TextureId = 0", "Error in ActivateTexture()! ");
        }
}
```

Textures in the Vertex Shader

Texture information in the vertex shader is basically passed through to the fragment shader for our basic lighting method. The aTextureCoord receives input vertex texture coordinates.

```
attribute vec2 aTextureCoord;
```

The vTextureCoord variable passes through this texture coordinate information into the fragment shader.

```
varying vec2 vTextureCoord;
vTextureCoord = aTextureCoord;
```

Textures in the Fragment Shader

Texture information in the fragment shader is used in combination with the diffuse, specular, ambient, and emissive light produced by the object to find its final color. The color from the currently active texture that matches the texture coordinates in vTextureCoord is found through the texture2D() function. That color is then modified by the total combined light from the diffuse, specular, ambient, and emissive light colors for that vertex.

```
uniform sampler2D sTexture;
varying vec2 vTextureCoord;
vec4 color = texture2D(sTexture, vTextureCoord);
vec4 tempColor = color * (vec4(DiffuseTerm,1) + vec4(SpecularTerm,1) + vec4(AmbientTerm,1) +
vec4(EmissiveTerm,1));
```

Summary

In this chapter, I covered the basic concepts of programming graphics in OpenGL ES 2.0 for Android. I started with a general description of how OpenGL rendering works, then moved on to a more specific explanation, discussing the 3D math, transformations, and vertex and fragment shaders involved. Next, I gave an overview of the shader language used for vertex and fragment shaders, as well as some examples. Then I covered core custom classes that relate to fundamental concepts in OpenGL ES 2.0 that are essential to creating not only 3D games but any OpenGL ES 2.0 graphics application.

Chapter 5

Motion and Collision

In this chapter, I will discuss motion and collision. In terms of motion, I will discuss the basics of an object's linear velocity, linear acceleration, angular velocity, and angular acceleration. I cover Newton's three laws of motion and the new Physics class we have created to implement these laws of motion. I then discuss a hands-on example that demonstrates with a bouncing and rotating cube how to apply linear and angular acceleration to objects. Next, I cover collision detection and collision response and add code to our Physics class to implement this collision detection and collision response. I then go through a hands-on example where we add in another cube on top of the one from the previous hands-on example and demonstrate our new collision detection and response code. Finally, we create a gravity grid and then demonstrate its use by adding this grid to the previous hands-on example with the two colliding cubes.

Overview of Motion

I will start by covering linear and angular velocity and acceleration, as well as Newton's three laws of motion, and then put all of these to use in the hands-on examples.

Linear Velocity and Linear Acceleration

Linear velocity is a vector quantity that has direction and magnitude. For example, let's say that a car is heading northeast on a street at 35 miles per hour. This can be represented by a vector pointing in the northeasterly direction and with a magnitude that would represent 35 miles an hour. Let's say the driver presses down on the brake. This produces an acceleration in the opposite direction to the velocity, which has the result of slowing the car down. After the car stops, the driver then puts the car into reverse and drives down the street backward. The resulting velocity and acceleration vectors are now back in alignment. (See Figure 5-1.)

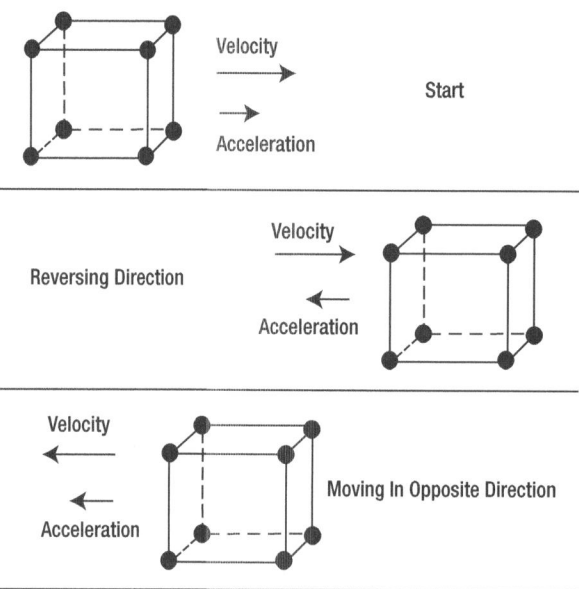

Figure 5-1. Car reversing then moving backward

Figure 5-1 shows the velocity and acceleration vectors for this car as it moves down the street, slows down, and then reverses direction.

Average velocity is the change in distance divided by the change in time. Figure 5-2 shows this definition, with the variable x representing the position and t representing time.

$$V_{Average} = \frac{\Delta x}{\Delta t} = \frac{x_{final} - x_{initial}}{t_{final} - t_{initial}}$$

Figure 5-2. Average velocity definition

The average velocity may be a good representation, if the car maintains a constant speed throughout the time interval. However, if during the time interval the car's speed is extremely high during one portion and extremely low during another, the average speed would not be a good representation of the car's behavior.

The instantaneous velocity is the change in position of an object divided by the change in time when the time interval approaches 0. If you are given a function that represents the position of an object with respect to time x(t), then the first derivative of that function, or x'(t), is the velocity function. (See Figure 5-3.)

$$v = \lim_{\Delta t \to 0} \frac{\Delta x}{\Delta t} = \frac{dx}{dt}$$

Figure 5-3. Instantaneous velocity definition

The average acceleration of an object is the change in velocity divided by the change in time. (See Figure 5-4.)

$$a_{Average} = \frac{\Delta v}{\Delta t} = \frac{v_{final} - v_{initial}}{t_{final} - t_{initial}}$$

Figure 5-4. Average acceleration definition

The instantaneous acceleration is the change in velocity divided by the change in time as the time interval goes to 0. (See Figure 5-5.)

$$a = \lim_{\Delta t \to 0} \frac{\Delta v}{\Delta t} = \frac{dv}{dt}$$

Figure 5-5. Instantaneous acceleration definition

Newton's Laws of Motion

In order for an object to change its velocity, an external force must be applied to that object. Newton's laws of motion are used to describe this force in terms of the object's mass and acceleration. There are three laws of motion, and they are described as follows:

Newton's first law: Consider a body on which no net force acts. If the body is at rest, it will remain at rest. If the body is moving with a constant velocity, it will continue to do so.

Newton's second law: The sum of the external forces acting on an object are equal to that object's mass, multiplied by its acceleration. Figure 5-6 shows the vector form of this law. Figure 5-7 shows the scalar version of this law, with forces grouped along the x, y, and z axes.

$$\sum \vec{F} = m\vec{a}$$

Figure 5-6. Newton's second law vector equation

$$\sum F_x = ma_x$$

$$\sum F_y = ma_y$$

$$\sum F_z = ma_z$$

Figure 5-7. Newton's second law scalar equations

Newton's third law: External forces on objects occur in pairs. That is, if body 1 collides with body 2, it exerts a force on body 2. Body 2 also exerts an equal but opposite force on body 1. (See Figure 5-8.)

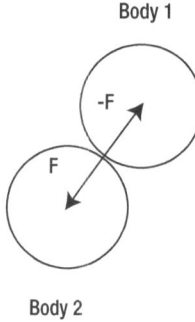

Figure 5-8. Newton's third law demonstrated by colliding spheres

Gravity

The earth's gravity is also a force that can act on objects. For example, Newton's second law can be altered to refer to the earth's gravity. The weight of an object on the earth is actually a force that is equal to the mass of the object, multiplied by the gravitational acceleration or free fall acceleration at the location of the object. (See Figure 5-9.)

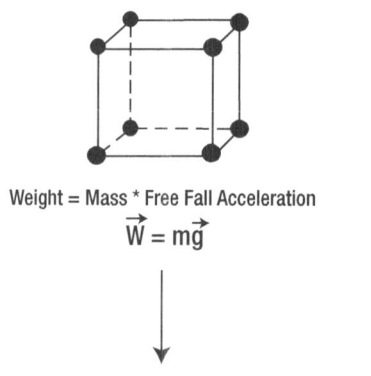

Weight = Mass * Free Fall Acceleration

$$\vec{W} = m\vec{g}$$

Figure 5-9. Force of gravity

Angular Velocity and Angular Acceleration

An object can also have a velocity and acceleration in terms of its angular movement. The distance an object rotates about its rotation axis is measured in radians or degrees. In Figure 5-10, you see an object that is rotating about its rotation axis. The initial starting point is the angle theta1, and the ending angle is theta2. The change, or delta angle amount, is theta2 - theta1.

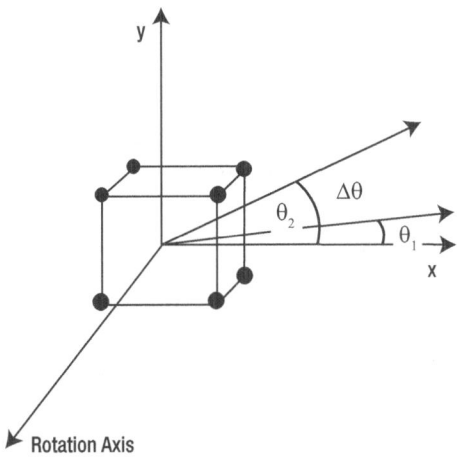

Figure 5-10. *Rotating an object*

The average angular velocity is the change in the angular position, divided by the change in time. The instantaneous angular velocity is the first derivative of the angular position with respect to time. (See Figure 5-11.)

$$\omega_{Average} = \frac{\Delta\Theta}{\Delta t} = \frac{\Theta_2 - \Theta_1}{t2 - t1}$$

$$\omega = \frac{d\Theta}{dt}$$

Figure 5-11. *Angular velocity*

The average angular acceleration is the change in the angular velocity, divided by the change in time. The instantaneous angular acceleration is the first derivative of the angular velocity. (See Figure 5-12.)

$$\alpha_{Average} = \frac{\Delta\omega}{\Delta t} = \frac{\omega 2 - \omega 1}{t2 - t1}$$

$$\alpha = \frac{d\omega}{dt}$$

Figure 5-12. *Angular acceleration*

Rotational Forces

A change in angular velocity of an object is caused by a force that, when applied to the object, changes its rate of rotation around its rotation axis. This force that causes a change in the rotation of the object is called torque. Torque can be measured as the product of the force that causes the torque and the perpendicular distance of the force to the rotational axis. (See Figure 5-13.)

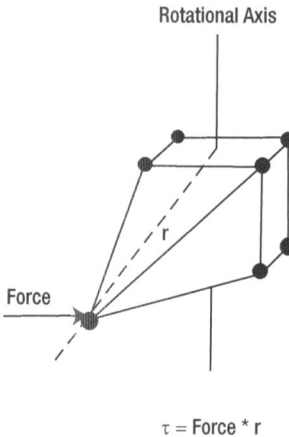

$$\tau = \text{Force} * r$$

Figure 5-13. *Torque*

There is also an angular equivalent of Newton's second law, which is that the torque force applied to the object is equal to the inertia of the object being rotated, times the angular acceleration of the object. (See Figure 5-14.)

$$\tau = I\alpha$$

Figure 5-14. *Torque equation*

The inertia of the object depends on the shape of the object and how exactly the rotation axis is oriented. For example, Figure 5-15 shows the rotational inertia for a hoop for a rotation axis that goes through the center of the hoop.

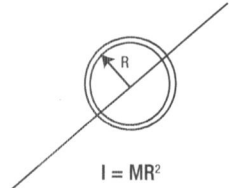

$$I = MR^2$$

Figure 5-15. *Inertia of a hoop*

The Physics Class

Our Physics class contains code related to updating the object's position and rotation, based on linear and rotational forces that are applied to the object. The Object3d class contains a Physics class variable called m_Physics. All of the object's physics-related data, such as velocity and acceleration, and physics functions, such as applying a force to the 3D object, are contained in this variable. This section covers our Physics class.

Listing 5-1 shows some useful constants that are used in our Physics class. PI is defined in radians and is equivalent to 180 degrees, or a half circle. TWO_PI is defined as twice the value of PI in radians and is 360 degrees, or a full circle. HALF_PI is defined as PI/2, which is 90 degrees, or a right angle. A QUARTER_PI is defined as PI/4, which is 45 degrees.

Listing 5-1. Static Physics Constants

```
static float PI        = (float)(3.1415926535897932384626433832795028841971693997511);
static float TWO_PI     = (float)(2.0*PI);
static float HALF_PI    = (float)(PI/2.0);
static float QUARTER_PI = (float)(PI/4.0);
```

In Listing 5-2, I define variables related to linear velocity and acceleration. The variable m_Velocity holds the object's linear velocity in the x, y, and z directions and is created and initialized to (0,0,0). The variable m_Acceleration holds the object's linear acceleration in the x, y, and z directions and is created and initialized to (0,0,0). The variable m_MaxVelocity holds the maximum absolute linear velocity in the x, y, z directions that the object can achieve. The m_MaxAcceleration variable holds the maximum absolute linear acceleration in the x, y, and z directions that the object can achieve.

Listing 5-2. Linear-Related Physics Variables

```
private Vector3 m_Velocity        = new Vector3(0,0,0);
private Vector3 m_Acceleration     = new Vector3(0,0,0);
private Vector3 m_MaxVelocity      = new Vector3(1.25f, 1.25f, 1.25f);
private Vector3 m_MaxAcceleration = new Vector3(1.0f,1.0f,1.0f);
```

The code in Listing 5-3 deals with angular velocity and acceleration. The m_AngularVelocity variable holds the angular velocity of an object around its rotational axis. The m_AngularAcceleration variable holds the angular acceleration of an object around its rotational axis. The m_MaxAngularVelocity variable holds the maximum absolute value of the angular velocity. The m_MaxAngularAcceleration variable holds the maximum absolute value of the angular acceleration for the object.

Listing 5-3. Angular Velocity and Acceleration Variables

```
private float   m_AngularVelocity        = 0;
private float   m_AngularAcceleration     = 0;
private float   m_MaxAngularVelocity      = 4 * PI;
private float   m_MaxAngularAcceleration  = HALF_PI;
```

Listing 5-4 shows the gravity-related variables in the Physics class. The m_ApplyGravity variable is true if gravity is to be applied to the object, and by default it is false. The variable m_Gravity specifies the gravitational acceleration acting on the object. The m_GroundLevel variable specifies the height of the ground. The m_JustHitGround variable is true if the object has just hit the ground level. The m_Mass variable holds the mass of the object.

Listing 5-4. Gravity-Related Variables

```
private boolean m_ApplyGravity   = false;
private float    m_Gravity        = 0.010f;
private float    m_GroundLevel    = 0;
private boolean m_JustHitGround   = false;
private float    m_Mass           = 100.0f;
```

The ApplyTranslationalForce() function takes a force vector as an input and converts this force into a linear acceleration value that is added to the total linear acceleration for the object. Basically, this function adds a new translational force to the object. The acceleration value is calculated from Newton's second law, which is F = ma. The acceleration based on this formula is a = F/m, or linear acceleration is equal to the force applied to the object divided by the mass of the object. (See Listing 5-5.)

Listing 5-5. Applying the Translational Force

```
void ApplyTranslationalForce(Vector3 Force)
{
        // Apply a force to the object
        // F = ma
        // F/m = a
        // 1. Calculate translational acceleration on object due to new force and add this
        // to the current acceleration for this object.
        Vector3 a = new Vector3(Force);
        if (m_Mass != 0)
        {
                a.Divide(m_Mass);
        }
        m_Acceleration.Add(a);
}
```

The ApplyRotationalForce() function applies a new rotational force to the object. The function takes a force and the perpendicular length from the application of this force to the object's rotation axis as input.

The force is converted to angular acceleration using the formula

$$AngularAcceleration = (Force * r) / Rotational Inertia.$$

This new angular acceleration is then added to the total angular acceleration to be applied to this object. The rotational inertia is simplified to a hoop with a radius of 1, so that the rotational inertia is just the mass of the object. (See Listing 5-6.)

Listing 5-6. Applying Rotation Forces to an Object

```
void ApplyRotationalForce(float Force, float r)
{
        // 1. Torque = r X F;
        //     T = I * AngularAcceleration;
        //     T/I = AngularAccleration;
        //
```

```
//    I = mr^2 = approximate with hoop inertia with r = 1 so that I = mass;
float Torque   = r * Force;
float aangular = 0;
float I        = m_Mass;

if (I != 0)
{
        aangular = Torque/I;
}
m_AngularAcceleration += aangular;
}
```

The function `UpdateValueWithinLimit` updates the input value by increment, according to the limit parameter. That is, the function returns the incremented value within the range -limit and limit. (See Listing 5-7.)

Listing 5-7. `UpdateValueWithinLimit` Function

```
float UpdateValueWithinLimit(float value, float increment, float limit)
{
        float retvalue = 0;

        // Increments the value by the increment if the result
        // is within +- limit value
        float tempv = value + increment;
        if (tempv > limit)
        {
                retvalue = limit;
        }
        else if (tempv < -limit)
        {
                retvalue = -limit;
        }
        else
        {
                retvalue += increment;
        }
        return retvalue;
}
```

The `TestSetLimitValue()` function clamps the value of the input parameter value to -limit to limit. (See Listing 5-8.)

Listing 5-8. `TestSetLimitValue` Function

```
float TestSetLimitValue(float value, float limit)
{
        float retvalue = value;

        // If value is greater than limit then set value = limit
        // If value is less than -limit then set value = -limit
        if (value > limit)
```

```
        {
                retvalue = limit;
        }
        else if (value < -limit)
        {
                retvalue = -limit;
        }
        return retvalue;
}
```

The `ApplyGravityToObject()` function applies the force of the gravitational acceleration to the y component of the object's acceleration. (See Listing 5-9.)

Listing 5-9. Applying Gravity to an Object

```
void ApplyGravityToObject()
{
        // Apply gravity to object - Assume standard OpenGL axis orientation of positive y being up
        m_Acceleration.y = m_Acceleration.y - m_Gravity;
}
```

The `UpdatePhysicsObject()` function is the main update function where the position, velocity, and acceleration of the object is updated, based on the linear and angular accelerations to the object caused by external forces that have been applied. (See Listing 5-10.)

The function does the following:

1. Adds the acceleration caused by the force of gravity to the object, if m_ApplyGravity is true

2. Updates the linear acceleration of the object and clamps the values to within the range -m_MaxAcceleration to m_MaxAcceleration. Updates the linear velocity of the object, based on the linear acceleration, and clamps the value to within the range -m_MaxVelocity to m_MaxVelocity

3. Updates the angular acceleration and clamps the value to within the range -m_MaxAngularAcceleration to m_MaxAngularAcceleration. Updates the angular velocity, based on the angular acceleration, and clamps the value to -m_MaxAngularVelocity to m_MaxAngularVelocity

4. Sets the linear and angular accelerations to 0. All linear and angular accelerations caused by external forces acting on this object have been accounted for and processed.

5. Updates the linear position and takes gravity and the height of the ground into account, if applicable. If the object has just hit the ground, m_JustHitGround is set to true. The y component of the object's velocity is set to 0, and the position of the object is set to the ground level specified by m_GroundLevel, if the object is below the ground level and falling.

6. Updates the angular position of the object

Listing 5-10.   Updating the Object's Physics

```
void UpdatePhysicsObject(Orientation orientation)
{
        // 0. Apply Gravity if needed
        if (m_ApplyGravity)
        {
                ApplyGravityToObject();
        }

        // 1. Update Linear Velocity
        ///////////////////////////////////////////////////////////////////////
        m_Acceleration.x  = TestSetLimitValue(m_Acceleration.x, m_MaxAcceleration.x);
        m_Acceleration.y  = TestSetLimitValue(m_Acceleration.y, m_MaxAcceleration.y);
        m_Acceleration.z  = TestSetLimitValue(m_Acceleration.z, m_MaxAcceleration.z);

        m_Velocity.Add(m_Acceleration);
        m_Velocity.x = TestSetLimitValue(m_Velocity.x, m_MaxVelocity.x);
        m_Velocity.y = TestSetLimitValue(m_Velocity.y, m_MaxVelocity.y);
        m_Velocity.z = TestSetLimitValue(m_Velocity.z, m_MaxVelocity.z);

        // 2. Update Angular Velocity
        ///////////////////////////////////////////////////////////////////////
        m_AngularAcceleration = TestSetLimitValue(m_AngularAcceleration, m_MaxAngularAcceleration);

        m_AngularVelocity += m_AngularAcceleration;
        m_AngularVelocity = TestSetLimitValue(m_AngularVelocity,m_MaxAngularVelocity);

        // 3. Reset Forces acting on Object
        //    Rebuild forces acting on object for each update
        ///////////////////////////////////////////////////////////////////////
        m_Acceleration.Clear();
        m_AngularAcceleration = 0;

        //4. Update Object Linear Position
        ///////////////////////////////////////////////////////////////////////
        Vector3 pos = orientation.GetPosition();
        pos.Add(m_Velocity);

        // Check for object hitting ground if gravity is on.
        if (m_ApplyGravity)
        {
                if ((pos.y < m_GroundLevel)&& (m_Velocity.y < 0))
                {
                        if (Math.abs(m_Velocity.y) > Math.abs(m_Gravity))
                        {
                                m_JustHitGround = true;
                        }
                        pos.y = m_GroundLevel;
                        m_Velocity.y = 0;
                }
        }
```

```
        //5. Update Object Angular Position
        ////////////////////////////////////////////////////////////////////////////
        // Add Rotation to Rotation Matrix
        orientation.AddRotation(m_AngularVelocity);
}
```

Hands-on Example: Linear Motion and Angular Motion Using Forces

Here I guide you through a hands-on example demonstrating the use of forces to create linear and angular movement in 3D objects. To follow along with this exercise, the best thing to do is to create a new work space on your development system, download the code for this chapter from apress.com, and then import the project into your new work space.

Creating a Four-Sided Textured Cube

In our previous examples, we used a cube with a texture on two sides. Because we are demonstrating angular rotation here, it would be easier to see the effect if the cube had a texture on all four sides facing the viewer. The code in Listing 5-11 has been added to the Cube class to produce a cube with a texture mapped on four sides.

Listing 5-11. Cube with Four-Sided Texture in Cube Class

```
static float CubeData4Sided[] =
{
        // x,       y,      z,      u,      v       nx,  ny,  nz
        -0.5f,   0.5f,   0.5f,  0.0f,   0.0f,    -1,   1,   1,   // front top left        0
        -0.5f,  -0.5f,   0.5f,  0.0f,   1.0f,    -1,  -1,   1,   // front bottom left     1
         0.5f,  -0.5f,   0.5f,  1.0f,   1.0f,     1,  -1,   1,   // front bottom right    2
         0.5f,   0.5f,   0.5f,  1.0f,   0.0f,     1,   1,   1,   // front top right       3

        -0.5f,   0.5f,  -0.5f,  1.0f,   0.0f,    -1,   1,  -1,   // back top left         4
        -0.5f,  -0.5f,  -0.5f,  1.0f,   1.0f,    -1,  -1,  -1,   // back bottom left      5
         0.5f,  -0.5f,  -0.5f,  0.0f,   1.0f,     1,  -1,  -1,   // back bottom right     6
         0.5f,   0.5f,  -0.5f,  0.0f,   0.0f,     1,   1,  -1    // back top right        7
};
```

Modifying the Object3d Class

The Object3d class has to be modified to add functionality from our Physics class. First, we have to add two new variables. The m_Physics variable is our interface to the object's physics properties.

```
private Physics m_Physics;
```

Another variable we must add is the m_Visible variable that is true if we want that object to be visible and thus drawn to the screen.

```
private boolean m_Visible = true;
```

In the Object3d class's constructor, we have to create a new Physics object.

```
m_Physics = new Physics(iContext);
```

We then have to add functions that set and test the visibility. The SetVisibility() function sets if the object is visible or not.

```
void SetVisibility(boolean value) { m_Visible = value; }
```

The IsVisible() function returns true if the object is visible false otherwise.

```
boolean IsVisible() { return m_Visible; }
```

We also have to add in the GetObjectPhysics() function that allows access to the m_Physics object from outside the class.

```
Physics GetObjectPhysics() { return m_Physics; }
```

The UpdateObjectPhysics() function calls the UpdatePhysicsObject() function with the orientation of the object to do the actual physics update.

The UpdateObject3d() is the main entry point for updating the physics of our object. If the object is visible, the physics of that object are updated. (See Listing 5-12.)

Listing 5-12. The Physics Update Entry Point in the Object3d Class

```
void UpdateObjectPhysics()
{
        m_Physics.UpdatePhysicsObject(m_Orientation);
}

void UpdateObject3d()
{
        if (m_Visible)
        {
                UpdateObjectPhysics();
        }
}
```

Finally, the DrawObject() function draws the object if the object is visible. The additions to this function from previous examples are shown in bold in Listing 5-13.

Listing 5-13. Drawing an Object If It's Visible

```
void DrawObject(Camera Cam, PointLight light)
{
        if (m_Visible)
        {
                DrawObject(Cam,
                        light,
```

```
                    m_Orientation.GetPosition(),
                    m_Orientation.GetRotationAxis(),
                    m_Orientation.GetScale());
        }
}
```

Modifying the MyGLRenderer Class

The MyGLRenderer class must also be modified, because the final program will consist of a cube that is affected by gravity. A linear upward force is applied to the cube, and a rotational force is applied to the cube when it hits the ground. The net result is that the cube appears to jump when it hits the ground and starts spinning faster and faster.

The variable m_Force1 is the linear force that is applied to the cube when it hits the ground level.

```
private Vector3 m_Force1 = new Vector3(0,20,0);
```

The m_RotationalForce variable is the rotational force that is applied to the cube every time it hits the ground.

```
private float m_RotationalForce = 3;
```

In the CreateCube() function, Cube.CubeData4Sided is used to provide texture mapping on four sides of the cube instead of two.

```
MeshEx CubeMesh = new MeshEx(8,0,3,5,Cube.CubeData4Sided, Cube.CubeDrawOrder);
```

The gravity for m_Cube is set to true, so that the cube will fall until it hits the ground level.

```
m_Cube.GetObjectPhysics().SetGravity(true);
```

In the onDrawFrame() function (see Listing 5-14), new code in bold has been added, which

1. Updates the cube physics by calling UpdateObject3d()

2. Tests if the cube has just hit the ground by calling GetHitGroundStatus()

3. Applies, the upward translational force m_Force1 and the rotational force m_RotationalForce to the cube, if the cube has just hit the ground

4. Resets the just-hit-ground status

This is the key code that creates the bouncing/rotating cube effect. (See Figure 5-16.)

Listing 5-14. Modifying the onDrawFrame() Function

```
@Override
public void onDrawFrame(GL10 unused)
{
        GLES20.glClearColor(0.0f, 0.0f, 0.0f, 1.0f);
        GLES20.glClear( GLES20.GL_DEPTH_BUFFER_BIT | GLES20.GL_COLOR_BUFFER_BIT);
```

```
m_Camera.UpdateCamera();
///////////////////////////// Update Object Physics
// Cube1
m_Cube.UpdateObject3d();
boolean HitGround = m_Cube.GetObjectPhysics().GetHitGroundStatus();
if (HitGround)
{
        m_Cube.GetObjectPhysics().ApplyTranslationalForce(m_Force1);
        m_Cube.GetObjectPhysics().ApplyRotationalForce(m_RotationalForce, 10.0f);
        m_Cube.GetObjectPhysics().ClearHitGroundStatus();
}
///////////////////////////// Draw Object
m_Cube.DrawObject(m_Camera, m_PointLight);
}
```

Figure 5-16. *The jumping rotating cube*

Overview of Collisions

In this overview of collisions, I will cover collision detection and the actual application of the action and reaction forces on both colliding objects.

Collision Detection

The type of collision detection we will use in this book is based on a sphere where the entire 3D object fits within the boundaries of a collision sphere. The two objects involved in a collision have initial velocities of V1Initial and V2Initial and final velocities after the collision of V1Final and V2Final. The center of mass of Body1 and Body2 are both assumed to be at the center of the bounding or collision sphere. The collision normal is a vector that passes through the center of masses of both objects and is key in determining the final collision velocities and direction of the objects. The forces acting on the objects when they collide will act along the collision normal and will be equal and opposite in direction, according to Newton's third law of motion. (See Figure 5-17.)

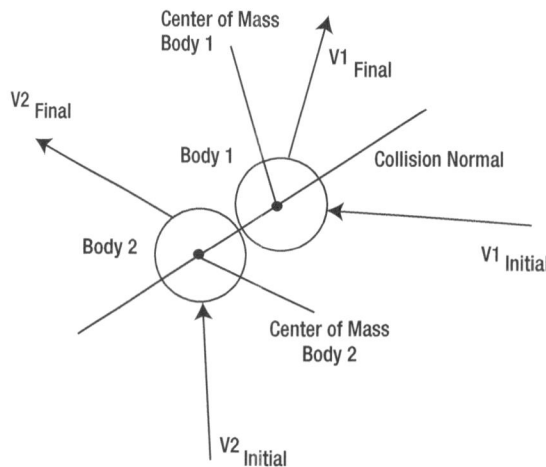

Figure 5-17. *Collision between two 3D objects represented by bounding spheres*

Modifying the MeshEx Class

First, we must be able to calculate the radius of the collision sphere for a 3D object. In order to do this, we have to add some code to our MeshEx class.

The following variables have been added. The m_Size variable measures the largest size of the 3D object mesh in the x, y, and z directions.

```
private Vector3 m_Size  = new Vector3(0,0,0);
```

The m_Radius variable holds the radius of the collision sphere that holds the entire object.

```
private float m_Radius = 0;
```

The m_RadiusAverage variable holds the average of the biggest parts of the object in the x, y, and z axes directions. This radius may not enclose the entire object and is not used for our collision detecti on method discussed later.

```
private float m_RadiusAverage = 0;
```

In the MeshEx constructor, we call the function that calculates the bounding sphere of the mesh being created, which is CalculateRadius(). (See Listing 5-15.)

The CalculateRadius() function calculates the bounding sphere radius for a 3D object by doing the following:

1. Searching through all the object's vertices and determining the smallest and largest x, y, and z coordinates.

2. Finding the size of the object along its x, y, and z axes, based on the minimum and maximum values of the x, y, and z coordinates found in the step above.

3. Calculating the collision radius from the largest portion of the object in the x, y, or z axes direction. Assuming the object is centered at the origin, the largest size values in either the x, y, or z direction would represent the object's diameter. The collision radius would be half of that diameter.

4. Calculating the average radius based on the average of the object's x, y, and z lengths as the diameter. The final average radius is half of this diameter.

Listing 5-15. Calculating the Radius of an Object's Mesh

```
void CalculateRadius()
{
        float XMin = 100000000;
        float YMin = 100000000;
        float ZMin = 100000000;

        float XMax = -100000000;
        float YMax = -100000000;
        float ZMax = -100000000;

        int ElementPos = m_MeshVerticesDataPosOffset;

        // Loop through all vertices and find min and max values of x,y,z
        for (int i = 0; i < m_VertexCount; i++)
        {
                float x = m_VertexBuffer.get(ElementPos);
                float y = m_VertexBuffer.get(ElementPos+1);
                float z = m_VertexBuffer.get(ElementPos+2);

                // Test for Min
                if (x < XMin)
                {
                        XMin = x;
                }

                if (y < YMin)
                {
                        YMin = y;
                }
```

```
                if (z < ZMin)
                {
                        ZMin = z;
                }

                // Test for Max
                if (x > XMax)
                {
                        XMax = x;
                }

                if (y > YMax)
                {
                        YMax = y;
                }

                if (z > ZMax)
                {
                        ZMax = z;
                }
                ElementPos = ElementPos + m_CoordsPerVertex;
        }

        // Calculate Size of Mesh in the x,y,z directions
        m_Size.x = Math.abs(XMax - XMin);
        m_Size.y = Math.abs(YMax - YMin);
        m_Size.z = Math.abs(ZMax - ZMin);

        // Calculate Radius
        float LargestSize = -1;
        if (m_Size.x > LargestSize)
        {
                LargestSize = m_Size.x;
        }

        if (m_Size.y > LargestSize)
        {
                LargestSize = m_Size.y;
        }

        if (m_Size.z > LargestSize)
        {
                LargestSize = m_Size.z;
        }

        m_Radius = LargestSize/2.0f;

        // Calculate Average Radius;
        m_RadiusAverage = (m_Size.x + m_Size.y + m_Size.z) / 3.0f;
        m_RadiusAverage = m_RadiusAverage/2.0f;
}
```

Modifying the Object3d Class

The GetRadius() function has been added to the Object3d class. This function returns the collision radius of the object's mesh. (See Listing 5-16.)

Listing 5-16. The GetRadius() Function

```
float GetRadius()
{
        if (m_MeshEx != null)
        {
                return m_MeshEx.GetRadius();
        }
        return -1;
}
```

The GetScaledRadius() function returns the radius of the bounding/collision sphere of the object scaled by the object's scale factor. (See Listing 5-17.) Thus, an object that has been scaled twice the size of the original mesh will have a radius twice that of the original mesh.

Listing 5-17. Getting the Scaled Object3d Mesh Radius

```
float GetScaledRadius()
{
        float LargestScaleFactor = 0;
        float ScaledRadius = 0;
        float RawRadius = GetRadius();

        Vector3 ObjectScale = m_Orientation.GetScale();

        if (ObjectScale.x > LargestScaleFactor)
        {
                LargestScaleFactor = ObjectScale.x;
        }

        if (ObjectScale.y > LargestScaleFactor)
        {
                LargestScaleFactor = ObjectScale.y;
        }

        if (ObjectScale.z > LargestScaleFactor)
        {
                LargestScaleFactor = ObjectScale.z;
        }
        ScaledRadius = RawRadius * LargestScaleFactor;
        return ScaledRadius;
}
```

Types of Collisions

The two types of collisions that we will consider in this section are a normal collision and a penetrating collision. A normal collision is where the two objects' bounding spheres collide at their boundaries within the collision tolerance level. (See Figure 5-18.)

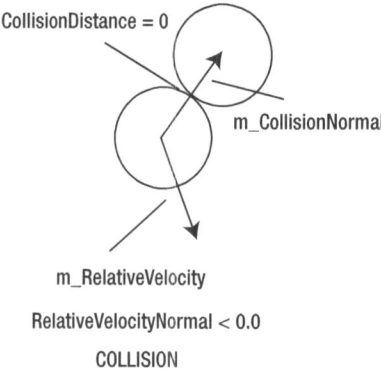

Figure 5-18. *Collision*

A penetrating collision is where the two objects' boundary spheres are overlapping and both spheres are moving toward each other. The two spheres are moving toward each other because the collision normal and the relative velocity between the two spheres point in different directions. That is, their dot product is less than 0. (See Figure 5-19.)

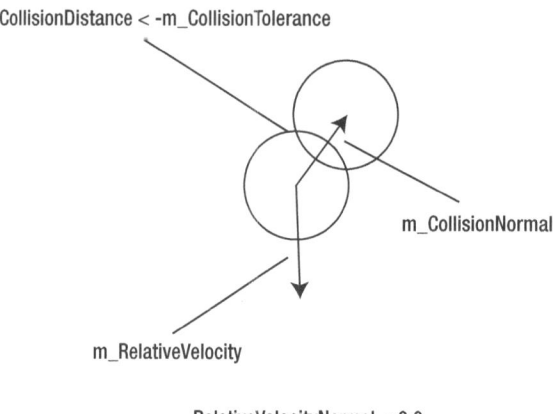

Figure 5-19. *Penetrating collision*

Another case that you have to consider is when the two bounding spheres overlap but are headed away from each other. When the objects are moving away from each other, the dot product between the collision normal and the relative velocity is greater or equal to 0. This case is not considered a collision, because the objects are heading away from each other (see Figure 5–20). We will get into more detail into these cases later in this section.

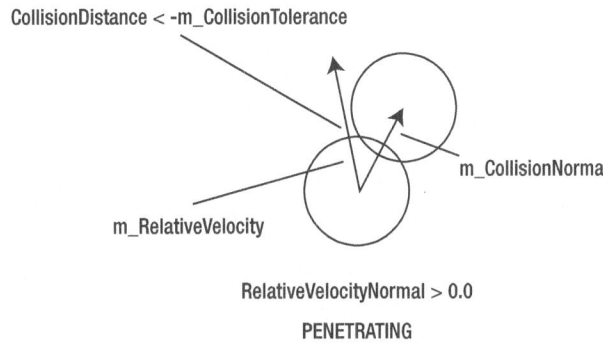

CollisionDistance < -m_CollisionTolerance

m_CollisionNormal

m_RelativeVelocity

RelativeVelocityNormal > 0.0

PENETRATING

Figure 5-20. Penetrating

Modifying the Physics Class

The Physics class holds the main implementation for collision detection. We have to add some variables and a function to this class.

We added an enumeration called CollisionStatus that holds the outcome of our collision detection testing. The values are the following:

>   COLLISION: A collision has occurred.

>   NOCOLLISION: The bodies tested are not touching at all.

>   PENETRATING: The bodies tested are penetrating each other but are moving away from each other and thus not colliding.

>   PENETRATING_COLLISION: The bodies tested are penetrating each other and are moving toward each other and, thus, are colliding.

```
enum CollisionStatus
{
      COLLISION,
      NOCOLLISION,
      PENETRATING,
      PENETRATING_COLLISION
}
```

Next, we add in variables for the collision tolerance. If the collision distance is within the range -COLLISIONTOLERANCE to COLLISIONTOLERANCE, the two bodies would be considered to be colliding with each other, and the value COLLISION is returned.

```
private float COLLISIONTOLERANCE = 0.1f;
private float m_CollisionTolerance = COLLISIONTOLERANCE;
```

The m_CollisionNormal vector is the vector from the center of mass of one object to the center of mass of the other object.

```
private Vector3 m_CollisionNormal;
```

The m_RelativeVelocity vector is a vector that represents the relative velocity of one object to the other object that is being tested for collision.

```
private Vector3 m_RelativeVelocity;
```

The CheckForCollisionSphereBounding() is the function that actually does the collision detection for two 3D objects. It accepts two objects as input and returns a value of the type CollisionStatus. The following steps are performed by the function in order to determine the collision status.

1. Calculate the collision distance between the two objects.

2. Calculate the collision normal between the two objects.

3. Calculate the relative velocity of the two objects along the collision normal.

4. Determine the collision status based on the collision distance and the relative velocity along the collision normal.

The ImpactRadiusSum variable is the sum of the radius of object1 and radius of object2. If the collision were exactly at the boundary of the collision spheres, then the CollisionDistance between the two objects would be zero, and the distance between the center of the masses of the objects would be just ImpactRadiusSum, as seen in Figure 5-21.

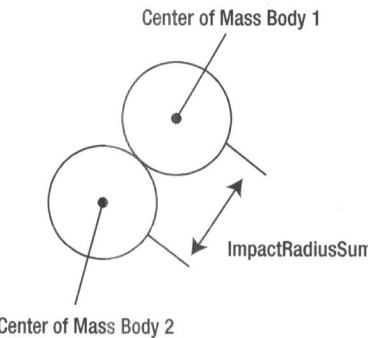

Figure 5-21. The perfect collision

The CollisionDistance measures the distance between the boundaries of the collision spheres of the two objects. It is calculated by subtracting the ImpactRadiusSum from the distance between the objects' center of mass. (See Figure 5-22.)

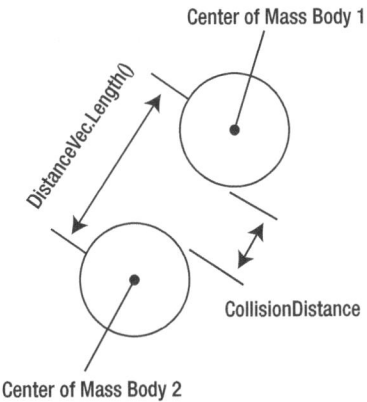

Figure 5-22. *The collision distance*

The CollisionNormal is calculated by normalizing the distance vector between the two centers of mass calculated in the previous step. The relative velocity between the two objects is also calculated. By taking the dot product between the relative velocity vector and the collision normal vector, the magnitude of the relative velocity along the collision normal is found. (See Figure 5-23.)

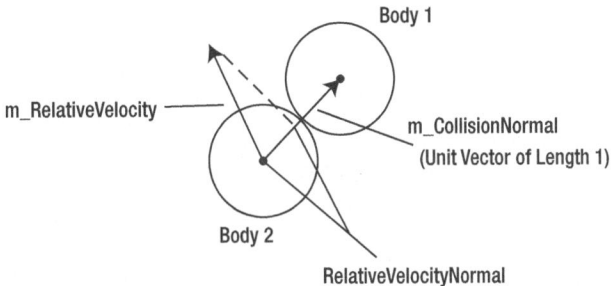

Figure 5-23. *Calculating the collision direction*

Now that you know the collision distance and the relative velocity along the collision normal, you have the all the information you need to find out if the objects are colliding. You know if they are touching each other, based on the collision distance, and if they are moving toward each other, based on the relative velocity of the objects along the collision normal. See Listing 5-18 to see the actual code.

Listing 5-18. *Collision Detection Function*

```
CollisionStatus CheckForCollisionSphereBounding(Object3d body1, Object3d body2)
{
        Float   ImpactRadiusSum        = 0;
        float   RelativeVelocityNormal = 0;
        float   CollisionDistance      = 0;
        Vector3 Body1Velocity;
        Vector3 Body2Velocity;
        CollisionStatus retval;
```

```
// 1. Calculate Separation
ImpactRadiusSum = body1.GetScaledRadius() + body2.GetScaledRadius();

Vector3 Position1 = body1.m_Orientation.GetPosition();
Vector3 Position2 =body2.m_Orientation.GetPosition();

Vector3 DistanceVec = Vector3.Subtract(Position1, Position2);
CollisionDistance = DistanceVec.Length() - ImpactRadiusSum;

// 2. Set Collision Normal Vector
DistanceVec.Normalize();
m_CollisionNormal = DistanceVec;

// 3. Calculate Relative Normal Velocity:
Body1Velocity = body1.GetObjectPhysics().GetVelocity();
Body2Velocity = body2.GetObjectPhysics().GetVelocity();

m_RelativeVelocity = Vector3.Subtract(Body1Velocity , Body2Velocity);
RelativeVelocityNormal = m_RelativeVelocity.DotProduct(m_CollisionNormal);

// 4. Test for collision
if((Math.abs(CollisionDistance) <= m_CollisionTolerance) && (RelativeVelocityNormal < 0.0))
{
        retval = CollisionStatus.COLLISION;
}
else
if ((CollisionDistance < -m_CollisionTolerance) && (RelativeVelocityNormal < 0.0))
{
        retval = CollisionStatus.PENETRATING_COLLISION;
}
else
if (CollisionDistance < -m_CollisionTolerance)
{
        retval = CollisionStatus.PENETRATING;
}
else
{
        retval = CollisionStatus.NOCOLLISION;
}

    return retval;
}
```

Calculating Collisions

Now we will derive the formula used to calculate the force acting on the two colliding objects after impact. The forces acting on the objects will be equal and opposite, according to Newton's third law of motion.

The three equations we will need are listed below. The first two equations are Newton's second law applied to both objects, with the force on one object the opposite of the other. The third equation is the coefficient of restitution or "e," which determines how elastic the collision is. A fully elastic collision will produce a collision where kinetic energy is preserved and e = 1. A fully inelastic collision will produce a collision where kinetic energy is completely lost with e = 0. The e in our equation is calculated from a ratio of the final relative velocities of our objects after collision to the relative velocities before the collision. This gives a good idea of how much kinetic energy is preserved after the collision.

```
Force = mass1 * acceleration1
-Force = mass2 * acceleration2
E = -(V1Final - V2Final) / (V1Initial - V2Initial);
```

Replace acceleration with variables for the change in velocity. We have three equations and three unknowns. The unknowns are V1Final, V2Final, and Force generated by the collision of these two objects.

```
Force = mass1 * ( V1Final - V1Initial )
-Force = mass2 * ( V2Final - V2Initial )
e = -(V1Final - V2Final) / (V1Initial - V2Initial);
```

Solve the first equation for V1Final.

```
Force/mass1 = mass1 * ( V1Final - V1Initial )/mass1
Force/mass1 = V1Final - V1Initial
Force/mass1 + V1Initial = V1Final
```

Solve the second equation for V2Final.

```
-Force/mass2 = mass2 * ( V2Final - V2Initial ) / mass2
-Force/mass2 = V2Final  - V2Initial
-Force/mass2 + V2Initial = V2Final
```

Plug V1Final and V2Final into the equation for the coefficient of restitution, or e.

```
 e = -(V1Final - V2Final) / (V1Initial - V2Initial);
-e(V1Initial - V2Initial) = V1Final - V2Final
-e(V1Initial - V2Initial) = Force/mass1 + V1Initial - (-Force/mass2 + V2Initial )
-e(V1Initial - V2Initial)  - V1Initial = Force/mass1 + Force/mass2 - V2Initial
 -e(V1Initial - V2Initial)  - V1Initial + V2Initial = Force/mass1 + Force/mass2
-e(V1Initial - V2Initial)  - V1Initial + V2Initial = (1/mass1 + 1/mass2) Force
```

Substitute VRelative = V1initial - V2Initial.

```
-e(VRelative)  - VRelative = (1/mass1 + 1/mass2) Force
VRelative( -e -1 ) = (1/mass1 + 1/mass2) Force
-VRelative( e + 1 ) = (1/mass1 + 1/mass2) Force
-VRelative( e + 1 ) /  (1/mass1 + 1/mass2) = Force
```

Thus, the final forces that act on the objects are

```
ForceAction = -VRelative( e + 1 ) /  (1/mass1 + 1/mass2)
ForceReaction = - ForceAction
```

Modifying the Physics Class

Next, we have to add code, to process a collision.

The `ApplyLinearImpulse()` function (see Listing 5-19) in the Physics class actually implements the collision action and reaction forces. This function has three main components, which

1. Calculate the force generated by the collision along the collision normal of the two objects.

2. Find the vector form of the action force, by taking that magnitude of the collision force found in step 1 and putting this along the collision normal between the objects. The reaction force is found by taking the negative of the action force.

3. Add the forces acting on both objects to each of the objects, by using the `ApplyTranslationalForce()` function.

Listing 5-19. The `ApplyLinearImpulse()` Function

```
void ApplyLinearImpulse(Object3d body1, Object3d body2)
{
        float m_Impulse;

        // 1. Calculate the impulse along the line of action of the Collision Normal
        m_Impulse = (-(1+m_CoefficientOfRestitution) * (m_RelativeVelocity.DotProduct
(m_CollisionNormal))) / ((1/body1.GetObjectPhysics().GetMass() + 1/body2.GetObjectPhysics().
GetMass()));

        // 2. Apply Translational Force to bodies
        // f = ma;
        // f/m = a;
        Vector3 Force1 =  Vector3.Multiply( m_Impulse, m_CollisionNormal);
        Vector3 Force2 =  Vector3.Multiply(-m_Impulse, m_CollisionNormal);

        body1.GetObjectPhysics().ApplyTranslationalForce(Force1);
        body2.GetObjectPhysics().ApplyTranslationalForce(Force2);
}
```

Hands-on Example: Collisions

In this exercise, we will create another cube above the cube created in the previous hands-on example. This cube will fall and collide with the first cube. The net effect will be to have two cubes continuously colliding with each other.

Modifying the MyGLRenderer Class

We need to make some modifications to the MyGLRenderer class in our project. The modifications involve adding code to create a second cube and adding code to process the collision between this new cube and old cube from the previous hands-on example.

First, we have to add the variable for the new cube, which is m_Cube2.

```
private Cube m_Cube2;
```

Next, we have to create the new cube. The creation of the new cube is similar to the creation of the first cube. (See Listing 5-20.)

Listing 5-20. Creating the New Cube

```
void CreateCube2(Context iContext)
{
        //Create Cube Shader
        Shader Shader = new Shader(iContext, R.raw.vsonelight, R.raw.fsonelight);        // ok

        MeshEx CubeMesh = new MeshEx(8,0,3,5,Cube.CubeData4Sided, Cube.CubeDrawOrder);

        // Create Material for this object
        Material Material1 = new Material();

        // Create Texture
        Texture TexAndroid = new Texture(iContext,R.drawable.ic_launcher);
        Texture[] CubeTex = new Texture[1];
        CubeTex[0] = TexAndroid;

        m_Cube2 = new Cube(iContext,
                           CubeMesh,
                                CubeTex,
                                Material1,
                                Shader);

        // Set Intial Position and Orientation
        Vector3 Axis = new Vector3(0,1,0);
        Vector3 Position = new Vector3(0.0f, 4.0f, 0.0f);
        Vector3 Scale = new Vector3(1.0f,1.0f,1.0f);

        m_Cube2.m_Orientation.SetPosition(Position);
        m_Cube2.m_Orientation.SetRotationAxis(Axis);
        m_Cube2.m_Orientation.SetScale(Scale);

        // Gravity
        m_Cube2.GetObjectPhysics().SetGravity(true);
}
```

The onSurfaceCreated() function has to be modified to create the new cube. The changes needed are listed in bold in Listing 5-21.

Listing 5-21. onSurfaceCreated() Function Changes

```
@Override
public void onSurfaceCreated(GL10 unused, EGLConfig config)
{
        m_PointLight = new PointLight(m_Context);
        SetupLights();
```

```
        // Create a 3d Cube
        CreateCube(m_Context);

        // Create a Second Cube
        CreateCube2(m_Context);
}
```

In the onDrawFrame() function, the physics properties of the second cube are updated through the UpdateObject3d() function. The two spheres are checked for a valid collision type, and if true, then appropriate linear forces are applied to each object. (See Listing 5-22.)

Listing 5-22. onDrawFrame() Modifications

```
@Override
public void onDrawFrame(GL10 unused)
{
        GLES20.glClearColor(0.0f, 0.0f, 0.0f, 1.0f);
        GLES20.glClear( GLES20.GL_DEPTH_BUFFER_BIT | GLES20.GL_COLOR_BUFFER_BIT);

        m_Camera.UpdateCamera();

        ///////////////////////////// Update Object Physics
        // Cube1
        m_Cube.UpdateObject3d();
        boolean HitGround = m_Cube.GetObjectPhysics().GetHitGroundStatus();
        if (HitGround)
        {
                m_Cube.GetObjectPhysics().ApplyTranslationalForce(m_Force1);
                m_Cube.GetObjectPhysics().ApplyRotationalForce(m_RotationalForce, 10.0f);
                m_Cube.GetObjectPhysics().ClearHitGroundStatus();
        }

        // Cube2
        m_Cube2.UpdateObject3d();

        // Process Collisions
        Physics.CollisionStatus TypeCollision = m_Cube.GetObjectPhysics().
CheckForCollisionSphereBounding(m_Cube, m_Cube2);

        if ((TypeCollision == Physics.CollisionStatus.COLLISION) ||
        (TypeCollision == Physics.CollisionStatus.PENETRATING_COLLISION))
        {
                m_Cube.GetObjectPhysics().ApplyLinearImpulse(m_Cube, m_Cube2);
        }

        ///////////////////////////// Draw Objects
        m_Cube.DrawObject(m_Camera, m_PointLight);
        m_Cube2.DrawObject(m_Camera, m_PointLight);
}
```

Figure 5-24 shows the final result, with two cubes continuously colliding with each other along the vertical y axis.

Figure 5-24. The two colliding cubes

Newton's Law of Gravity

Newton's law of gravity states that every body in the universe attracts every other body in the universe. For example, assume you have two masses that are separated by the distance R. Mass2 exerts a force of Force on Mass1, and Mass1 exerts a force -Force on Mass2 that is equal and opposite. (See Figure 5-25.)

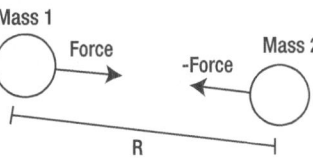

Figure 5-25. Two masses attracting each other

Force is equal to the product of the masses of the two objects, divided by the distance between them, squared and multiplied by the gravitational constant. (See Figure 5-26.)

$$\text{Force} = G \; \frac{\text{Mass1} * \text{Mass2}}{R^2}$$

Figure 5-26. Newton's law of gravity

For the purposes of this book, we will use a general modified form of the equation in Figure 5-26 for a gravity grid that will react in a similar way according to that of Newton's law of gravity. The main purpose of the gravity grid will be to produce some visually impressive effects.

Drone Grid Case Study: Creating a Gravity Grid Using a Vertex Shader

In this case study, we will add a gravity grid to the previous hands-on example of the two colliding cubes bouncing on top of each other. The gravity grid consists of a grid of points that behave according to Newton's law of gravity. That is, the grid points are simulated as masses that are attracted to other masses that are placed on the grid. The purpose of our gravity grid in this example will be to illustrate how the movement of the cubes can change the shape of the gravity grid to produce some visually interesting effects. In addition, a spotlight will be placed on the grid under the masses that are added onto it. The purpose of the spotlight is to highlight and enhance the deformations caused by the masses on the gravity grid.

Modifying the Physics Class

The Physics class needs to be modified to hold the radius of the spotlight that is shown on the grid below an object that is added on to the grid. The variable m_MassEffectiveRadius is the radius for the spotlight on the grid.

```
private float m_MassEffectiveRadius = 10;  // Radius for mass effect on gravity grid
```

The functions GetMassEffectiveRadius() and SetMassEffectiveRadius() retrieve and set the radius for the spotlight.

```
float GetMassEffectiveRadius() {return m_MassEffectiveRadius;}
void SetMassEffectiveRadius(float value) { m_MassEffectiveRadius = value;}
```

Modifying the MeshEx Class

Next, we have to add some code to the MeshEx class. This new code will be used to draw lines instead of triangles.

MeshType is a new enumeration that has the values of Triangles and Lines.

```
enum MeshType
{
        Triangles,
        Lines
}
```

A new variable is also added of MeshType called m_MeshType.

```
private MeshType m_MeshType;
```

In the MeshEx() constructor, the type of mesh to draw is defaulted to triangles.

```
m_MeshType = MeshType.Triangles;
```

Functions to set and retrieve the type of mesh being drawn are also added.

```
void SetMeshType(MeshType Type){m_MeshType = Type;}
MeshType GetMeshType() {return m_MeshType;}
```

In the DrawMesh() function, we add and change the code to actually draw the mesh. We draw either triangles or lines, based on the value of m_MeshType. (See Listing 5-23.)

Listing 5-23. Code to Draw Either Triangles or Lines

```
if (m_MeshType == MeshType.Triangles)
{
        GLES20.glDrawElements(GLES20.GL_TRIANGLES,
                                m_DrawListBuffer.capacity(),
                                GLES20.GL_UNSIGNED_SHORT,
                                m_DrawListBuffer);
}
else
if (m_MeshType == MeshType.Lines)
{
        GLES20.glDrawElements(GLES20.GL_LINES,
                                m_DrawListBuffer.capacity(),
                                GLES20.GL_UNSIGNED_SHORT,
                                m_DrawListBuffer);
}
```

The GravityGridEx Class

Next, we create a new class called GravityGridEx. This is the class that represents the gravity grid that our objects will be placed on.

The actual grid object is of type MeshEx and is called m_LineMeshGrid.

```
private MeshEx m_LineMeshGrid;
```

The vertex data for the grid has to be defined. The number of coordinates per vertex is 3, which are the x, y, and z values of the point location for the grid.

```
private int m_CoordsPerVertex = 3;
```

The offset into the vertex array to the vertex position data is 0.

```
private int m_MeshVerticesDataPosOffset = 0;
```

The offset into the vertex array to the uv texture is -1, meaning there is no texture for this grid.

```
private int m_MeshVerticesUVOffset = -1;
```

The offset into the vertex array to the vertex normal is -1, meaning there are no vertex normals for this grid.

```
private int m_MeshVerticesNormalOffset = -1;
```

The m_Vertices array holds the vertex data for the grid.

```
private float[] m_Vertices;
```

The m_DrawOrder array stores the order that the vertices held in m_Vertices are rendered.

```
private short[] m_DrawOrder;
```

The number of masses on the grid is held in m_NumberMasses.

```
private int m_NumberMasses = 0;
```

The index into the array data for the masses is held in MassesIndex.

```
private int MassesIndex = 0;
```

The maximum number of masses allowed on the grid is held in MAX_MASSES.

```
private int MAX_MASSES = 30;
```

The values of each of the masses on the grid are held in an array called m_MassValues.

```
private float[] m_MassValues = new float[MAX_MASSES];
```

The values of the locations of the masses on the grid are held in an array called m_MassLocations in the format x, y, z for each mass. So every three float elements in the array represent one mass.

```
private float[] m_MassLocations = new float[MAX_MASSES*3];
```

The values for the radius in which to draw the spotlight are stored in m_MassEffectiveRadius.

```
private float[] m_MassEffectiveRadius = new float[MAX_MASSES];
```

The spotlight color to place on the grid for each of the masses is stored in m_MassSpotLightColor in r, g, b format. So, every three float array elements represent data for a single mass.

```
private float[] m_MassSpotLightColor  = new float[MAX_MASSES*3]; // 3 r,g,b values per mass
```

The shader for this grid is held in m_Shader.

```
private Shader m_Shader;
```

The link to the vertex position variable in the shader is held in m_PositionHandle.

```
private int m_PositionHandle;
```

The color of the grid is held in m_GridColor.

```
private Vector3 m_GridColor;
```

The value to send to the modelviewprojection matrix in the shader is located in m_MVPMatrix.

```
private float[] m_MVPMatrix = new float[16];
```

The grid location boundaries along the x axis are held in the following variables:

```
private float m_XMinBoundary;
private float m_XMaxBoundary;
```

The grid location boundaries along the z axis are held in the following variables:

```
private float m_ZMinBoundary;
private float m_ZMaxBoundary;
```

The GravityGridEx constructor creates the grid, based on the input parameters to the constructor.

The following steps are performed in the creation of the gravity grid in the constructor:

1. The array that holds the vertex data m_Vertices is initialized by allocating enough memory to hold all the vertex data.

2. Two nested for loops, the outer loop running the length of the z axis of the grid and the other inner loop running the length of the x axis of the grid, create the points of the grid and puts them into the m_Vertices array.

3. The array m_DrawOrder, which holds the indices into m_Vertices for the actual method of how to draw the mesh, is initialized by allocating enough memory to hold each line to be drawn.

4. The m_DrawOrder array is filled with the indices of the vertices that need to have lines drawn between them. Because we are drawing lines, each two entries of m_DrawOrder would represent one line. This is done in two loops, one for the horizontal lines, and the other for vertical lines of the grid.

5. The actual grid is created using the m_Vertices and the m_DrawOrder arrays created in the previous steps and stored in the m_LineMeshGrid variable.

6. The mesh type to draw is set to lines.

7. The call to the ClearMasses() function sets the mass value of all the objects on the grid to 0.

See Listing 5-24 for details of the code that executes the preceding steps.

Listing 5-24. GravityGridEx Constructor

```
// Creates a grid of lines on the XZ plane at GridHeight height
// of size GridSizeZ by GridSizeX in number of grid points
GravityGridEx(Context iContext,
        Vector3 GridColor,
        float GridHeight,
        float GridStartZValue, float GridStartXValue,
        float GridSpacing,
        int GridSizeZ, int GridSizeX,
        Shader iShader)
{
        m_Context = iContext;
        m_Shader = iShader;
        m_GridColor = GridColor;

        // Set Grid Boundaries
        float NumberCellsX = GridSizeX - 1;
        float NumberCellsZ = GridSizeZ - 1;

        m_XMinBoundary = GridStartXValue;
        m_XMaxBoundary = GridStartXValue + (NumberCellsX * GridSpacing);

        m_ZMinBoundary = GridStartZValue;
        m_ZMaxBoundary = GridStartZValue + (NumberCellsZ * GridSpacing);
        int NumberVertices = GridSizeZ * GridSizeX;
        int TotalNumberCoords = m_CoordsPerVertex * NumberVertices;

        Log.e("GRAVITYGRIDEX" , "TotalNumberCoords = " + TotalNumberCoords);
        m_Vertices = new float[TotalNumberCoords];

        // Create Vertices for Grid
        int index = 0;
        for (float z = 0; z < GridSizeZ; z++)
        {
                for (float x = 0; x < GridSizeX; x++)
                {
                        // Determine World Position of Vertex
                        float xpos = GridStartXValue + (x * GridSpacing);
                        float zpos = GridStartZValue + (z * GridSpacing);

                        if (index >= TotalNumberCoords)
                        {
                                Log.e("GRAVITYGRIDEX" , "Array Out of Bounds ERRROR, Index >=
TotalNumberCoords");
                        }
                        // Assign Vertex to array
                        m_Vertices[index]     = xpos;       //x coord
                        m_Vertices[index + 1] = GridHeight;  // y coord
                        m_Vertices[index + 2] = zpos;
                // z coord
```

```
                        // Increment index counter for next vertex
                        index = index + 3;
                }
        }

        // Create DrawList for Grid
        int DrawListEntriesX = (GridSizeX-1) * 2;
        int TotalDrawListEntriesX = GridSizeZ * DrawListEntriesX;

        int DrawListEntriesZ = (GridSizeZ-1) * 2;
        int TotalDrawListEntriesZ = GridSizeX * DrawListEntriesZ;

        int TotalDrawListEntries = TotalDrawListEntriesX + TotalDrawListEntriesZ;

        Log.e("GRAVITYGRIDEX" , "TotalDrawListEntries = " + TotalDrawListEntries);
        m_DrawOrder = new short[TotalDrawListEntries];

        index = 0;
        for (int z = 0; z < GridSizeZ; z++)
        {
                // Create Draw List for Horizontal Lines
                for (int x = 0; x < (GridSizeX-1);x++)
                {
                        if (index >= TotalDrawListEntries)
                        {
                                Log.e("GRAVITYGRIDEX" , "Array Out of Bounds ERRROR- Horizontal,
Index >= TotalDrawListEntries");
                        }

                        int CurrentVertexIndex = (z*GridSizeX) + x;
                        m_DrawOrder[index] = (short)CurrentVertexIndex;
                        m_DrawOrder[index + 1]= (short)(CurrentVertexIndex + 1);

                        index = index + 2;
                }
        }

        for (int z = 0; z < (GridSizeZ-1); z++)
        {
                // Create Draw List for Vertical Lines
                for (int x = 0; x < (GridSizeX);x++)
                {
                        if (index >= TotalDrawListEntries)
                        {
                                Log.e("GRAVITYGRIDEX" , "Array Out of Bounds ERRROR-Vertical, Index
>= TotalDrawListEntries");
                        }

                        int CurrentVertexIndex      = (z*GridSizeX) + x;
                        int VertexIndexBelowCurrent = CurrentVertexIndex + GridSizeX;

                        m_DrawOrder[index]           = (short)CurrentVertexIndex;
```

```
                m_DrawOrder[index + 1]        = (short)VertexIndexBelowCurrent;

                index = index + 2;
            }
        }
        // Create Mesh
        m_LineMeshGrid = new MeshEx(m_CoordsPerVertex, m_MeshVerticesDataPosOffset, m_
MeshVerticesUVOffset, m_MeshVerticesNormalOffset,m_Vertices,m_DrawOrder);
        m_LineMeshGrid.SetMeshType(MeshType.Lines);

        // Clear Value of Masses
        ClearMasses();
}
```

The function ClearMasses() does the actual job of clearing the grid of all the masses from the m_MassValues array. Clearing all the masses from the grid has to be done for each frame update, because a mass such as an enemy object can be destroyed and thus will need to be removed from the gravity grid. For every frame update, only masses that are currently active will be added to the gravity grid. (See Listing 5-25.)

Listing 5-25. Clearing the Grid

```
void ClearMasses()
{
        for (int i = 0; i < MAX_MASSES; i++)
        {
                m_MassValues[i] = 0;
        }
}
```

The ResetGrid() function clears the grid of all masses and all other related variables that are needed to keep track of the number of masses. (See Listing 5-26.)

Listing 5-26. Resetting the Grid

```
void ResetGrid()
{
        // Clears Grid of All Masses
        MassesIndex = 0;
        m_NumberMasses = 0;
        ClearMasses();
}
```

The code in Listing 5-27 provides access to key grid data, including the following:

1. The maximum number of masses allowed on the gravity grid

2. The current number of masses on the gravity grid

3. The x boundaries of the gravity grid

4. The z boundaries of the gravity grid

Listing 5-27. Accessing Key Grid Data

```
int GetMaxMasses(){return MAX_MASSES;}
int GetNumberMassesOnGrid(){return m_NumberMasses;}
float GetXMinBoundary(){return m_XMinBoundary;}
float GetXMaxBoundary() {return m_XMaxBoundary;}
float GetZMinBoundary(){return m_ZMinBoundary;}
float GetZMaxBoundary(){return m_ZMaxBoundary;}
```

The function AddMass() in Listing 5-28 adds an object to the gravity grid. This function has to be used with the ClearMasses() function, in order to make sure all the masses currently on the gravity grid are up to date.

The function does the following:

1. Calculates the indices for the location and spotlight arrays for this new object

2. Checks to see if the gravity grid is already full. If it is, then it returns with a false value

3. Places the value of the new object's mass in the m_MassValues array

4. Places the x, y, z position of the new object in the m_MassLocations array using the index calculated from step 1

5. Places the radius of the spotlight for the object in the m_MassEffectiveRadius array

6. Places the spotlight color for the object in the m_MassSpotLightColor array using the index calculated in step 1

Listing 5-28. Adding a Mass to the Gravity Grid

```
boolean AddMass(Object3d Mass)
{
        boolean result = true;

        int MassLocationIndex        = MassesIndex * 3; // each mass has 3 components x,y,z
        int SpotLightLocationIndex   = MassesIndex * 3; // each spotlight has 3 components r,g,b

        if (m_NumberMasses >= MAX_MASSES)
        {
                result = false;
                return result;
        }

        float[] Color;

        // Add Value of the Mass
        m_MassValues[MassesIndex] = Mass.GetObjectPhysics().GetMass();

        // Add the x,y,z location of the Mass
        m_MassLocations[MassLocationIndex]    = Mass.m_Orientation.GetPosition().x;
        m_MassLocations[MassLocationIndex + 1]= Mass.m_Orientation.GetPosition().y;
```

```
        m_MassLocations[MassLocationIndex + 2]= Mass.m_Orientation.GetPosition().z;
        MassLocationIndex = MassLocationIndex + 3;

        // Add the Radius of the Spotlight for the Mass
        m_MassEffectiveRadius[MassesIndex] = Mass.GetObjectPhysics().GetMassEffectiveRadius();

        // Add the SpotLight Color for the mass
        Color = Mass.GetGridSpotLightColor();
        m_MassSpotLightColor[SpotLightLocationIndex] = Color[0];
        m_MassSpotLightColor[SpotLightLocationIndex + 1] = Color[1];
        m_MassSpotLightColor[SpotLightLocationIndex + 2] = Color[2];
        SpotLightLocationIndex = SpotLightLocationIndex + 3;

        MassesIndex++;
        m_NumberMasses++;

        return result;
}
```

The AddMasses() function shown in Listing 5-29 adds in iNumberMasses objects from the Masses array to the gravity grid. This is basically the same function as in Listing 5-28, except that instead of reading in the data for one object, multiple objects are read in from an array.

Listing 5-29. Adding Multiple Masses from an Array

```
boolean AddMasses(int iNumberMasses, Object3d[] Masses)
{
        boolean result = true;

        int MassLocationIndex = MassesIndex * 3; // each mass has 3 components x,y,z
        int SpotLightLocationIndex = MassesIndex * 3; // each spotlight has 3 components r,g,b

        float[] Color;
        for (int i = 0; i < iNumberMasses; i++)
        {
                if (m_NumberMasses >= MAX_MASSES)
                {
                        return false;
                }

                // Add Value of the Mass
                m_MassValues[MassesIndex] = Masses[i].GetObjectPhysics().GetMass();

                // Add the x,y,z location of the Mass
                m_MassLocations[MassLocationIndex]    = Masses[i].m_Orientation.GetPosition().x;
                m_MassLocations[MassLocationIndex + 1]= Masses[i].m_Orientation.GetPosition().y;
                m_MassLocations[MassLocationIndex + 2]= Masses[i].m_Orientation.GetPosition().z;
                MassLocationIndex = MassLocationIndex + 3;

                // Add the Radius of the Spotlight for the Mass
                m_MassEffectiveRadius[MassesIndex] = Masses[i].GetObjectPhysics().
GetMassEffectiveRadius();
```

```
            // Add the SpotLight Color for the mass
            Color = Masses[i].GetGridSpotLightColor();
            m_MassSpotLightColor[SpotLightLocationIndex] = Color[0];
            m_MassSpotLightColor[SpotLightLocationIndex + 1] = Color[1];
            m_MassSpotLightColor[SpotLightLocationIndex + 2] = Color[2];
            SpotLightLocationIndex = SpotLightLocationIndex + 3;

            MassesIndex++;
            m_NumberMasses++;
        }

        return result;
}
```

The SetUpShader() function prepares the vertex shader to render the gravity grid.

The function does the following:

1. Activates the shader

2. Gets the position handle that serves to link the vertex position variable in the shader to the main program that will send the vertex data to the shader

3. Sets the specific values of the masses that are on the grid, such as the mass value, radius of the spotlight, mass location, and the color of the spotlight

4. Sets the color of the gravity grid in the shader

5. Sets the value of the modelviewprojection matrix in the shader

See Listing 5-30 for the actual code details.

Listing 5-30. Setting Up the Vertex Shader for the Gravity Grid

```
void SetUpShader()
{
        // Add program to OpenGL environment
        m_Shader.ActivateShader();

        // get handle to vertex shader's vPosition member
        m_PositionHandle = m_Shader.GetShaderVertexAttributeVariableLocation("aPosition");

        // Set Gravity Line Variables
        m_Shader.SetShaderUniformVariableValueInt("NumberMasses", m_NumberMasses);
        m_Shader.SetShaderVariableValueFloatVector1Array("MassValues", MAX_MASSES, m_MassValues, 0);
        m_Shader.SetShaderVariableValueFloatVector3Array("MassLocations", MAX_MASSES,
m_MassLocations, 0);
        m_Shader.SetShaderVariableValueFloatVector1Array("MassEffectiveRadius", MAX_MASSES,
m_MassEffectiveRadius, 0);
        m_Shader.SetShaderVariableValueFloatVector3Array("SpotLightColor",MAX_MASSES,
m_MassSpotLightColor, 0);
```

```
        // Set Color of Line
        m_Shader.SetShaderUniformVariableValue("vColor", m_GridColor);

        // Set View Proj Matrix
        m_Shader.SetShaderVariableValueFloatMatrix4Array("uMVPMatrix", 1, false, m_MVPMatrix, 0);
}
```

The function GenerateMatrices() builds the modelviewprojection matrix from the view matrix and the projection matrix. The grid does not need to be moved or rotated anywhere, so we can skip the step where the model is translated and rotated into the world space. (See Listing 5-31.)

Listing 5-31. Generating the modelviewprojection Matrix

```
void GenerateMatrices(Camera Cam)
{
        Matrix.multiplyMM(m_MVPMatrix, 0, Cam.GetProjectionMatrix(), 0, Cam.GetViewMatrix(), 0);
}
```

The DrawGrid() function creates the needed matrices, sets up the vertex shader for rendering, and then draws the actual gravity grid mesh. (See Listing 5-32.)

Listing 5-32. Drawing the Gravity Grid Mesh

```
void DrawGrid(Camera Cam)
{
        // Set up Shader
        GenerateMatrices(Cam);
        SetUpShader();

        // Draw Mesh
        m_LineMeshGrid.DrawMesh(m_PositionHandle, -1, -1);
}
```

Creating the New Vertex Shader

A new vertex shader needs to be created in order to change the way vertices are placed in the 3D world for the grid mesh object. The basic idea of this new vertex shader is that each vertex or point on the gravity grid will have all the attractive gravitational forces from all the masses on the grid calculated, and the sum of these forces will help determine the final position of each grid point. The spotlight color contribution from all the objects on the grid are also calculated for each grid point and added to the original color. (See Figure 5-27.)

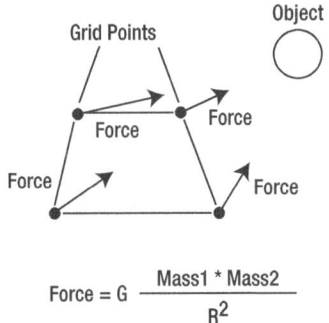

$$Force = G \; \frac{Mass1 * Mass2}{R^2}$$

Figure 5-27. The gravity grid

Next, I'll discuss what code is in the actual gravity grid vertex shader. The modelviewprojection matrix is held in uMVPMatrix.

```
uniform mat4 uMVPMatrix;
```

The grid vertex locations are sent from the main program to the aPosition variable.

```
attribute vec3 aPosition;
```

The code in Listing 5-33 creates shader variables that deal with the following:

1. The number of masses currently on the grid

2. The maximum number of masses on the grid

3. The mass values of all the objects on the grid

4. The locations of all the objects on the grid

5. The radius of the grid spotlight under the object

6. The spotlight color related to the object that is displayed on the grid

Listing 5-33. Grid Object Information

```
uniform int NumberMasses;
const int MAX_MASSES = 30;
uniform float MassValues[MAX_MASSES];
uniform vec3  MassLocations[MAX_MASSES];
uniform float MassEffectiveRadius[MAX_MASSES];
uniform vec3  SpotLightColor[MAX_MASSES];
```

The color of the vertex that is input to the shader is held in vColor.

```
uniform vec3 vColor;
```

The final color of the vertex, including colors contributed from spotlights, is sent to the fragment shader by the Color variable.

```
varying vec3 Color;
```

The function IntensityCircle() returns an intensity value from 0 to 1 that is most intense in the center of the circle and 0 when Radius = MaxRadius. (See Listing 5-34.)

Listing 5-34. The Intensity Circle for the Spotlights

```
float IntensityCircle(float Radius, float MaxRadius)
{
        float retval;
        retval = 1.0 - (Radius/MaxRadius);
        return retval;
}
```

Main() Function of Shader

In the main() function of the shader, where the actual shader code starts to execute, we first create a NewPos vector variable, to hold the incoming vertex locations.

```
vec3  NewPos;
NewPos = aPosition;
```

The part of the shader code shown in Listing 5-35 is the main loop that processes all the objects on the grid and determines the net force acting on the current grid vertex by all the objects on the grid. It also determines the final color of the vertex, based on the original color and the total of the spotlight colors from the objects.

The code in Listing 5-35 does the following:

1. Sets a maximum force through the ForceMax variable

2. Initializes the cumulative spotlight color from all the objects on the grid to black (0,0,0)

3. Initializes the cumulative spotlight color for each of the active objects on the grid (Mass > 0)

4. Calculates the direction to the object from the vertex

5. Calculates the distance from the object to the vertex

6. Calculates the gravitational force attraction, using the formula Force = (MassValues[i] * (2.0)) / (R * R), which roughly approximates Newton's law of gravity, where both objects' masses are the same and the gravitational constant is 1

7. Uses the IntensityCircle function to determine the vertex's spotlight color for that object, if the distance to the object from the vertex is within the object's spotlight distance

8. Choses from Newton's law of gravity and ForceMax the minimum force from the current force calculated

9. Adds the displacement that this force causes the grid vertex by adding this force to the current vertex position

Listing 5-35. Calculating the Forces and Colors for a Grid Point

```
// F = G *( M1 * M2)/ (R*R)
// F = m * a
// F/m = a
// Force = (MassOnGravityGrid * MassVertex) / (RadiusBetweenMasses * RadiusBetweenMasses);
float Force;
float ForceMax = 0.6; //0.5;
vec3 VertexPos = NewPos;

vec3 MassSpotLightColor = vec3(0,0,0);

for (int i = 0; i < MAX_MASSES; i++)
{
        // If mass value is valid then process this mass for the grid
        if (MassValues[i] > 0.0)
        {
                vec3 Mass2Vertex = VertexPos - MassLocations[i];
                vec3 DirectionToVertex = normalize(Mass2Vertex);
                vec3 DirectionToMass = -DirectionToVertex;

                float R = length(Mass2Vertex);

                Force =  (MassValues[i] * (2.0)) / (R * R);

                if (R < MassEffectiveRadius[i])
                {
                        float Intensity = IntensityCircle(R, MassEffectiveRadius[i]);
                        MassSpotLightColor = MassSpotLightColor + (SpotLightColor[i] * Intensity);
                }

                Force = min(Force, ForceMax);

                VertexPos = VertexPos + (Force * DirectionToMass);
        }
}
```

The final vertex position saved in gl_Position is calculated by multiplying the modelviewprojection matrix by the vertex location in VertexPos.

```
gl_Position = uMVPMatrix * vec4(VertexPos,1);
```

The final color of the vertex Color is derived from the original color vColor of the vertex, added to the sum of the spotlight colors from all the objects on the gravity grid.

```
Color = vColor + MassSpotLightColor;
```

Modifying the MyGLRenderer Class

Next, we have to add more code to our MyGLRenderer class. This code creates and updates the gravity grid.

The actual gravity grid is held in the variable m_Grid, which is a GravityGridEx class.

```
private GravityGridEx m_Grid;
```

The CreateGrid() function actually creates the gravity grid of size 33 by 33 blocks, with grid lines that are dark blue in color. (See Listing 5-36.)

Listing 5-36. Creating the Gravity Grid

```
void CreateGrid(Context iContext)
  {
        Vector3 GridColor      = new Vector3(0,0.0f,0.3f);
        float   GridHeight     = -0.5f;
        float   GridStartZValue = -15;
        float   GridStartXValue = -15;
        float   GridSpacing    = 1.0f;
        int     GridSizeZ      = 33;  // grid vertex points in the z direction
        int     GridSizeX      = 33;  // grid vertex point in the x direction

        Shader  iShader = new Shader(iContext, R.raw.vsgrid, R.raw.fslocalaxis);

        m_Grid = new GravityGridEx(iContext,
                          GridColor,
                          GridHeight,
                          GridStartZValue,
                          GridStartXValue,
                          GridSpacing,
                          GridSizeZ,
                          GridSizeX,
                          iShader);
}
```

In the CreateCube() function, you have to set the grid spotlight color that your object will produce and set the spotlight radius or mass effective radius for this spotlight. In the case of our first cube, the spotlight color will be red, and the spotlight radius will be 6. (See Listing 5-37.)

Listing 5-37. Adding to the CreateCube() Function

```
Vector3 GridColor = new Vector3(1,0,0);
m_Cube.SetGridSpotLightColor(GridColor);
m_Cube.GetObjectPhysics().SetMassEffectiveRadius(6);
```

In the CreateCube2() function, we add code to set the grid spotlight color to green and the spotlight radius to 6 of object m_Cube2. (See Listing 5-38.)

Listing 5-38. Adding to the CreateCube2( ) Function

```
Vector3 GridColor = new Vector3(0,1,0);
m_Cube2.SetGridSpotLightColor(GridColor);
m_Cube2.GetObjectPhysics().SetMassEffectiveRadius(6);
```

We add a function UpdateGravityGrid(), which updates our gravity grid by resetting the grid to clear out all the masses. Then we add the masses we want to appear on the grid. Let's add our first cube with the red spotlight. (See Listing 5-39.)

Listing 5-39. Updating the Gravity Grid

```
void UpdateGravityGrid()
{
        // Clear Masses from Grid from Previous Update
        m_Grid.ResetGrid();

        // Add Cubes to Grid
         m_Grid.AddMass(m_Cube);
}
```

In the onSurfaceCreated() function, we add a call to the CreateGrid() function to create our new gravity grid when our GL surface has been created. (See Listing 5-40.)

Listing 5-40. Modifying the onSurfaceCreated( ) Function

```
@Override
public void onSurfaceCreated(GL10 unused, EGLConfig config)
{
        m_PointLight = new PointLight(m_Context);
        SetupLights();

        // Create a 3d Cube
        CreateCube(m_Context);

        // Create a Second Cube
        CreateCube2(m_Context);

        // Create a new gravity grid
        CreateGrid(m_Context);
}
```

The onDrawFrame() function has to be modified to update and draw the gravity grid. The changes are in bold print. (See Listing 5-41.)

Listing 5-41. Modifying the onDrawFrame( ) Function

```
@Override
public void onDrawFrame(GL10 unused)
{
        GLES20.glClearColor(0.0f, 0.0f, 0.0f, 1.0f);
        GLES20.glClear( GLES20.GL_DEPTH_BUFFER_BIT | GLES20.GL_COLOR_BUFFER_BIT);
```

```
        m_Camera.UpdateCamera();

        ///////////////////////////// Update Object Physics
        // Cube1
        m_Cube.UpdateObject3d();
        boolean HitGround = m_Cube.GetObjectPhysics().GetHitGroundStatus();
        if (HitGround)
        {
                m_Cube.GetObjectPhysics().ApplyTranslationalForce(m_Force1);
                m_Cube.GetObjectPhysics().ApplyRotationalForce(m_RotationalForce, 10.0f);
                m_Cube.GetObjectPhysics().ClearHitGroundStatus();
        }
        // Cube2
        m_Cube2.UpdateObject3d();

        // Process Collisions
        Physics.CollisionStatus TypeCollision = m_Cube.GetObjectPhysics()
.CheckForCollisionSphereBounding(m_Cube, m_Cube2);

        if ((TypeCollision == Physics.CollisionStatus.COLLISION) ||
        (TypeCollision == Physics.CollisionStatus.PENETRATING_COLLISION))
        {
                m_Cube.GetObjectPhysics().ApplyLinearImpulse(m_Cube, m_Cube2);
        }

        ///////////////////////////// Draw Objects
        m_Cube.DrawObject(m_Camera, m_PointLight);
        m_Cube2.DrawObject(m_Camera, m_PointLight);

        ///////////////////////////// Update and Draw Grid
        UpdateGravityGrid();
        m_Grid.DrawGrid(m_Camera);
}
```

Now run the application. You should see something that resembles Figure 5-28.

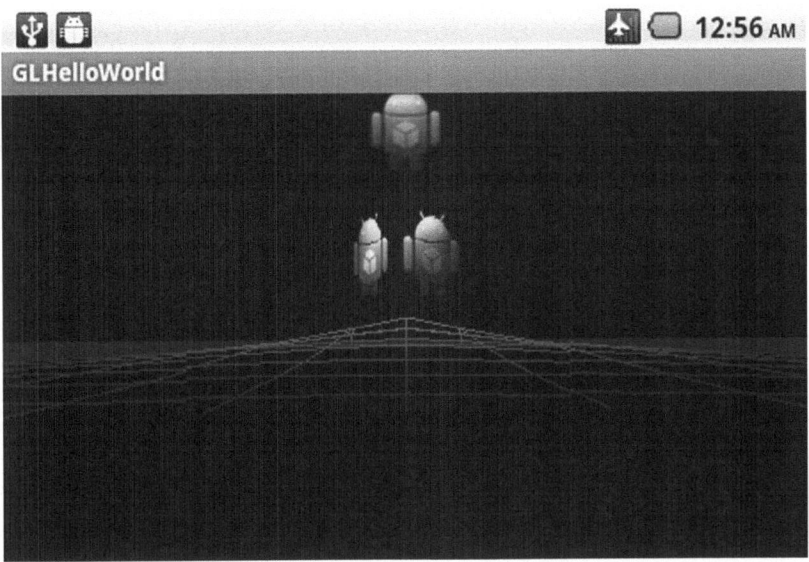

Figure 5-28. Gravity grid with one cube added

Now, let's add the other cube to the gravity grid. Go to the UpdateGravityGrid() function and add the following line:

```
m_Grid.AddMass(m_Cube2);
```

This adds the second cube to the gravity grid. You should see something similar to Figure 5-29. Note that the color of the spotlight under the grid has changed, and the grid generally seems to be higher because of the addition of the new mass.

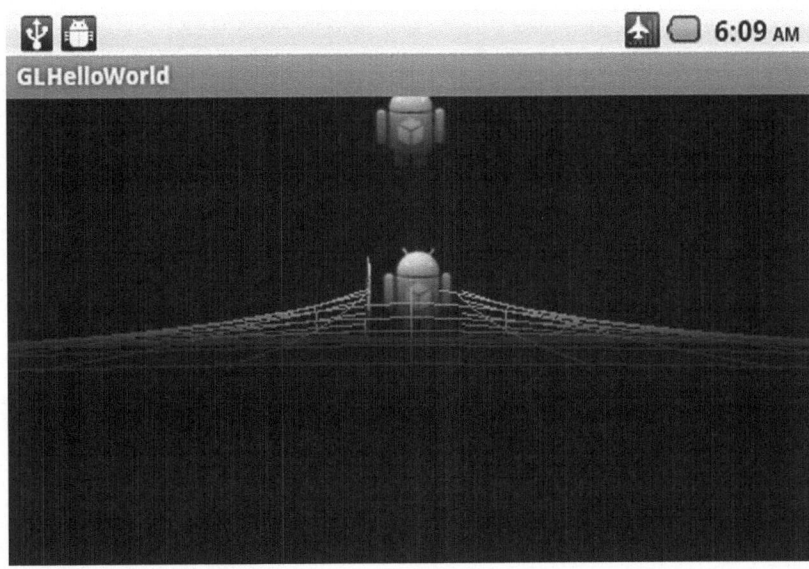

Figure 5-29. Gravity grid with two cubes added

Summary

In this chapter, I discussed motion and collision. I started with linear and angular velocity and acceleration and Newton's three laws of motion. Then I covered our custom Physics class that implemented motion for our objects. I then presented a hands-on example in which we demonstrated the application of linear and angular force on an object. Next, I covered collision detection and collision response. We added to our previous hands-on example by adding another cube that continuously collided with the first cube. Then we designed a gravity grid that obeyed Newton's law of gravity. Finally, we implemented and demonstrated this gravity grid in a hands-on example.

Chapter 6

Game Environment

In this chapter, I will cover the game environment. I start off by covering the creation of sound on Android. I cover our custom Sound class and then modifications we will need to make to other class objects to integrate sounds into them. I then go through a hands-on example in which we play explosion sounds when two cubes collide with each other. I then cover how to create a heads-up display, or HUD. The new classes needed to implement the HUD are covered, followed by a case study in which a heads-up display is created for a game. Next, I cover saving and loading data and show how you can add code into your classes to save and load class data. Finally, I discuss a hands-on example in which we modify our case study to preserve the HUD item data and the orientation and physics state of the two cubes that are colliding with each other.

Overview of Sounds on Android

In this section, I discuss the Sound class and other changes in the Object3d class that have to be made in order to add sounds to our 3D objects.

The Sound Class

The Sound class uses the existing Android SoundPool class to play and manage sounds. A reference to the existing SoundPool object that holds the main pool of sounds is held in m_SoundPool.

```
private SoundPool m_SoundPool;
```

The index to the specific sound in the sound pool is held in m_SoundIndex.

```
private int m_SoundIndex = -1;
```

The Sound constructor creates a new sound. The sound is loaded from the resource ResourceID into the SoundPool object Pool. An index is returned to the newly created sound within this sound pool. (See Listing 6-1.)

Listing 6-1. Sound Constructor

```
Sound(Context iContext, SoundPool Pool, int ResourceID)
{
        m_SoundPool = Pool;
        m_SoundIndex = m_SoundPool.load(iContext, ResourceID, 1);
}
```

In terms of playing back a sound, the left speaker volume output m_LeftVolume accepts the range 0 to 1.

```
float   m_LeftVolume = 1;
```

The right speaker volume level m_RightVolume also accepts the range 0 to 1.

```
float   m_RightVolume = 1;
```

In terms of the priority level for playback m_Priority (required if resources are limited), the higher the number, the greater the priority.

```
int     m_Priority = 1;
```

The variable m_Loop determines if the sound is looped or not. A negative value means the sound will be looped forever. A positive number specifies the number of times to loop the sound. A 0 indicates that there is no looping.

```
int     m_Loop = 0;
```

The variable m_Rate determines the rate at which to play back the sound. A 1.0 would play the sound back normally. A 2.0 would play back the sound at twice the rate as normal. The range is 0.5 to 2.0.

```
float   m_Rate = 1;
```

The PlaySound() function plays back the sound by calling the associated sound pool's play() function with the index of the sound m_SoundIndex, along with parameters describing how you should play the sound. By default, we play back the sound at full volume on the left and right speakers at the normal rate, without any looping of the sound. (See Listing 6-2.)

Listing 6-2. Playing Back a Sound

```
void PlaySound()
{
        /*
        *       soundID         a soundID returned by the load() function
                leftVolume      left volume value (range = 0.0 to 1.0)
                rightVolume     right volume value (range = 0.0 to 1.0)
                priority        stream priority (0 = lowest priority)
```

```
                loop              loop mode (0 = no loop, -1 = loop forever)
                rate              playback rate (1.0 = normal playback, range 0.5 to 2.0)

    *
    */
        m_SoundPool.play(m_SoundIndex, m_LeftVolume, m_RightVolume, m_Priority, m_Loop, m_Rate);
}
```

Modifying the Object3d Class

Next, the Object3d class has to be modified to use our new Sound class.

First, new sound-related variables have to be added.

The maximum number of sounds for a single Object3d class is held in MAX_SOUNDS.

```
private int MAX_SOUNDS = 5;
```

The current number of sounds available is held in m_NumberSounds.

```
private int m_NumberSounds = 0;
```

The sound effects are actually held in the array m_SoundEffects, and each element is of type Sound. Refer to "The Sound Class" section (preceding) for more information on the Sound class.

```
private Sound[] m_SoundEffects = new Sound[MAX_SOUNDS];
```

The m_SoundEffectsOn array holds booleans that allow you to turn the sound effects on or off.

```
private boolean[] m_SoundEffectsOn = new boolean[MAX_SOUNDS];
```

The AddSound() function adds a Sound object to the m_SoundEffects array at the next available slot. The slot number the object is stored in is returned if the operation is successful. If there are no available slots, a -1 is returned. (See Listing 6-3.)

Listing 6-3. Adding a Sound Object

```
int AddSound(Sound iSound)
{
        int Index = m_NumberSounds;

        if (m_NumberSounds >= MAX_SOUNDS)
        {
                return -1;
        }

        m_SoundEffects[Index] = iSound;
        m_NumberSounds++;

        return Index;
}
```

The SetSFXOnOff() function turns on or off all the sounds associated with this Object3d class. (See Listing 6-4.)

Listing 6-4. Turning the SFX On or Off

```
void SetSFXOnOff(boolean Value)
{
        for (int i = 0; i < m_NumberSounds;i++)
        {
                m_SoundEffectsOn[i] = Value;
        }
}
```

The AddSound() function creates a new Sound class object from a resource ResourceID and the sound pool Pool and adds the sound to the m_SoundEffects array that holds the sound effects for this class. (See Listing 6-5.)

Listing 6-5. Creating a New Sound from a Resource

```
int AddSound(SoundPool Pool, int ResourceID)
{
        int SoundIndex = -1;
        Sound SFX = new Sound(m_Context, Pool, ResourceID);
        SoundIndex = AddSound(SFX);

        return SoundIndex;
}
```

The PlaySound() function plays the sound effect that is associated with the SoundIndex input parameter for this class. Recall that each time a new sound is added to this Object3d class, an index handle is returned. You must use this index handle as input to the PlaySound() function, if you want to play the sound back. (See Listing 6-6.)

Listing 6-6. Playing the Sound

```
void PlaySound(int SoundIndex)
{
        if ((SoundIndex < m_NumberSounds) &&
        (m_SoundEffectsOn[SoundIndex]))
        {
                // Play Sound
                m_SoundEffects[SoundIndex].PlaySound();
        }
        else
        {
                Log.e("OBJECT3D", "ERROR IN PLAYING SOUND, SOUNDINDEX = " + SoundIndex);
        }
}
```

Hands-on Example: Sounds

In this section, we will add the playing of explosive sounds every time our two cubes from our previous chapters collide. Each cube will have its own explosive sound, which will be played every time the cubes collide. For this hands-on example, you will have to download the source code from the Source Code/Download area of apress.com and install it on your development system to a new work space. Two sound effects in the form of .wav files have been added to the project and are located in the res/raw directory.

Modifying the MyGLRenderer Class

For this hands-on example, we will need to add some code to the MyGLRenderer class.

The sound pool we will use to store and play back the sound from is located in m_SoundPool.

```
private SoundPool     m_SoundPool;
```

The sound index of the explosive sound for our first cube is stored in m_SoundIndex1.

```
private int m_SoundIndex1;
```

The sound index of the explosive sound for our second cube is stored in m_SoundIndex2.

```
private int m_SoundIndex2;
```

The m_SFXOn variable determines if the sound effects are set to on or off.

```
private boolean m_SFXOn = true;
```

The CreateSoundPool() function creates the sound pool that is used to create and store sounds for our cube collisions. (See Listing 6-7.)

The SoundPool constructor accepts the following parameters:

- **MaxStreams:** This is the maximum number of simultaneous streams for this SoundPool object, which is set to 10.
- **StreamType:** The audio stream type and that for games will normally be STREAM_MUSIC.
- **SrcQuality:** This is the sample-rate converter quality that currently has no effect and is set to 0 for the default.

Listing 6-7. Creating the Sound Pool

```
void CreateSoundPool()
{
        int maxStreams = 10;
        int streamType = AudioManager.STREAM_MUSIC;
        int srcQuality = 0;
```

```
        m_SoundPool = new SoundPool(maxStreams, streamType, srcQuality);

        if (m_SoundPool == null)
        {
                Log.e("RENDERER " , "m_SoundPool creation
failure!!!!!!!!!!!!!!!!!!!!!!!!!!!!!!!!!!!!!!!!!!!!!!!!!!!!!!!!!!!!");
        }
}
```

The `CreateSound()` function creates and adds a sound to our cubes, using as input the sound pool and the resource id of the specific sound effect. The sound effects for each cube are also turned on. (See Listing 6-8.)

Listing 6-8. Creating the Sounds for Our Cubes

```
void CreateSound(Context iContext)
{
        m_SoundIndex1 = m_Cube.AddSound(m_SoundPool, R.raw.explosion2);
        m_Cube.SetSFXOnOff(m_SFXOn);

        m_SoundIndex2 = m_Cube2.AddSound(m_SoundPool, R.raw.explosion5);
        m_Cube2.SetSFXOnOff(m_SFXOn);
}
```

We create the sound pool and the sound effects for each of the cubes by calling `CreateSoundPool()` and `CreateSound()` from the `onSurfaceCreated()` function, which is called when our OpenGL surface is created. (See Listing 6-9.)

Listing 6-9. Creating the Sound Pool and Sound Effects for Our Cubes

```
@Override
public void onSurfaceCreated(GL10 unused, EGLConfig config)
{
        m_PointLight = new PointLight(m_Context);
        SetupLights();

        // Create a 3d Cube
        CreateCube(m_Context);

        // Create a Second Cube
        CreateCube2(m_Context);

        // Create a new gravity grid
        CreateGrid(m_Context);

        // Create SFX
        CreateSoundPool();
        CreateSound(m_Context);
}
```

Next, we must modify the onDrawFrame() function, to play the collision sounds associated with each of the cubes. Each cube plays its own explosion sound through the PlaySound() function, using the sound index associated with the sound. See the highlighted code in Listing 6-10.

Listing 6-10. Modifying the onDrawFrame() function

```
@Override
public void onDrawFrame(GL10 unused)
{
        GLES20.glClearColor(0.0f, 0.0f, 0.0f, 1.0f);
        GLES20.glClear( GLES20.GL_DEPTH_BUFFER_BIT | GLES20.GL_COLOR_BUFFER_BIT);

        m_Camera.UpdateCamera();

        ///////////////////////////// Update Object Physics
        // Cube1
        m_Cube.UpdateObject3d();
        boolean HitGround = m_Cube.GetObjectPhysics().GetHitGroundStatus();
        if (HitGround)
        {
                m_Cube.GetObjectPhysics().ApplyTranslationalForce(m_Force1);
                m_Cube.GetObjectPhysics().ApplyRotationalForce(m_RotationalForce, 10.0f);
                m_Cube.GetObjectPhysics().ClearHitGroundStatus();
         }

        // Cube2
        m_Cube2.UpdateObject3d();

        // Process Collisions
        Physics.CollisionStatus TypeCollision =
        m_Cube.GetObjectPhysics().CheckForCollisionSphereBounding(m_Cube, m_Cube2);

        if ((TypeCollision == Physics.CollisionStatus.COLLISION) ||
            (TypeCollision == Physics.CollisionStatus.PENETRATING_COLLISION))
        {
                m_Cube.GetObjectPhysics().ApplyLinearImpulse(m_Cube, m_Cube2);
                // SFX
                m_Cube.PlaySound(m_SoundIndex1);
                m_Cube2.PlaySound(m_SoundIndex2);
        }

        ///////////////////////////// Draw Objects
        m_Cube.DrawObject(m_Camera, m_PointLight);
        m_Cube2.DrawObject(m_Camera, m_PointLight);

        ///////////////////////////// Update and Draw Grid
        UpdateGravityGrid();
        m_Grid.DrawGrid(m_Camera);
}
```

The final task is to run our project. You should hear the collision sounds play each time the cubes hit each other.

Overview of a Heads-Up Display

In this section, I will cover the basic features of our heads-up display, as well as the necessary classes that we will need to create to support the HUD.

Overview of Our HUD

Our HUD is composed of components of the HUDItem class. The actual graphic images for each HUD item are a BillBoard class that implements a 2D billboarding system. In billboarding, a flat rectangle with the image of the items we want to display on the HUD, such as scores and the player's health, are placed in front of the camera and turned toward the camera, so that the images appear flat (see Figure 6-1). HUD items are updated by directly copying the new graphics data to the texture associated with the HUD item.

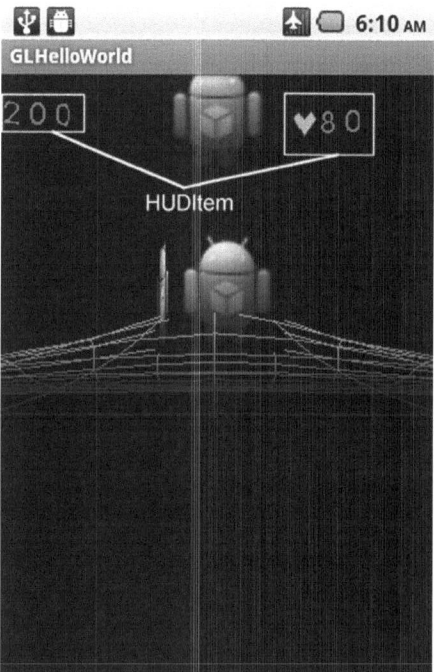

Figure 6-1. *HUD diagram*

Creating the BillBoard Class

In billboarding, the basic idea is to take a 2D rectangular image and turn it, so that it faces the camera. Generally, this is used as a cheap way to make a 2D image look three-dimensional. Billboarding is accomplished in our BillBoard class, which I cover in this section.

The BillBoard class is derived from the Cube class, which is derived from the Object3d class.

```
public class BillBoard extends Cube
```

The constructor for the BillBoard class is shown in Listing 6-11. The constructor first calls the constructor for the superclass Cube. Next, the scale for the billboard is set to normal along the x and y local axes and minimized on the z axis, to make the billboard as thin as possible.

Listing 6-11. The BillBoard Class Constructor

```
BillBoard(Context iContext,
        MeshEx iMeshEx,
        Texture[] iTextures,
        Material iMaterial,
        Shader iShader )
{
        super(iContext, iMeshEx, iTextures, iMaterial, iShader );
        Vector3 Scale = new Vector3(1.0f,1.0f,0.1f);
        m_Orientation.SetScale(Scale);
}
```

The SetBillBoardTowardCamera() function is where the billboard is actually turned toward the camera or viewer.

The process of implementing billboarding is as follows:

1. Get the front vector of the billboard object and project it onto the xz plane, which is ForwardVecProj.

2. Get the billboard position and project it onto the xz plane, which is BillBoardPositionProj.

3. Get the position of the camera and project it onto the xz plane, which is CameraPositionProj.

4. Calculate the vector from the billboard to the camera, which is Bill2CameraVecProj.

5. Find the angle between the forward or front vector of the billboard object and the camera, which is Theta.

6. Calculate the rotation axis by calculating the cross-product of the billboard's front vector and the billboard to camera vector to form the rotation axis.

7. Rotate the billboard toward the camera.

See Figure 6-2 for a visual depiction of the preceding steps, and see Listing 6-12 for the code that implements the billboarding procedure.

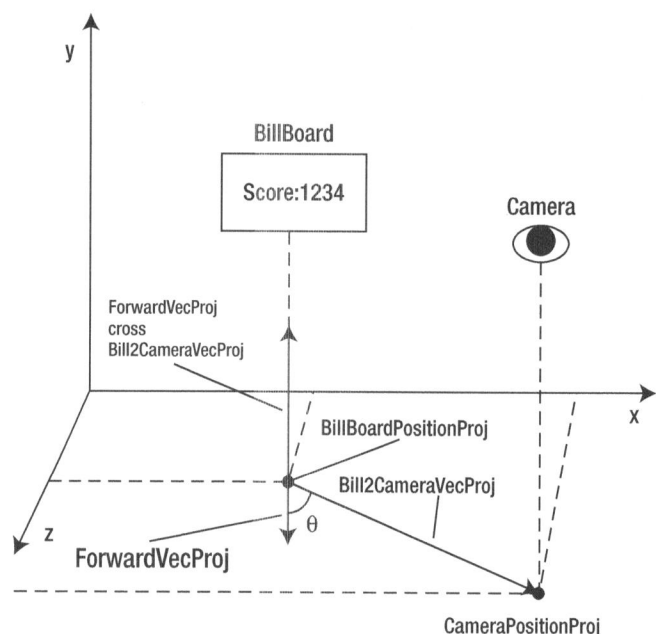

Figure 6-2. *Billboarding*

Listing 6-12. *Billboarding Procedure*

```
void SetBillBoardTowardCamera(Camera Cam)
{
        // 1. Get Front Vector of Billboard Object projected on xz plane
        Vector3 ForwardVecProj = new Vector3(m_Orientation.GetForwardWorldCoords().x,  0,
m_Orientation.GetForwardWorldCoords().z);
        // 2. Get The BillBoard Position projected on xz plane
        Vector3 BillBoardPositionProj = new Vector3(m_Orientation.GetPosition().x, 0,
m_Orientation.GetPosition().z);
        // 3. Get Position of Camera on 2d XZ Plane
        Vector3 CameraPositionProj = new Vector3(Cam.GetCameraEye().x, 0, Cam.GetCameraEye().z);

        // 4. Calculate Vector from Billboard to Camera
        Vector3 Bill2CameraVecProj = Vector3.Subtract(CameraPositionProj , BillBoardPositionProj);
        Bill2CameraVecProj.Normalize();

        // 5. Find Angle between forward of Billboard object and camera
        // P = forwardxy
        // Q = Vec_Bill_Camera
        // P and Q are normalized Vectors
        // P.Q = P*Q*cos(theta)
        // P.Q/P*Q = cos(theta)
        // acos(P.Q/P*Q) = theta;

        // P.Q > 0 then angle between vectors is less than 90 deg
        // P.Q < 0 then angle between vectors is greater than 90 deg.
        // P.Q = 0 then angle between vector is exactly 90 degs.
```

```
        // Get current theta
        // returns 0-PI radians
        float Theta = (float)Math.acos(ForwardVecProj.DotProduct(Bill2CameraVecProj));
        float DegreeTheta = Theta * 180.0f/Physics.PI;

        // 6. Cross Product to form rotation axis
        Vector3 RotAxis = Vector3.CrossProduct(ForwardVecProj, Bill2CameraVecProj);

        // 7. Rotate BillBoard Toward Camera
        // cos in radians
        if ((Math.cos(Theta) < 0.9999) && (Math.cos(Theta) > -0.9999))
        {
                m_Orientation.SetRotationAxis(RotAxis);
                m_Orientation.AddRotation(DegreeTheta);
        }
        else
        {
                //Log.e( "BILLBOARD", "No Cylindrical Rotation!! , Theta = " + Theta);
        }
}
```

Finally, the UpdateObject3d() function is called continuously to update the orientation of the billboard object by calling the SetBillBoardTowardCamera() function discussed in Listing 6-13.

Listing 6-13. Updating the Billboard

```
void UpdateObject3d(Camera Cam)
{
        super.UpdateObject3d();
        SetBillBoardTowardCamera(Cam);
}
```

Creating the BillBoardFont Class

The BillBoardFont class is used to associate a specific character with the billboard texture image.

The BillBoardFont class is derived from the BillBoard class.

```
public class BillBoardFont extends BillBoard
```

The variable m_Character is used to hold the alphanumeric value that represents the billboard texture for this class.

```
private char m_Character;
```

The BillBoardFont() constructor is shown in Listing 6-14. First, the constructor of the superclass is called, which would be the constructor for the BillBoard class. Then, the character that this billboard represents is set in the variable m_Character.

Listing 6-14. BillBoardFont() Constructor

```
BillBoardFont(Context iContext,  MeshEx iMeshEx, Texture[] iTextures, Material iMaterial, Shader
iShader, char Character)
{
        super(iContext, iMeshEx, iTextures, iMaterial, iShader );
        m_Character = Character;
}
```

The GetCharacter() function returns the character that this billboard is associated with.

```
char GetCharacter() {return m_Character;}
```

The SetCharacter() function sets the character that will be associated with this billboard.

```
void SetCharacter(char value) { m_Character = value;}
```

The IsFontCharacter() function returns true if the input parameter value is the alphanumeric character that this billboard texture represents and false otherwise. (See Listing 6-15.)

Listing 6-15. Testing for a Character Value

```
boolean IsFontCharacter(char value)
{
        if (m_Character == value)
        {
                return true;
        }
        else
        {
                return false;
        }
}
```

Modifying the Texture Class

Next, we have to modify the Texture class to add a new function.

The CopySubTextureToTexture() function copies the texture in the input parameter BitmapImage to the bitmap associated with the Texture object. The main purpose in using this function is to update the data on the HUD, such as score, health, etc. Each letter and number that is displayed on the HUD has a separate graphic bitmap associated with it. For example, the HUD item for score has a billboard texture associated with it. When the score has to be updated, individual bitmaps are copied to the billboard texture using the CopySubTextureToTexture() function and placed at XOffset, YOffset location on the bitmap. (See Listing 6-16.)

Listing 6-16. Copying Over a Texture

```
void CopySubTextureToTexture(int Level, int XOffset, int YOffset, Bitmap BitmapImage)
{
        // Copies the texture in BitmapImage to the bitmap associated with this Texture object
        /*
        public static void texSubImage2D (int target, int level, int xoffset, int yoffset,
Bitmap bitmap)
        Added in API level 1
        Calls glTexSubImage2D() on the current OpenGL context. If no context is current the
        behavior is the same as calling glTexSubImage2D() with no current context, that is,
eglGetError()
        will return the appropriate error. Unlike glTexSubImage2D() bitmap cannot be null and will
raise
        an exception in that case. All other parameters are identical to those used for
glTexSubImage2D().
        */
        ActivateTexture();
        GLUtils.texSubImage2D(GLES20.GL_TEXTURE_2D, Level, XOffset, YOffset, BitmapImage);
        CheckGLError("GLUtils.texSubImage2D");
}
```

Creating the BillBoardCharacterSet Class

The BillBoardCharacterSet class holds the character fonts for use with the HUD. The Settext() function sets the text you want to display, then you use the RenderToBillBoard() function to put this text on the input BillBoard object's texture.

The maximum number of characters in this set is specified by MAX_CHARACTERS.

```
static int MAX_CHARACTERS = 50;
```

The number of characters actually in the character set is held in m_NumberCharacters.

```
private int m_NumberCharacters = 0; // Number characters in the character set
```

The character set itself is made up of an array of BillBoardFonts.

```
private BillBoardFont[] m_CharacterSet = new BillBoardFont[MAX_CHARACTERS];
```

The text to be placed on the billboard when the Settext() function is called is stored in m_Text, which is a character array of MAX_CHARACTERS_TEXT in length.

```
private int MAX_CHARACTERS_TEXT = 100;
private char[] m_Text = new char[MAX_CHARACTERS_TEXT];
```

The BillBoardFont objects that correspond to the characters in m_Text are stored in m_TextBillBoard.

```
private BillBoardFont[] m_TextBillBoard = new BillBoardFont[MAX_CHARACTERS_TEXT];
```

The GetNumberCharactersInSet() function returns the current number of characters in the character set.

```
int GetNumberCharactersInSet() {return m_NumberCharacters;}
```

The GetCharacter() function returns the BillBoardFont object associated with this character set located at place index. (See Listing 6-17.)

Listing 6-17. GetCharacter Function

```
BillBoardFont GetCharacter(int index)
{
        BillBoardFont Font = null;

        if (index < m_NumberCharacters)
        {
                Font = m_CharacterSet[index];
        }

        return Font;
}
```

The GetFontWidth() function gets the width of the bitmap that represents a character for this character set. (See Listing 6-18.)

Listing 6-18. Getting the Width of the Font

```
int GetFontWidth()
{
        int Width = 0;
        if (m_NumberCharacters > 0)
        {
                BillBoardFont Character = m_CharacterSet[0];
                Texture Tex = Character.GetTexture(0);
                Bitmap Image = Tex.GetTextureBitMap();
                Width = Image.getWidth();
        }

        return Width;
}
```

The GetFontHeight() function returns the height of the bitmap that represents a character for this character set. (See Listing 6-19.)

Listing 6-19. Getting the Font Height

```
int GetFontHeight()
{
        int Height = 0;
        if (m_NumberCharacters > 0)
        {
                BillBoardFont Character = m_CharacterSet[0];
                Texture Tex = Character.GetTexture(0);
```

```
                    Bitmap Image = Tex.GetTextureBitMap();
                    Height = Image.getHeight();
        }

        return Height;
}
```

The AddToCharacterSet() function adds a BillBoardFont object to the character set. The function returns true if it is successfully added or false if there is not enough room. (See Listing 6-20.)

Listing 6-20. Adding BillBoardFont Object to the Character Set

```
boolean AddToCharacterSet(BillBoardFont Character)
{
        if (m_NumberCharacters < MAX_CHARACTERS)
        {
                m_CharacterSet[m_NumberCharacters] = Character;
                m_NumberCharacters++;
                return true;
        }
        else
        {
                Log.e("BILLBOARD CHARACTER SET" , "NOT ENOUGH ROOM TO ADD ANOTHER CHARACTER TO
CHARACTER SET");
                return false;
        }
}
```

The FindBillBoardCharacter() function searches for the input character within the character set. If it is found, then the corresponding BillBoardFont object is returned. Null is returned otherwise. (See Listing 6-21.)

Listing 6-21. Searching the Character Set

```
BillBoardFont FindBillBoardCharacter(char character)
{
        BillBoardFont Font = null;
        for (int i = 0; i < m_NumberCharacters; i++)
        {
                if (m_CharacterSet[i].IsFontCharacter(character))
                {
                        Font = m_CharacterSet[i];
                }
        }
        return Font;
}
```

The SetText() function converts an array of characters to the corresponding array of BillBoardFont objects stored in the m_TextBillBoard array. (See Listing 6-22.)

Listing 6-22. Setting the Text for Rendering

```
void SetText(char[] Text)
{
        String TextStr = new String(Text);
        TextStr = TextStr.toLowerCase();
        m_Text = TextStr.toCharArray();

        for (int i = 0; i < m_Text.length; i++)
        {
                BillBoardFont Character = FindBillBoardCharacter(m_Text[i]);
                if (Character != null)
                {
                        m_TextBillBoard[i] = Character;
                }
                else
                {
                        Log.e("CHARACTER SET ERROR" , "SETTEXT ERROR , " + m_Text[i] + "
NOT FOUND!!!!!");
                }
        }
}
```

The `DrawFontToComposite()` function copies the bitmap image in the BillBoardFont object `Obj` into the bitmap image on the BillBoard object `Composite`, starting at location X , Y. The width of the destination texture in the `Composite` variable is also tested to make sure that the source texture fits into the destination texture. (See Listing 6-23.)

Listing 6-23. Drawing a Font from the Character Set to a BillBoard Object

```
void DrawFontToComposite(BillBoardFont Obj, int X, int Y, BillBoard Composite)
{
        Texture TexSource = Obj.GetTexture(0);
        Bitmap BitmapSource = TexSource.GetTextureBitMap();
        int BitmapSourceWidth = BitmapSource.getWidth();

        Texture TexDest = Composite.GetTexture(0);
        Bitmap BitmapDest = TexDest.GetTextureBitMap();
        int BitmapDestWidth = BitmapDest.getWidth();

        // Put Sub Image on Composite
        int XEndTexture = X + BitmapSourceWidth;
        if (XEndTexture >= BitmapDestWidth)
        {
                Log.e("BillBoardCharacterSet::DrawFontToComposite" , "ERROR Overwriting Dest Texture,
Last X Position To Write = " + XEndTexture + ", Max Destination Width = " + BitmapDestWidth);
        }
        else
        {
                TexDest.CopySubTextureToTexture(0, X, Y, BitmapSource);
        }
}
```

The RenderToBillBoard() function renders the text that is set by the SetText() function to the bitmap texture in the Composite input variable at location XOffset, YOffset on the bitmap, with 0,0 indicating the upper left-hand corner of the texture. Each character graphic is drawn on the Composite using the DrawFontToComposite() function. (See Listing 6-24.)

Listing 6-24. Rendering the Text to a BillBoard

```
void RenderToBillBoard(BillBoard Composite, int XOffset, int YOffset)
{
        int Length = m_Text.length;
        for (int i = 0; i < Length; i++)
        {
                BillBoardFont Character = m_TextBillBoard[i];
                if (Character != null)
                {
                        // Draw this font to the composite by copying the bitmap image data
                        Texture Tex = Character.GetTexture(0);
                        Bitmap Image = Tex.GetTextureBitMap();
                        int Width = Image.getWidth();
                        int XCompositeOffset = XOffset + (Width * i);

                        DrawFontToComposite(Character, XCompositeOffset, YOffset, Composite);
                }
        }
}
```

Creating the HUDItem Class

The HUDItem class holds the data for an individual HUD item, such as a score or health statistic.

If this HUDItem is in use and valid, then m_ItemValid is true; otherwise, it is false.

```
private boolean m_ItemValid;
```

The name that is used to reference this HUD item is held in m_ItemName.

```
private String m_ItemName;
```

The numeric value associated with this HUD item, if there is one, is held in m_NumericalValue.

```
private int m_NumericalValue;
```

The text value associated with this HUD item, if any, is held in m_TextValue.

```
private String m_TextValue = null;
```

The position of the HUD item in local HUD coordinates with x = 0 and y = 0 being the center of the camera view.

```
private Vector3 m_ScreenPosition;
```

The m_Text variable holds the text and text character graphics associated with the HUD item, if any.

```
private BillBoardCharacterSet m_Text;
```

The m_Icon variable holds an icon associated with the HUD item, if any. A heart graphic for health statistics is an example.

```
private Texture m_Icon;
```

The m_HUDImage variable stores the actual full graphic image for a HUD item. Alphanumeric characters and graphic icons are copied to this billboard for final display on the HUD.

```
private BillBoard m_HUDImage;
```

If m_Dirty is true, then the m_HUDImage billboard must be updated, because the item has changed value. For example, the player's score has changed.

```
private boolean m_Dirty = false;
```

If the HUD item is visible, then m_IsVisible is true. It is false otherwise.

```
private boolean m_IsVisible = true;
```

There are also functions to provide access to the preceding private variables from outside the class. Please refer to the actual code from the Source Code/Download area of apress.com for more information.

The constructor for the HUDItem class is shown in Listing 6-25.

Listing 6-25. HUDItem Constructor

```
HUDItem(String ItemName,
        int NumericalValue,
        Vector3 ScreenPosition,
        BillBoardCharacterSet Text,
        Texture Icon,
        BillBoard HUDImage)
{
        m_ItemName = ItemName;
        m_NumericalValue= NumericalValue;
        m_ScreenPosition= ScreenPosition;
        m_Text = Text;
        m_Icon = Icon;
        m_HUDImage= HUDImage;
}
```

Creating the HUD Class

Now, we need to create the actual HUD class that will represent our HUD.

The variable MAX_HUDITEMS holds the maximum number of items that can be in the HUD, which is set to 10.

```
private int MAX_HUDITEMS = 10;
```

The m_HUDItems array holds the items for this HUD.

```
private HUDItem[] m_HUDItems = new HUDItem[MAX_HUDITEMS];
```

A blank texture consisting of a black background is held in m_BlankTexture.

```
private Texture m_BlankTexture;
```

The HUD constructor is shown in Listing 6-26. The constructor creates and loads a new blank texture from the R.drawable.blankhud resource and assigns it to m_BlankTexture. All the slots for HUD items are initialized with an empty item and set to invalid.

Listing 6-26. HUD Constructor

```
HUD(Context iContext)
{
        m_BlankTexture = new Texture(iContext, R.drawable.blankhud);

        String  ItemName = "NONE";
        int     NumericalValue= 0;
        Vector3 ScreenPosition= null;

        BillBoardCharacterSet CharacterSet = null;
        Texture Icon = null;
        BillBoard HUDImage = null;

        // Initialize m_HUDItems
        for (int i = 0; i < MAX_HUDITEMS; i++)
        {
                m_HUDItems[i] = new HUDItem(ItemName, NumericalValue, ScreenPosition,
CharacterSet,Icon, HUDImage);
                m_HUDItems[i].SetItemValidState(false);
        }
}
```

The FindEmptyHUDItemSlot() function finds and returns the index of an empty HUD item slot or -1 if no slots are available. (See Listing 6-27.)

Listing 6-27. Finding an Empty HUD Item Slot

```
int FindEmptyHUDItemSlot()
{
        int EmptySlot = -1;
        for (int i = 0; i < MAX_HUDITEMS; i++)
        {
                if (m_HUDItems[i].IsValid() == false)
                {
                        return i;
                }
        }
        return EmptySlot;
}
```

The AddHUDItem() function adds a new item into the HUD. The item is set to be a valid HUD item and is also set to be dirty, because we must have this new item rendered onto the HUD after it is added. (See Listing 6-28.)

Listing 6-28. Adding a New HUD Item

```
boolean AddHUDItem(HUDItem Item)
{
        boolean result = false;

        int EmptySlot =  FindEmptyHUDItemSlot();
        if (EmptySlot >= 0)
        {
                m_HUDItems[EmptySlot] = Item;
                m_HUDItems[EmptySlot].SetItemValidState(true);
                m_HUDItems[EmptySlot].SetDirty(true);
                result = true;
        }
        return result;
}
```

The FindHUDItem() function returns the index of the HUD item that has ID as a name or -1, if there is no such item. (See Listing 6-29.)

Listing 6-29. Finding an Item on the HUD Using an ID

```
int FindHUDItem(String ID)
{
        int Slot = -1;
        for (int i = 0; i < MAX_HUDITEMS; i++)
        {
                if ((m_HUDItems[i].GetName() == ID) &&
                    (m_HUDItems[i].IsValid()))
                {
                        Slot = i;
                }
        }
        return Slot;
}
```

The GetHUDItem() function returns a HUDItem object if there is an item with the name ItemID in the HUD. If there is no such item in the HUD, then a null pointer is returned. (See Listing 6-30.)

Listing 6-30. Retrieving a HUD Item by the Item's ID

```
HUDItem GetHUDItem(String ItemID)
{
        HUDItem Item = null;
        int Slot = FindHUDItem(ItemID);
```

```
        if (Slot >= 0)
        {
                Item = m_HUDItems[Slot];
        }
        return Item;
}
```

The DeleteHUDItem() function deletes a HUD item named ItemName from the HUD, if it exists, by setting its state to invalid. It then returns true. If the HUD item could not be found, the function returns false. (See Listing 6-31.)

Listing 6-31. Deleting an Item in the HUD

```
boolean DeleteHUDItem(String ItemName)
{
        boolean result = false;
        int Slot =  FindHUDItem(ItemName);
        if (Slot >= 0)
        {
                m_HUDItems[Slot].SetItemValidState(false);
                result = true;
        }
        return result;
}
```

The UpdateHUDItemNumericalValue() function finds and updates the numerical value of the HUD item that matches ID. It also sets the dirty status to true, so that the updated graphics data will be copied to the m_HUDImage billboard texture associated with the HUD item. (See Listing 6-32.)

Listing 6-32. Updating Numerical HUD Items

```
void UpdateHUDItemNumericalValue(String ID, int NumericalValue)
{
        int Slot =  FindHUDItem(ID);
        HUDItem HItem = m_HUDItems[Slot];
        if (HItem != null)
        {
                // Update Key fields in HUDItem
                HItem.SetNumericalValue(NumericalValue);
                HItem.SetDirty(true);
        }
}
```

The UpdateHUDItem() function updates the HUD item Item using the camera Cam. What this means is that

1. The position in the world for this HUD item is calculated based on the camera's position, orientation, and the local HUD coordinates of the HUD item. In local HUD coordinates, the x = 0 and y = 0 means the center of the camera viewpoint. So, local coordinates of (1,2) mean that the HUD item is placed 1 unit to the right of center and 2 units up above the center. When z = 0, this means the HUD item is placed at the position of the near projection plane, which would probably make the item unviewable, so you require a positive value here, such as 0.5.

2. If the HUD item is dirty, its billboard texture is updated. That is the billboard texture associated with the HUDItem object, which is m_HUDImage updated.

3. First m_HUDImage is cleared by copying a blank texture over it.

4. If there is an icon associated with this HUD item, it is copied to m_HUDImage.

5. The numerical value of the HUD item is rendered to m_HUDImage.

6. If there is a string value associated with the HUD item, then it is rendered to m_HUDImage.

7. The HUD item has been updated as "cleaned," so it is no longer set to "dirty."

8. The HUD item is positioned in the world using the location calculated in the first step.

9. The m_HUDImage's UpdateObject3d() function is then called, so that the billboard is turned to face the camera.

See Listing 6-33 for the source code.

Listing 6-33. Updating the HUD Item

```
void UpdateHUDItem(Camera Cam, HUDItem Item)
{
        // Update HUDItem position and rotation in the 3d world
        // to face the camera.
        Vector3 PositionLocal = Item.GetLocalScreenPosition();
        Vector3 PositionWorld = new Vector3(0,0,0);

        Vector3 CamPos = new Vector3(Cam.GetCameraEye().x, Cam.GetCameraEye().y, Cam.GetCameraEye().z);
        Vector3 CameraForward = Cam.GetOrientation().GetForwardWorldCoords();
        Vector3 CameraUp = Cam.GetOrientation().GetUpWorldCoords();
        Vector3 CameraRight = Cam.GetOrientation().GetRightWorldCoords();
```

```
// Local Camera Offsets
Vector3 CamHorizontalOffset = Vector3.Multiply(PositionLocal.x, CameraRight);
Vector3 CamVerticalOffset = Vector3.Multiply(PositionLocal.y, CameraUp);

float ZOffset = Cam.GetProjNear() + PositionLocal.z;
Vector3 CamDepthOffset = Vector3.Multiply(ZOffset, CameraForward);

// Create Final PositionWorld Vector
PositionWorld = Vector3.Add(CamPos, CamHorizontalOffset);
PositionWorld = Vector3.Add(PositionWorld, CamVerticalOffset);
PositionWorld = Vector3.Add(PositionWorld, CamDepthOffset);

// Put images from icon and numerical data onto the composite hud texture
BillBoard HUDComposite = Item.GetHUDImage();
Texture HUDCompositeTexture = HUDComposite.GetTexture(0);
Bitmap  HUDCompositeBitmap = HUDCompositeTexture.GetTextureBitMap();

BillBoardCharacterSet Text = Item.GetText();

int FontWidth = Text.GetFontWidth();
Texture Icon = Item.GetIcon();
int IconWidth = 0;

if (Item.IsDirty())
{
        // Clear Composite Texture;
        Bitmap BlankBitmap = m_BlankTexture.GetTextureBitMap();
        HUDCompositeTexture.CopySubTextureToTexture(0, 0, 0, BlankBitmap);

        if (Icon != null)
        {
                // Draw Icon on composite
                Bitmap HealthBitmap = Icon.GetTextureBitMap();
                IconWidth = HealthBitmap.getWidth();
                HUDCompositeTexture.CopySubTextureToTexture(0,0,0, HealthBitmap);
        }

        // Update Numerical Value and render to composite billboard
        String text = String.valueOf(Item.GetNumbericalValue());
        Text.SetText(text.toCharArray());
        Text.RenderToBillBoard(HUDComposite, IconWidth, 0);

        // Update Text Value and render to composite billboard
        String TextValue = Item.GetTextValue();
        if (TextValue != null)
        {
                int XPosText = IconWidth + (text.length() * FontWidth);
                Text.SetText(TextValue.toCharArray());
                Text.RenderToBillBoard(HUDComposite, XPosText, 0);
        }

                Item.SetDirty(false);
}
```

```
        HUDComposite.m_Orientation.GetPosition().Set(PositionWorld.x, PositionWorld.y,
PositionWorld.z);

                // Update BillBoard orientation
        HUDComposite.UpdateObject3d(Cam);
}
```

The UpdateHUD() function updates every HUD item that is visible and valid by calling the
UpdateHUDItem() function. (See Listing 6-34.)

Listing 6-34. Updating the HUD

```
void UpdateHUD(Camera Cam)
{
        for (int i = 0; i < MAX_HUDITEMS; i++)
        {
                if (m_HUDItems[i].IsValid() && m_HUDItems[i].IsVisible())
                {
                        UpdateHUDItem(Cam,m_HUDItems[i]);
                }
        }
}
```

The RenderHUD() function renders the m_HUDImage BillBoard object for each HUD item that is visible
and valid. (See Listing 6-35.)

Listing 6-35. Rendering the HUD

```
void RenderHUD(Camera Cam, PointLight light)
{
        for (int i = 0; i < MAX_HUDITEMS; i++)
        {
                if (m_HUDItems[i].IsValid()&& m_HUDItems[i].IsVisible())
                {
                        HUDItem Item = m_HUDItems[i];
                        BillBoard HUDComposite = Item.GetHUDImage();
                        HUDComposite.DrawObject(Cam, light);
                }
        }
}
```

Modifying the Object3d Class

Next, we need to add some code to the Object3d class, so that the black portions of our HUD items
are transparent when displayed on the screen with other objects.

The m_Blend variable is set to true if we intend to combine colors from an object being rendered with
colors already in the background and false otherwise.

```
private boolean m_Blend = false;
```

The GetMaterial() function returns a reference to the object's material.

```
Material GetMaterial() {return m_Material;}
```

The SetBlend() function sets the m_Blend variable.

```
void SetBlend(boolean value) { m_Blend = value; }
```

The DrawObject() function code is added to enable blending if m_Blend is true and disable blending after the object is rendered. See the code in bold in Listing 6-36.

Listing 6-36. Modifying the DrawObject() Function

```
void DrawObject(Camera Cam, PointLight light)
{
        if (m_Blend)
        {
                GLES20.glEnable(GLES20.GL_BLEND);
                GLES20.glBlendFunc(GLES20.GL_SRC_ALPHA, GLES20.GL_ONE);
        }

        if (m_Visible)
        {
                DrawObject(Cam, light, m_Orientation.GetPosition(), m_Orientation.GetRotationAxis(),
m_Orientation.GetScale());
        }

        if (m_Blend)
        {
                GLES20.glDisable(GLES20.GL_BLEND);
        }
}
```

Drone Grid Case Study: Creating the HUD

In this section, we will create the HUD for our Drone Grid case-study game. The HUD will consist of two HUD items. One item will be the player's score, and the other item will be the player's health. You will need to download the Android project for this chapter, if you haven't done so already, and install it into a new work space. The project will contain graphics for the fonts and an icon for the player's health.

Modifying the MyGLRenderer Class

We will have to alter the MyGLRenderer class from the previous hands-on example dealing with sound.

The m_CharacterSetTextures array holds the character set textures for our HUD, which consist of letters, numbers, and extra characters.

```
private Texture[] m_CharacterSetTextures = new Texture[BillBoardCharacterSet.MAX_CHARACTERS];
```

The m_CharacterSet variable holds the character set we will use for the HUD.

```
private BillBoardCharacterSet m_CharacterSet = null;
```

The m_HUDTexture holds a texture used for a HUD item.

```
private Texture m_HUDTexture = null;
```

The m_HUDComposite holds a reference to a BillBoard that will be used for a HUD item.

```
private BillBoard m_HUDComposite = null;
```

The HUD is m_HUD.

```
private HUD m_HUD = null;
```

The player's health is stored in m_Health.

```
private int m_Health = 100;
```

The player's score is stored in m_Score.

```
private int m_Score = 0;
```

The textures needed for m_CharacterSet are initialized in the CreateCharacterSetTextures() function and placed in m_CharacterSetTextures. (See Listing 6-37.)

Listing 6-37. Creating the Textures for the Character Set

```
void CreateCharacterSetTextures(Context iContext)
{
        // Numeric
        m_CharacterSetTextures[0] = new Texture(iContext, R.drawable.charset1);
        m_CharacterSetTextures[1] = new Texture(iContext, R.drawable.charset2);
        m_CharacterSetTextures[2] = new Texture(iContext, R.drawable.charset3);
        m_CharacterSetTextures[3] = new Texture(iContext, R.drawable.charset4);
        m_CharacterSetTextures[4] = new Texture(iContext, R.drawable.charset5);
        m_CharacterSetTextures[5] = new Texture(iContext, R.drawable.charset6);
        m_CharacterSetTextures[6] = new Texture(iContext, R.drawable.charset7);
        m_CharacterSetTextures[7] = new Texture(iContext, R.drawable.charset8);
        m_CharacterSetTextures[8] = new Texture(iContext, R.drawable.charset9);
        m_CharacterSetTextures[9] = new Texture(iContext, R.drawable.charset0);

        // Alphabet
        m_CharacterSetTextures[10] = new Texture(iContext, R.drawable.charseta);
        m_CharacterSetTextures[11] = new Texture(iContext, R.drawable.charsetb);
        m_CharacterSetTextures[12] = new Texture(iContext, R.drawable.charsetc);
        m_CharacterSetTextures[13] = new Texture(iContext, R.drawable.charsetd);
        m_CharacterSetTextures[14] = new Texture(iContext, R.drawable.charsete);
        m_CharacterSetTextures[15] = new Texture(iContext, R.drawable.charsetf);
        m_CharacterSetTextures[16] = new Texture(iContext, R.drawable.charsetg);
```

```
        m_CharacterSetTextures[17] = new Texture(iContext, R.drawable.charseth);
        m_CharacterSetTextures[18] = new Texture(iContext, R.drawable.charseti);
        m_CharacterSetTextures[19] = new Texture(iContext, R.drawable.charsetj);
        m_CharacterSetTextures[20] = new Texture(iContext, R.drawable.charsetk);
        m_CharacterSetTextures[21] = new Texture(iContext, R.drawable.charsetl);
        m_CharacterSetTextures[22] = new Texture(iContext, R.drawable.charsetm);
        m_CharacterSetTextures[23] = new Texture(iContext, R.drawable.charsetn);
        m_CharacterSetTextures[24] = new Texture(iContext, R.drawable.charseto);
        m_CharacterSetTextures[25] = new Texture(iContext, R.drawable.charsetp);
        m_CharacterSetTextures[26] = new Texture(iContext, R.drawable.charsetq);
        m_CharacterSetTextures[27] = new Texture(iContext, R.drawable.charsetr);
        m_CharacterSetTextures[28] = new Texture(iContext, R.drawable.charsets);
        m_CharacterSetTextures[29] = new Texture(iContext, R.drawable.charsett);
        m_CharacterSetTextures[30] = new Texture(iContext, R.drawable.charsetu);
        m_CharacterSetTextures[31] = new Texture(iContext, R.drawable.charsetv);
        m_CharacterSetTextures[32] = new Texture(iContext, R.drawable.charsetw);
        m_CharacterSetTextures[33] = new Texture(iContext, R.drawable.charsetx);
        m_CharacterSetTextures[34] = new Texture(iContext, R.drawable.charsety);
        m_CharacterSetTextures[35] = new Texture(iContext, R.drawable.charsetz);

        // Debug Symbols
        m_CharacterSetTextures[36] = new Texture(iContext, R.drawable.charsetcolon);
        m_CharacterSetTextures[37] = new Texture(iContext, R.drawable.charsetsemicolon);
        m_CharacterSetTextures[38] = new Texture(iContext, R.drawable.charsetcomma);
        m_CharacterSetTextures[39] = new Texture(iContext, R.drawable.charsetequals);
        m_CharacterSetTextures[40] = new Texture(iContext, R.drawable.charsetleftparen);
        m_CharacterSetTextures[41] = new Texture(iContext, R.drawable.charsetrightparen);
        m_CharacterSetTextures[42] = new Texture(iContext, R.drawable.charsetdot);
}
```

The SetUpHUDComposite() function initializes m_HUDComposite as a BillBoard object that will be used in the creation of a HUD item. The m_HUDComposite will be displaying the actual score or health level of the player. (See Listing 6-38.)

Listing 6-38. Setting Up a BillBoard Object for Use with a HUD Item

```
void SetUpHUDComposite(Context iContext)
{
        m_HUDTexture = new Texture(iContext, R.drawable.hud);

        Shader Shader = new Shader(iContext, R.raw.vsonelight, R.raw.fsonelight);    // ok
        MeshEx Mesh = new MeshEx(8,0,3,5,Cube.CubeData, Cube.CubeDrawOrder);

        // Create Material for this object
        Material Material1 = new Material();
        Material1.SetEmissive(1.0f, 1.0f, 1.0f);

        Texture[] Tex = new Texture[1];
        Tex[0] = m_HUDTexture;

        m_HUDComposite = new BillBoard(iContext, Mesh, Tex, Material1, Shader);
```

```
        // Set Intial Position and Orientation
        Vector3 Position = new Vector3(0.0f, 3.0f, 0.0f);
        Vector3 Scale = new Vector3(1.0f,0.1f,0.01f);

        m_HUDComposite.m_Orientation.SetPosition(Position);
        m_HUDComposite.m_Orientation.SetScale(Scale);
        m_HUDComposite.GetObjectPhysics().SetGravity(false);

        // Set black portion of HUD to transparent
        m_HUDComposite.GetMaterial().SetAlpha(1.0f);
        m_HUDComposite.SetBlend(true);
}
```

The CreateCharacterSet() function creates a new BillBoardCharacterSet object for use in our HUD. For each character, number, or other symbol we want to add to our character set, we create a new BillBoardFont object and initialize it with the texture and text value associated with it. Then we call AddToCharacterSet() with the newly created font to add this font into our character set. Also note that we use a different fragment shader than normal, which is R.raw.fsonelightnodiffuse. This fragment shader only uses the color of the texture to determine the final color of the fragment. (See Listing 6-39.)

Listing 6-39. Creating the Character Set

```
 void CreateCharacterSet(Context iContext)
{
        //Create Shader
        Shader Shader = new Shader(iContext, R.raw.vsonelight, R.raw.fsonelightnodiffuse);
        // ok

        // Create Debug Local Axis Shader
        MeshEx Mesh = new MeshEx(8,0,3,5,Cube.CubeData, Cube.CubeDrawOrder);

         // Create Material for this object
        Material Material1 = new Material();
        Material1.SetEmissive(1.0f, 1.0f, 1.0f);

        // Create Texture
        CreateCharacterSetTextures(iContext);

        // Setup HUD
        SetUpHUDComposite(iContext);

        m_CharacterSet = new BillBoardCharacterSet();

        int NumberCharacters = 43;
        char[] Characters = new char[BillBoardCharacterSet.MAX_CHARACTERS];
        Characters[0] = '1';
        Characters[1] = '2';
        Characters[2] = '3';
        Characters[3] = '4';
        Characters[4] = '5';
        Characters[5] = '6';
```

```
Characters[6] = '7';
Characters[7] = '8';
Characters[8] = '9';
Characters[9] = '0';

// AlphaBets
Characters[10] = 'a';
Characters[11] = 'b';
Characters[12] = 'c';
Characters[13] = 'd';
Characters[14] = 'e';
Characters[15] = 'f';
Characters[16] = 'g';
Characters[17] = 'h';
Characters[18] = 'i';
Characters[19] = 'j';
Characters[20] = 'k';
Characters[21] = 'l';
Characters[22] = 'm';
Characters[23] = 'n';
Characters[24] = 'o';
Characters[25] = 'p';
Characters[26] = 'q';
Characters[27] = 'r';
Characters[28] = 's';
Characters[29] = 't';
Characters[30] = 'u';
Characters[31] = 'v';
Characters[32] = 'w';
Characters[33] = 'x';
Characters[34] = 'y';
Characters[35] = 'z';

// Debug
Characters[36] = ':';
Characters[37] = ';';
Characters[38] = ',';
Characters[39] = '=';
Characters[40] = '(';
Characters[41] = ')';
Characters[42] = '.';

for (int i = 0; i < NumberCharacters; i++)
{
        Texture[] Tex = new Texture[1];
                Tex[0] = m_CharacterSetTextures[i];

                BillBoardFont Font = new BillBoardFont(iContext,  Mesh, Tex, Material1,
Shader, Characters[i]);
                Font.GetObjectPhysics().SetGravity(false);
                m_CharacterSet.AddToCharacterSet(Font);
        }
}
```

The CreateHealthItem() function creates a new HUDItem for our player's health and adds it to the HUD. (See Listing 6-40.)

Listing 6-40. Creating and Adding a Health Item to the HUD

```
 void CreateHealthItem()
{
        Texture HUDTexture = new Texture(m_Context, R.drawable.hud);

        Shader Shader = new Shader(m_Context, R.raw.vsonelight, R.raw.fsonelightnodiffuse);
        // ok
        MeshEx Mesh = new MeshEx(8,0,3,5,Cube.CubeData, Cube.CubeDrawOrder);

        // Create Material for this object
        Material Material1 = new Material();
        Material1.SetEmissive(1.0f, 1.0f, 1.0f);

        Texture[] Tex = new Texture[1];
        Tex[0] = HUDTexture;

        BillBoard HUDHealthComposite = new BillBoard(m_Context,  Mesh, Tex, Material1, Shader);

        Vector3 Scale = new Vector3(1.0f,0.1f,0.01f);
        HUDHealthComposite.m_Orientation.SetScale(Scale);
        HUDHealthComposite.GetObjectPhysics().SetGravity(false);

         // Set Black portion of HUD to transparent
        HUDHealthComposite.GetMaterial().SetAlpha(1.0f);
        HUDHealthComposite.SetBlend(true);

        // Create Health HUD
        Texture HealthTexture = new Texture(m_Context, R.drawable.health);
        Vector3 ScreenPosition = new Vector3(0.8f, m_Camera.GetCameraViewportHeight()/2, 0.5f);

        HUDItem HUDHealth = new HUDItem("health", m_Health,
ScreenPosition,m_CharacterSet,HealthTexture, HUDHealthComposite);

        if (m_HUD.AddHUDItem(HUDHealth) == false)
        {
                Log.e("ADDHUDITEM" , "CANNOT ADD IN NEW HUD HEALTH ITEM");
        }
}
```

The CreateHUD() function creates the HUD and the two items in the HUD, which are the player's health and the player's score. (See Listing 6-41.)

Listing 6-41. Creating the HUD

```
void CreateHUD()
{
        // Create HUD
        m_HUD = new HUD(m_Context);

        // Create Score HUD
        Vector3 ScreenPosition = new Vector3(-m_Camera.GetCameraViewportWidth()/2 + 0.3f,
m_Camera.GetCameraViewportHeight()/2, 0.5f);

        // Create Score Item for HUD
        HUDItem HUDScore = new HUDItem("score", 0, ScreenPosition,m_CharacterSet,null,m_HUDComposite);
        if (m_HUD.AddHUDItem(HUDScore) == false)
        {
                Log.e("ADDHUDITEM" , "CANNOT ADD IN NEW HUD ITEM");
        }
        CreateHealthItem();
}
```

The UpdateHUD() function updates the numerical values of the player's health and the player's score in the HUD. (See Listing 6-42.)

Listing 6-42. Updating the HUD

```
void UpdateHUD()
{
        m_HUD.UpdateHUDItemNumericalValue("health", m_Health);
        m_HUD.UpdateHUDItemNumericalValue("score",m_Score);
}
```

The onSurfaceCreated() function requires some modifications to initialize our HUD. First, the viewing height and width on the actual Android device is retrieved. The camera is then modified with these values. Next, the character set that will be used with the HUD is initialized. Finally, the HUD is created. (See Listing 6-43.)

Listing 6-43. Modifying the onSurfaceCreated() Function

```
@Override
public void onSurfaceCreated(GL10 unused, EGLConfig config)
{
        m_PointLight = new PointLight(m_Context);
        SetupLights();

        // Create a 3d Cube
        CreateCube(m_Context);

        // Create a Second Cube
        CreateCube2(m_Context);

        // Create a new gravity grid
        CreateGrid(m_Context);
```

```
// Create SFX
CreateSoundPool();
CreateSound(m_Context);

// Create HUD
// Get Width and Height of surface
        m_ViewPortHeight = m_Context.getResources().getDisplayMetrics().heightPixels;
        m_ViewPortWidth = m_Context.getResources().getDisplayMetrics().widthPixels;

SetupCamera();

// Create Character Set
CreateCharacterSet(m_Context);

CreateHUD
}
```

The onDrawFrame() function has to be modified to accommodate our HUD. We have added code that will decrease the health and increase the score on the HUD every time the two cubes collide with each other. We have also added code to update the numeric values on the HUD, update the HUD, and render the HUD to the screen. (See Listing 6-44.)

Listing 6-44. Modifying the onDrawFrame() Function

```
if ((TypeCollision == Physics.CollisionStatus.COLLISION) || (TypeCollision ==
Physics.CollisionStatus.PENETRATING_COLLISION))
{
        m_Cube.GetObjectPhysics().ApplyLinearImpulse(m_Cube, m_Cube2);
        // SFX
        m_Cube.PlaySound(m_SoundIndex1);
        m_Cube2.PlaySound(m_SoundIndex2);

        // HUD
        m_Health = m_Health - 1;
        if (m_Health < 0)
        {
                m_Health = 0;
        }
        m_Score = m_Score + 10;
}
//////////////////////////// HUD
UpdateHUD();
m_HUD.UpdateHUD(m_Camera);
m_HUD.RenderHUD(m_Camera, m_PointLight);
```

Now, run the application, and you should see the HUD with the score increasing and the health decreasing every time the two cubes collide. You should see something similar to Figure 6-1, shown earlier in this chapter.

Overview of Persistent Data

You may have noticed that when you change the orientation of your Android device, the program in our previous example restarts, and the two cubes' previous orientation and physics is lost. The score and health items in the HUD are also lost and reset. In the Android system, you can use SharedPreferences to save and load data to preserve the environment of your game.

To save the state of a class object, you can add in a function such as SaveState(), shown in Listing 6-45, which:

1. Creates a SharedPreferences variable by calling getSharedPreferences() on the context with the name of the record you want to save the data in

2. Creates a SharedPreferences.Editor editor variable that is used to store the data from the class object

3. Puts the data into the record by calling putXXXX("name", value) on the editor variable to associate the value with the "name." The XXXX is a data type such as Float, Int, etc.

4. Saves the data by calling the commit() function on the editor

Listing 6-45. Saving the State of an Object

```
void SaveState(String handle)
{
        SharedPreferences settings = m_Context.getSharedPreferences(handle, 0);
        SharedPreferences.Editor editor = settings.edit();

        editor.putFloat("x", m_Position.x);

        // Commit the edits!
        editor.commit();
}
```

To load the state of a class object, you could add a class such as LoadState(), shown in Listing 6-46, which:

1. Creates a SharedPreferences variable by calling getSharedPreferences() on the context with the name of the record you want to load the data from

2. Gets the data from the record by calling getXXXX("name", default value), in which XXXX is the data type, such as Float, Int, etc. If "name" does not exist, then the default value is returned.

Listing 6-46. Loading the State of an Object

```
void LoadState(String handle)
{
        // Restore preferences
        SharedPreferences settings = m_Context.getSharedPreferences(handle, 0);
        float x = settings.getFloat("x", 0);
}
```

Modifying the Orientation Class

We have added a SaveState() and LoadState() function to the Orientation class. These follow the format of the examples in the preceding Listings 6-45 and 6-46. In order to save space, I have decided not to include the code here. Please download the code for this chapter from apress.com.

Modifying the Physics Class

We have added a SaveState() and LoadState() function to the Physics class. These follow the format of the examples in the preceding Listings 6-45 and 6-46. In order to save space, I have decided not to include the code here. Please download the code for this chapter from apress.com.

Modifying the Object3d Class

We have added in a SaveObjectState() and LoadObjectState() function to the Object3d class. These follow the format of the examples in the preceding Listings 6-45 and 6-46. In order to save space, I have decided not to include the code here. Please download the code for this chapter from apress.com.

Hands-on Example: Saving Persistent Data

In this section, I cover a hands-on example in which we save the state of the two cubes' orientation and physics state. We also save the score and health items on the HUD. New code is added, so that the health value rolls over back to 100 when reaching 0. This makes it easier to see it being saved when the program is exited or the orientation of the screen changes.

Modifying the MyGLRenderer Class

The MyGLRenderer class has to be modified.

The state of the two colliding cubes are saved using the SaveCubes() function. (See Listing 6-47.)

Listing 6-47. Saving the Two Cubes' State

```
void SaveCubes()
{
        m_Cube.SaveObjectState("Cube1Data");
        m_Cube2.SaveObjectState("Cube2Data");
}
```

The state of the two colliding cubes is loaded back in, using the LoadCubes() function. (See Listing 6-48.)

Listing 6-48. Loading the Two Cubes' State

```
void LoadCubes()
{
        m_Cube.LoadObjectState("Cube1Data");
        m_Cube2.LoadObjectState("Cube2Data");
}
```

The LoadGameState() function loads the score and health HUD items, as well as the state of the two cubes, if their state was previously saved. (See Listing 6-49.)

Listing 6-49. Loading the Game State

```
void LoadGameState()
{
        SharedPreferences settings = m_Context.getSharedPreferences("gamestate", 0);
        int StatePreviouslySaved = settings.getInt("previouslysaved", 0);
        if (StatePreviouslySaved != 0)
        {
                // Load in previously saved state
                m_Score = settings.getInt("score", 0);
                m_Health = settings.getInt("health", 100);
                LoadCubes();
        }
}
```

The SaveGameState() function saves the values of the score, health, state of the cubes, and saves a 1 to "previouslysaved." (See Listing 6-50.)

Listing 6-50. Saving the Game State

```
void SaveGameState()
{
        // We need an Editor object to make preference changes.
        SharedPreferences settings = m_Context.getSharedPreferences("gamestate", 0);
        SharedPreferences.Editor editor = settings.edit();

        editor.putInt("score", m_Score);
        editor.putInt("health", m_Health);

        SaveCubes();
        editor.putInt("previouslysaved", 1);

         editor.commit();
}
```

In the onSurfaceCreated() function, we add code to call the function to load the saved game state, if there exists one.

```
LoadGameState();
```

Next, we need to change the code to roll over the health back to 100 when it reaches 0. (See Listing 6-51.)

Listing 6-51. Rolling over the Health

```
if ((TypeCollision == Physics.CollisionStatus.COLLISION) || (TypeCollision ==
Physics.CollisionStatus.PENETRATING_COLLISION))
{
        m_Cube.GetObjectPhysics().ApplyLinearImpulse(m_Cube, m_Cube2);

        // SFX
        m_Cube.PlaySound(m_SoundIndex1);
        m_Cube2.PlaySound(m_SoundIndex2);

        // HUD
        m_Health = m_Health - 1;
        if (m_Health < 0)
        {
                m_Health = 100;
        }
        m_Score = m_Score + 10;
}
```

Modifying the MyGLSurfaceView Class

We must also modify the MyGLSurfaceView class.

Add a line that holds a reference to the Custom GLRenderer.

```
public MyGLRenderer CustomGLRenderer = null;
```

Next, we have to make some changes to the constructor. (See Listing 6-52.)

Listing 6-52. MyGLSurfaceView Constructor Modifications

```
public MyGLSurfaceView(Context context)
{
        super(context);

        // Create an OpenGL ES 2.0 context.
        setEGLContextClientVersion(2);

        // Set the Renderer for drawing on the GLSurfaceView
        //setRenderer(new MyGLRenderer(context));

        CustomGLRenderer = new MyGLRenderer(context);
        setRenderer(CustomGLRenderer);
}
```

Modifying the MainActivity Class

Next, we have to make modifications to our MainActivity class.

The m_GLView needs to be changed to:

```
private MyGLSurfaceView m_GLView;
```

The onPause() function has to be modified to save the state of the game when onPause() is called. (See Listing 6-53.)

Listing 6-53. Modifying the onPause() Function

```
@Override
protected void onPause()
{
        super.onPause();
        m_GLView.onPause();

        // Save State
        m_GLView.CustomGLRenderer.SaveGameState();
}
```

Finally, run the project, change orientations, and exit and reenter the program repeatedly. You should now have a score and health that are saved when you exit the program or change orientations. You should also see that the two cubes' orientation and physics are saved as well.

Summary

In this chapter, I covered the game environment. We started off by creating sounds. Code was presented that supported the creation and playback of sounds. Next, we went through a hands-on example in which we added sound effects, applying this code to two colliding cubes. A HUD was discussed. Then the code required to implement this HUD was discussed, followed by a case study that implemented the HUD using the code. Finally, a way was presented to save the state of your game environment. A hands-on example was also presented, which demonstrated how you would save the state of your HUD as well as the state of objects in your game.

Chapter 7

Drone Grid Case Study: Creating the Player

This chapter covers a case study for a 3D game called Drone Grid. The objective of the game will be to protect your pyramid, which is located in the center of the playfield, from attacking enemies. You will do this by touching the screen and launching projectiles toward the enemies to destroy them. You will be in a fixed location on the edge of the playfield, and you can turn left or right 90 degrees. First, the creation of the classes representing the player's graphic in the game is presented. Next, we create classes relating to the player's viewpoint and the player's input. This is followed by the creation of the classes dealing with the player's weapons and ammunition. Next, we create classes dealing with explosions and a class dealing with game object properties. Finally, we present a hands-on example illustrating the use of our new classes, with which you will use your weapon to hit cubes.

Creating the Player Graphic

The player graphic for the Drone Grid case study will be a pyramid. In order to create this pyramid, several classes will have to be constructed, including a new Mesh class that will draw an object's vertices in a slightly different manner than in the MeshEx class that we have created previously. Modifications to other classes, including the Object3d class, are also needed.

Creating the Mesh Class

The Mesh class is very similar to the MeshEx class you learned about in Chapter 4. The key differences are in how an object's vertices are defined and how they are drawn. In the MeshEx class, an object's vertices are listed in one array, and a list of vertices from which to draw triangles or lines are listed in another array. In the Mesh class, there is only one array, which consists of the vertices that make up the triangles to be drawn. Listing 7-1 shows the vertex data for a single triangle consisting of three vertices to be used with a Mesh class. If you want to draw an additional triangle, you will have to add three more vertex data entries for this new triangle. Because more complex graphics are made up of many triangles, you will obviously have to add more triangles to the vertex data for a complex object.

Listing 7-1. A Triangle Defined for a Mesh Class

```
// Left Side              u    v    nx, ny, nz
-1.5f, -1.5f, -1.5f,      0,   1,   -1, -1, -1,        // v0 =  left, bottom, back
-1.5f, -1.5f,  1.5f,      1,   1,   -1, -1,  1,        // v1 =  left, bottom, front
 0.0f,  3.2f,  0.0f,  0.5f,   0,    0,  1,  0,        // v2 =  top point
```

The DrawMesh() function draws the mesh and is similar to the corresponding function in the MeshEx class. The key difference is that the OpenGL function glDrawArrays() is used to draw the mesh instead of glDrawElements(). (See Listing 7-2.)

The glDrawArrays() function takes as parameters in sequence:

1. The graphics primitive to draw, in this case GL_TRIANGLES

2. The number of the starting vertex to draw

3. The number of vertices to draw

Listing 7-2. The DrawMesh Function

```
void DrawMesh(int PosHandle, int TexHandle, int NormalHandle)
{
        SetUpMeshArrays(PosHandle, TexHandle, NormalHandle);

        // Draw the triangle
        //glDrawArrays (int mode, int first, int count)
        GLES20.glDrawArrays(GLES20.GL_TRIANGLES, 0, m_VertexCount);

        // Disable vertex array
        GLES20.glDisableVertexAttribArray(PosHandle);
        if (m_MeshHasUV)
        {
                GLES20.glDisableVertexAttribArray(TexHandle);
        }
        if (m_MeshHasNormals)
        {
                GLES20.glDisableVertexAttribArray(NormalHandle);
        }
}
```

Modifying the Object3d Class

Next, the Object3d class needs to be modified to use the Mesh class.

A new Mesh class variable, m_Mesh, is added.

```
private Mesh m_Mesh = null;
```

The constructor for the Object3d class must also be modified to account for the addition of the new Mesh variable. See changes in bold in Listing 7-3.

Listing 7-3. Object3d Constructor

```
Object3d(Context iContext, Mesh iMesh, MeshEx iMeshEx, Texture[] iTextures, Material iMaterial,
Shader iShader)
{
        m_Context          = iContext;
        m_Mesh             = iMesh;

        // REst of Code
}
```

The GetRadius() function is modified to return the radius of the object's m_Mesh variable, if it exists. (See Listing 7-4.) The object's mesh can now be held in either a Mesh or a MeshEx.

Listing 7-4. GetRadius() Function Modifications

```
float GetRadius()
{
        if (m_Mesh != null)
        {
                return m_Mesh.GetRadius();
        }

        if (m_MeshEx != null)
        {
                return m_MeshEx.GetRadius();
        }
        return -1;
}
```

The DrawObject() function, which draws the actual object's mesh, is modified to test to see if the m_Mesh variable contains a valid object. If it does, the Mesh's DrawMesh() function is called to do the actual rendering of the object. (See Listing 7-5.)

Listing 7-5. Modifying the DrawObject() Function

```
void DrawObject(Camera Cam,PointLight light,Vector3 iPosition,Vector3 iRotationAxis,Vector3 iScale)
{
        // Activate and set up the Shader and Draw Object's mesh

        // Generate Needed Matrices for Object
        GenerateMatrices(Cam, iPosition,iRotationAxis,iScale);

        // Add program to OpenGL environment
        m_Shader.ActivateShader();

        // Get Vertex Attribute Info in preparation for drawing the mesh
        GetVertexAttribInfo();
```

```
// Sets up the lighting parameters for this object
SetLighting(Cam, light, m_ModelMatrix, Cam.GetViewMatrix(), m_ModelViewMatrix,
m_NormalMatrix);

// Apply the projection and view transformation matrix to the shader
m_Shader.SetShaderVariableValueFloatMatrix4Array("uMVPMatrix", 1, false, m_MVPMatrix, 0);

// Activates texture for this object
ActivateTexture();

// Enable Hidden surface removal
GLES20.glEnable(GLES20.GL_DEPTH_TEST);

// Draw Mesh for this Object
if (m_Mesh != null)
{
        m_Mesh.DrawMesh(m_PositionHandle, m_TextureHandle, m_NormalHandle);
}
else
if (m_MeshEx != null)
{
        m_MeshEx.DrawMesh(m_PositionHandle, m_TextureHandle, m_NormalHandle);
}
else
{
        Log.d("class Object3d :", "No MESH in Object3d");
}
}
```

Modifying Other Classes That Use the Object3d Class

Other classes that use the Object3d class also will have to be modified. For example, the Cube class that is derived from the Object3d class will have to be modified in terms of adding a Mesh input parameter into the constructor and the call to the Object3d constructor. (See Listing 7-6.)

Listing 7-6. Modifying the Cube Class

```
public class Cube extends Object3d
Cube(Context iContext, Mesh iMesh, MeshEx iMeshEx, Texture[] iTextures, Material iMaterial,
Shader iShader)
{
        super(iContext, iMesh, iMeshEx, iTextures, iMaterial, iShader);
}
```

Please download the sample project for this chapter from the Source Code/Download area of apress.com to see other changes that have to be made to other classes.

Creating the Pyramid Class

The Pyramid class contains the actual vertex data for our Player graphic, which will be a 3D object in the shape of a pyramid. The Pyramid class vertex data is contained in the `PyramidVertices` array and is designed to be used with a Mesh class. (See Listing 7-7.)

Listing 7-7. The Pyramid Class

```
public class Pyramid extends Object3d
{
        static float[] PyramidVertices =
        {
          // Triangle Shape
          // Left Side            u   v  nx, ny, nz
          -1.5f, -1.5f, -1.5f,    0,  1, -1, -1, -1,       // v0 = left, bottom, back
          -1.5f, -1.5f,  1.5f,    1,  1, -1, -1,  1,       // v1 = left, bottom, front
           0.0f,  3.2f,  0.0f,  .5f,  0,  0,  1,  0,       // v2 = top point

          // Right Side
           1.5f, -1.5f,  1.5f,    0,  1,  1, -1,  1,       // v3 = right, bottom, front
           1.5f, -1.5f, -1.5f,    1,  1,  1, -1, -1,       // v4 = right, bottom, back
           0.0f,  3.2f,  0.0f, 0.5f,  0,  0,  1,  0,       // v2 = top point

          // Front
          -1.5f, -1.5f,  1.5f,    1,  1, -1, -1,  1,       // v1 = left, bottom, front
           1.5f, -1.5f,  1.5f,    0,  1,  1, -1,  1,       // v3 = right, bottom, front
           0.0f,  3.2f,  0.0f, 0.5f,  0,  0,  1,  0,       // v2 = top point

          // Back
          -1.5f, -1.5f, -1.5f,    0,  1, -1, -1, -1,       // v0 = left, bottom, back
           1.5f, -1.5f, -1.5f,    1,  1,  1, -1, -1,       // v4 = right, bottom, back
           0.0f,  3.2f,  0.0f, 0.5f,  0,  0,  1,  0,       // v2 = top point

          // Bottom
          -1.5f, -1.5f, -1.5f,    0,  0, -1, -1, -1,       // v0 = left, bottom, back
           1.5f, -1.5f, -1.5f,    0,  1,  1, -1, -1,       // v4 = right, bottom, back
           1.5f, -1.5f,  1.5f,    1,  1,  1, -1,  1,       // v3 = right, bottom, front

          // Bottom 2
          -1.5f, -1.5f, -1.5f,    0,  0, -1, -1, -1,       // v0 = left, bottom, back
          -1.5f, -1.5f,  1.5f,    1,  0, -1, -1,  1,       // v1 = left, bottom, front
           1.5f, -1.5f,  1.5f,    1,  1,  1, -1,  1        // v3 = right, bottom, front
        };
        Pyramid(Context iContext,  Mesh iMesh, MeshEx iMeshEx, Texture[] iTextures, Material
        iMaterial, Shader iShader, Shader LocalAxisShader)
        {
                super(iContext, iMesh, iMeshEx, iTextures, iMaterial, iShader);
        }
}
```

Creating the PowerPyramid Class

The PowerPyramid class is the class you will use to represent the player's physical presence in the game.

The m_ExplosionSFXIndex variable holds the index to the explosive sound played when the pyramid is hit.

The CreateExplosionSFX() function creates the explosive sound effect for the power pyramid and saves the index to this sound in the m_ExplosionSFXIndex variable.

The PlayExplosionSFX() function plays back the explosion that was created with the CreateExplosionSFX() function. (See Listing 7-8.)

Listing 7-8. The Player's PowerPyramid Class

```
public class PowerPyramid extends Object3d
{
        private int m_ExplosionSFXIndex = -1;

        PowerPyramid(Context iContext, Mesh iMesh, MeshEx iMeshEx, Texture[] iTextures, Material
        iMaterial, Shader iShader)
        {
                super(iContext, iMesh, iMeshEx, iTextures, iMaterial, iShader);
        }

        // Sound Effects
        void CreateExplosionSFX(SoundPool Pool, int ResourceID)
        {
                m_ExplosionSFXIndex = AddSound(Pool, ResourceID);
        }

        void PlayExplosionSFX()
        {
                if (m_ExplosionSFXIndex >= 0)
                {
                        PlaySound(m_ExplosionSFXIndex);
                }
        }
}
```

The player's pyramid, which is an instance of the PowerPyramid class, is shown in Figure 7-1.

Figure 7-1. *Player's power pyramid*

Creating the Player's Viewpoint and Input

The player's viewpoint for this game will be a first-person viewpoint, in which the player can turn left and right up to 90 degrees. The player will be able to fire 3D projectiles by touching the screen. Creating the player's viewpoint and the player's input requires modifications to the MyGLRenderer class and the MyGLSurfaceview class.

Modifying the MyGLRenderer Class

The MyGLRenderer class must be modified to add code to calculate the player's viewpoint and player's input.

The CameraMoved() function takes as input the change, or delta, in the rotation of the player's view around the x and y axes. The x and y change in angular position is altered by the ScaleFactor variable, of which you can make the rotation greater or smaller. (See Listing 7-9.)

Listing 7-9. Calculating the Camera Movement Delta

```
void CameraMoved(float DeltaXAxisRotation , float DeltaYAxisRotation)
{
        m_CameraMoved = true;
        float ScaleFactor = 3;
        m_DeltaXAxisRotation = DeltaXAxisRotation/ScaleFactor;
        m_DeltaYAxisRotation = DeltaYAxisRotation/ScaleFactor;
}
```

The ProcessCameraMove() function updates the camera's left/right rotation around the y axis, based on the value in m_DeltaYAxisRotation and limited by the values m_MaxCameraAngle and m_MinCameraAngle. (See Listing 7-10.)

Listing 7-10. Processing Player's Camera Movement

```
void ProcessCameraMove()
{
        Vector3 Axis = new Vector3(0,1,0);

        // Test Limits
        float CameraRotation = m_Camera.GetOrientation().GetRotationAngle();
        float NextRotationAngle = CameraRotation + m_DeltaYAxisRotation;
        if (NextRotationAngle > m_MaxCameraAngle)
        {
                m_DeltaYAxisRotation = m_MaxCameraAngle - CameraRotation;
        }
        else
        if (NextRotationAngle < m_MinCameraAngle)
        {
        m_DeltaYAxisRotation = m_MinCameraAngle - CameraRotation;
        }

        // Camera Test
        // Rotate Camera Around Y Axis
        m_Camera.GetOrientation().SetRotationAxis(Axis);
        m_Camera.GetOrientation().AddRotation(m_DeltaYAxisRotation);
        m_CameraMoved = false;
}
```

The onDrawFrame() function has to be modified to process a change in the player's view by calling ProcessCameraMove() if the m_CameraMoved has been set to true. (See Listing 7-11.)

Listing 7-11. Modification to onDrawFrame()

```
@Override
public void onDrawFrame(GL10 unused)
{
        GLES20.glClearColor(0.0f, 0.0f, 0.0f, 1.0f);
        GLES20.glClear( GLES20.GL_DEPTH_BUFFER_BIT | GLES20.GL_COLOR_BUFFER_BIT);

        // Player Update
        if (m_CameraMoved)
        {
                ProcessCameraMove();
        }
        m_Camera.UpdateCamera();
        // Rest of code
}
```

The ProcessTouch() function processes a user's touch. The starting point (Startx, Starty), at which the user touches the screen, and the ending point, at which the user lifts his or her finger from the screen (x,y), are input parameters.

The ProcessTouch() function (see Listing 7-12) does the following:

1. Finds the distance between the point at which the user touches the screen and the point at which the user lifts his or her finger

2. Sets the variable that keeps track of screen touches to true and sets the x, y screen coordinates of the touch location (m_TouchX, m_TouchY) if this distance is less than 10 (which means the intention of the user was to touch the screen to fire a weapon, instead of moving the view)

Listing 7-12. Processing the User's Touch

```
void ProcessTouch(float Startx, float Starty, float x, float y)
{
        Vector3 DiffVec = new Vector3(Startx - x, Starty - y, 0);
        float length = DiffVec.Length();
        if (length < 10)
        {
                // Player weapon has been fired
                m_ScreenTouched = true;
                m_TouchX = x;
                m_TouchY = y;
        }
}
```

Modifying the MyGLSurfaceView Class

The MyGLSurfaceView class also has to be modified to provide support for the player's view and player's touch input.

The m_PreviousX and m_PreviousY variables track the x and y screen locations of the last user's touch in the onTouchEvent() function.

```
private float m_PreviousX = 0;
private float m_PreviousY = 0;
```

The m_dx and m_dy variables hold the changes in x and y screen positions occurring when the user touches the screen.

```
private float m_dx = 0;
private float m_dy = 0;
```

The m_Startx and m_Starty variables hold the x and y screen positions when the user first touches the screen.

```
private float m_Startx = 0;
private float m_Starty = 0;
```

The onTouchEvent() function is the main hook into where the player's view and player's input are updated.

The onTouchEvent() function (see Listing 7-13) does the following:

1. Gets the current x and y screen position and stores it into variables x and y

2. Based on the user's action, one of the following occurs:

   ■ If the user starts a new touch, then this x, y starting location is saved in m_Startx and m_Starty.

   ■ If the user exits a touch by lifting his/her finger, then the ProcessTouch() function in the MyGLRenderer class is called.

   ■ If the user moves his/her finger across the screen, the CameraMoved() function in the MyGLRenderer class is called to update the player's view.

   ■ The current x, y screen location is saved as the previous location in anticipation of the next onTouchEvent() call.

Listing 7-13. Adding and Modifying the onTouchEvent() Function

```
@Override
public boolean onTouchEvent(MotionEvent e)
{
        // MotionEvent reports input details from the touch screen
                // and other input controls. In this case, you are only
                // interested in events where the touch position changed.

                float x = e.getX();
                float y = e.getY();

                switch (e.getAction())
                {
                        case MotionEvent.ACTION_DOWN:
                                m_Startx = x;
                                m_Starty = y;
                        break;

                        case MotionEvent.ACTION_UP:
                                CustomGLRenderer.ProcessTouch(m_Startx, m_Starty, x, y);
                        break;

                            case MotionEvent.ACTION_MOVE:
                                m_dx = x - m_PreviousX;
                                m_dy = y - m_PreviousY;

                                CustomGLRenderer.CameraMoved(m_dy, m_dx);
                        break;
                }
                m_PreviousX = x;
        m_PreviousY = y;
                return true;
}
```

Creating Player Weapons and Ammunition

We need to create new classes for the player's weapon and the ammunition that the weapon uses.
The Ammunition class derives from the Object3d class.

```
public class Ammunition extends Object3d
```

The m_FireStatus variable is true if this piece of ammunition has been fired.

```
private boolean        m_FireStatus = false;
```

The m_AmmunitionSpent variable allows you to keep track if this piece of ammunition has been used up.

```
private boolean        m_AmmunitionSpent = false;
```

The m_AmmunitionRange variable holds the maximum range of the ammunition, which is defaulted to
50 units in the OpenGL world.

```
private float          m_AmmunitionRange = 50;
```

The m_AmmunitionStartPosition variable holds the position the ammo is fired from and is initialized
to 0,0,0.

```
private Vector3 m_AmmunitionStartPosition = new Vector3(0,0,0);
```

The speed of the ammo in terms of OpenGL world units per update is held in m_AmmoSpeed and
defaults to 0.5.

```
private float m_AmmoSpeed = 0.5f;
```

The SFX index for the sound effect associated with this piece of ammo, if it exists, is m_FireSFXIndex.
The default, which is -1, means that there is no sound effect associated with this piece of
ammunition.

```
private int m_FireSFXIndex = -1;
```

The Ammunition constructor initializes the object by calling its base constructor and setting the
ammunition range, ammunition speed, and the mass of the ammunition, which is defaulted to 1.
(See Listing 7-14.)

Listing 7-14. Ammunition Constructor

```
Ammunition(Context iContext, Mesh iMesh, MeshEx iMeshEx, Texture[] iTextures, Material iMaterial,
Shader iShader, float AmmunitionRange,float AmmunitionSpeed)
{
        super(iContext, iMesh, iMeshEx, iTextures, iMaterial, iShader );
        m_AmmunitionRange = AmmunitionRange;
        m_AmmoSpeed = AmmunitionSpeed;
        GetObjectPhysics().SetMass(1.0f);
}
```

The CreateFiringSFX() function creates a new sound effect associated with an Ammunition object. (See Listing 7-15.)

Listing 7-15. Creating Sound Effects

```
void CreateFiringSFX(SoundPool Pool, int ResourceID)
{
        m_FireSFXIndex = AddSound(Pool, ResourceID);
}
```

The PlayFiringSFX() function plays the sound effect created by the CreateFiringSFX() function. (See Listing 7-16.)

Listing 7-16. Playing Back the Ammunition Sound Effect

```
void PlayFiringSFX()
{
        if (m_FireSFXIndex >= 0)
        {
                PlaySound(m_FireSFXIndex);
        }
}
```

The Reset() function sets the firing status of the ammunition and ammunition spent status to false, to indicate that the ammunition has not been fired and is ready for use. It also sets the velocity to 0. (See Listing 7-17.)

Listing 7-17. Resetting the Ammunition

```
void Reset()
{
        m_FireStatus = false;
        m_AmmunitionSpent = false;
        GetObjectPhysics().GetVelocity().Set(0, 0, 0);
}
```

The RenderAmmunition() function draws the ammunition to the screen by calling the DrawObject() function in the parent Object3d class. (See Listing 7-18.)

Listing 7-18. Rendering the Ammunition

```
void RenderAmmunition(Camera Cam, PointLight light, boolean DebugOn)
{
        DrawObject(Cam, light);
}
```

The UpdateAmmunition() function updates the ammunition object by calling the UpdateObject3d() function in the superclass. (See Listing 7-19.)

Listing 7-19. Updating the Ammunition

```
void UpdateAmmunition()
{
        // 1. Update Ammunition Physics, Position, Rotation
        UpdateObject3d();
}
```

The Fire() function is the key function in terms of actually firing the ammunition this class represents.

The Fire() function (see Listing 7-20) does the following:

1. Sets the m_FireStatus to true to indicate that this piece of ammunition has been fired and is currently moving through the 3D world

2. Calculates the velocity of the ammunition based on the input Direction vector, the m_AmmoSpeed ammunition speed, and the OffsetVelocity vector, which represents the movement of the object firing the ammunition

3. Sets the position of the ammunition based on the input AmmoPosition parameter

4. Sets the m_AmmunitionStartPosition class member variable

Listing 7-20. Firing the Ammunition

```
void Fire(Vector3 Direction,
        Vector3 AmmoPosition,
        Vector3 OffSetVelocity)
{
        // 1. Set Fire Status to true
        m_FireStatus = true;

        // 2. Set direction and speed of Ammunition
        // Velocity of Ammo
        Vector3 DirectionAmmo = new Vector3(Direction.x, Direction.y, Direction.z);
        DirectionAmmo.Normalize();

        Vector3 VelocityAmmo = Vector3.Multiply(m_AmmoSpeed, DirectionAmmo);

        // Velocity of Object with Weapon that has fired Ammo
        // Total Velocity
        Vector3 VelocityTotal = Vector3.Add(OffSetVelocity , VelocityAmmo);

        GetObjectPhysics().SetVelocity(VelocityTotal);
        m_Orientation.GetPosition().Set(AmmoPosition.x, AmmoPosition.y, AmmoPosition.z);

        // 3. Set Ammunition Initial World Position
        m_AmmunitionStartPosition.Set(AmmoPosition.x, AmmoPosition.y, AmmoPosition.z);
}
```

Next, we have to create the player's Weapon class that uses the ammunition discussed previously. The Weapon class is derived from the Object3d class.

```
public class Weapon extends Object3d
```

The MAX_DEFAULTAMMO variable indicates the maximum number of projectiles that this weapon can hold at one time.

```
private int MAX_DEFAULTAMMO = 20;
```

The m_WeaponClip variable array actually holds the weapon's ammunition.

```
private Ammunition[] m_WeaponClip = new Ammunition[MAX_DEFAULTAMMO];
```

The m_TimeLastFired variable holds the time this weapon was last fired in milliseconds.

```
private long m_TimeLastFired = 0;
```

The m_TimeReadyToFire variable is the time that this weapon will next be able to fire ammunition in milliseconds.

```
private long m_TimeReadyToFire = 0;
```

The m_FireDelay variable is the minimum time in milliseconds between the firing of projectiles.

```
private long m_FireDelay = 500;
```

The Weapon constructor calls the Object3d's constructor. (See Listing 7-21.)

Listing 7-21. Weapon Constructor

```
Weapon(Context iContext, Mesh iMesh, MeshEx iMeshEx, Texture[] iTextures, Material iMaterial,
Shader iShader)
{
        super(iContext, iMesh, iMeshEx, iTextures, iMaterial, iShader );
}
```

The TurnOnOffSFX() function turns on or off the sound effects associated with the weapon's ammunition. (See Listing 7-22.)

Listing 7-22. Turning On/Off Ammunition Sound Effects

```
void TurnOnOffSFX(boolean value)
{
        for (int i = 0; i < MAX_DEFAULTAMMO; i++)
        {
                m_WeaponClip[i].SetSFXOnOff(value);
        }
}
```

The ResetWeapon() function resets all of the weapon's ammunition. (See Listing 7-23.)

Listing 7-23. Resetting the Weapon

```
void ResetWeapon()
{
        // Reset All the Ammunition in the Weapon's Magazine
        for (int i = 0; i < MAX_DEFAULTAMMO; i++)
        {
                m_WeaponClip[i].Reset();
        }
}
```

The LoadAmmunition() function puts the Ammunition Ammo in the slot AmmoSlot in the m_WeaponClip array. (See Listing 7-24.)

Listing 7-24. Loading the Ammunition into the Weapon

```
void LoadAmmunition(Ammunition Ammo, int AmmoSlot)
{
        if (AmmoSlot >= MAX_DEFAULTAMMO)
        {
                AmmoSlot = MAX_DEFAULTAMMO - 1;
        }
        m_WeaponClip[AmmoSlot] = Ammo;
}
```

The FindReadyAmmo() function returns the index number of the first available piece of ammunition available for firing or -1, if there is no available ammunition. (See Listing 7-25.)

Listing 7-25. Finding Spare Ammunition

```
int FindReadyAmmo()
{
        for (int i = 0; i < MAX_DEFAULTAMMO; i++)
        {
                // If Ammo is not Fired
                if (m_WeaponClip[i].IsFired() == false)
                {
                        return i;
                }
        }
        return -1; // No More Ammo Available
}
```

The CheckAmmoCollision() function tests if the input object obj is colliding with any of the weapon's fired ammunition. It returns a reference to the object if there is a collision or null if there is no collision. (See Listing 7-26.)

Listing 7-26. Checking the Weapon's Ammunition Collision with an Object

```
Object3d CheckAmmoCollision(Object3d obj)
{
        Object3d ObjectCollided = null;
        for (int i = 0; i < MAX_DEFAULTAMMO; i++)
        {
                if (m_WeaponClip[i].IsFired() == true)
                {
                        //Check Collision
                        Physics.CollisionStatus result = m_WeaponClip[i].CheckCollision(obj);
                        if ((result == Physics.CollisionStatus.COLLISION) ||
                        (result == Physics.CollisionStatus.PENETRATING_COLLISION))
                        {
                                ObjectCollided = m_WeaponClip[i];
                        }
                }
        }
        return ObjectCollided;
}
```

The GetActiveAmmo() function puts references to all of the active fired ammunition from the weapon in the input ActiveAmmo array and returns the number of active ammunition. (See Listing 7-27.)

Listing 7-27. Getting All the Active Ammunition

```
int GetActiveAmmo(int StartIndex, Object3d[] ActiveAmmo)
{
        // Put all active fired ammunition in ActiveAmmo array
        // and return the number of fired ammunition
        int AmmoNumber = StartIndex;
        for (int i = 0; i < MAX_DEFAULTAMMO; i++)
        {
                if (m_WeaponClip[i].IsFired() == true)
                {
                        ActiveAmmo[AmmoNumber] = m_WeaponClip[i];
                        AmmoNumber++;
                }
        }
        return (AmmoNumber - StartIndex);
}
```

The Fire() function fires a projectile associated with the weapon in the direction held in the input parameter Direction and starting at the WeaponPosition position.

The Fire() function (see Listing 7-28) does the following:

1. Continues if the weapon is ready to fire, based on a minimum time delay value between firings; otherwise, it returns from the function

2. Finds a piece of ammunition that has not been fired yet

3. Calls an ammunition's `Fire()` function and plays any associated sound effects associated with that ammunition, if that ammunition exists

4. Calculates the time that the weapon can be fired again and puts this in the `m_TimeReadyToFire` variable

5. Returns true if the weapon could be fired and false otherwise

Listing 7-28. Firing the Weapon

```
boolean Fire(Vector3 Direction, Vector3 WeaponPosition)
{
        boolean WeaponFired = false;

        // 0. Test if this weapon is ready to fire
        long CurrentTime = System.currentTimeMillis();
        if (CurrentTime < m_TimeReadyToFire)
        {
                return false;
        }

        // 1. Find Ammo That is not spent
        int AmmoSlot = FindReadyAmmo();

        // 2. If Ammo Found then Fire Ammunition
        if (AmmoSlot >= 0)
        {
                WeaponFired = true;
                m_WeaponClip[AmmoSlot].Fire(Direction,WeaponPosition,
                GetObjectPhysics().GetVelocity());

                // Play SFX if available
                m_WeaponClip[AmmoSlot].PlayFiringSFX();
        }
        else
        {
                Log.e("AMMUNITION ", "AMMUNITION NOT FOUND");
                WeaponFired = false;
        }

        // 3. Firing Delay
        m_TimeLastFired = System.currentTimeMillis();
        m_TimeReadyToFire = m_TimeLastFired + m_FireDelay;

        return WeaponFired;
}
```

The RenderWeapon() function renders all of the weapon's ammunition that has been fired and is currently active. (See Listing 7-29.)

Listing 7-29. Rendering the Weapon's Ammunition

```
void RenderWeapon(Camera Cam, PointLight light, boolean DebugOn)
{
        // 1. Render Each Fired Ammunition in Weapon
        for (int i = 0; i < MAX_DEFAULTAMMO; i++)
        {
                if (m_WeaponClip[i].IsFired() == true)
                {
                        m_WeaponClip[i].RenderAmmunition(Cam, light, DebugOn);
                }
        }
}
```

The UpdateWeapon() function updates the ammunition for the weapon. (See Listing 7-30.)

For each piece of ammunition in the weapon, the UpdateWeapon() function does the following:

1. If the ammunition has been fired, it continues updating it. If not, then it checks to see if the next piece of ammunition has been fired.

2. It adds a rotational force to the ammunition and updates the object's physics.

3. If the distance the ammunition has traveled since it was fired is greater than the ammunition's range, it destroys the ammunition by calling the ammunition's Reset() function.

Listing 7-30. Updating the Weapon

```
void UpdateWeapon()
{
        // 1. Update Each Ammunition in Weapon
        for (int i = 0; i < MAX_DEFAULTAMMO; i++)
        {
                // If Ammunition is fired then Update Ammunition and Emit More AmmoDust Trail
particles
                if (m_WeaponClip[i].IsFired() == true)
                {
                        // Add Spin to Ammunition
                        m_WeaponClip[i].GetObjectPhysics().ApplyRotationalForce(30, 1);
                        m_WeaponClip[i].UpdateAmmunition();

                        // 2. Check if Ammunition is spent
                        float AmmoRange       = m_WeaponClip[i].GetAmmunitionRange();
                        Vector3 AmmoCurrentPos = m_WeaponClip[i].m_Orientation.GetPosition();
                        Vector3 AmmoInitPos    = m_WeaponClip[i].GetAmmunitionStartPosition();
                        Vector3 DistanceVector = Vector3.Subtract(AmmoCurrentPos, AmmoInitPos);

                        float DistanceMag = DistanceVector.Length();
```

```
                        if (DistanceMag > AmmoRange)
                        {
                                // Ammo is Spent so Reset Ammunition to ready to use status.
                                m_WeaponClip[i].Reset();
                        }
                }
        }
}
```

Figure 7-2 shows the player's weapon being fired. The green cube in the center of the picture is the player's ammunition from the weapon.

Figure 7-2. *Player's weapon fired*

Creating the Explosions

The explosions for our game will consist of many triangular polygons. The key classes for our explosion are the PolyParticleEx class, which represents the particles, and the SphericalPolygonExplosion class, which contains the particles that will make up our explosion.

Creating the PolyParticleEx Class

The PolyParticleEx class represents the particles of the explosion we want to create, along with functions to manage and manipulate the particle.

The PolyParticleEx class is derived from the Object3d class.

```
public class PolyParticleEx extends Object3d
```

The PolyParticleVertices variable array holds the mesh data for a PolyParticleEx particle. The particle is a triangle that has no texture coordinates but has lighting data in the form of vertex normals. (See Listing 7-31.)

Listing 7-31. Particle Mesh Definition

```
static float[] PolyParticleVertices =
{
        // Triangle Shape
        // Left Side              nx, ny,  nz
        0.0f,  0.0f, -0.5f,        0,  0,  -1,         // v0 =  bottom, back
        0.0f,  0.0f,  0.5f,        0,  0,   1,         // v1 =  bottom, front
        0.0f,  0.5f,  0.0f,        0,  1,   0,         // v2 =  top point
};
```

The m_Color variable holds the current color of the polygon particle.

```
private Vector3      m_Color = new Vector3(0,0,0);
```

The m_TimeStamp variable holds the time in milliseconds that the particle is created.

```
private long         m_TimeStamp;    // Time in milliseconds that Particle is created
```

The m_TimeDelay variable holds the life span of the particle in milliseconds.

```
private float        m_TimeDelay;    // LifeSpan of Particle in milliseconds
```

The m_Locked variable is true if set to launch or in use, false if available for use.

```
private boolean      m_Locked;       // true if set to launch or in use, false if available for use
```

The m_Active variable is true if the particle is onscreen and has to be rendered. It is false otherwise.

```
private boolean      m_Active;       // Onscreen = Render particle if Active
```

The m_ColorBrightness variable holds the current brightness level of the particle color.

```
private float        m_ColorBrightness;
```

The m_FadeDelta variable is the rate at which the particle fades out.

```
private float        m_FadeDelta;
```

The m_OriginalColor variable is the original color of the particle when created.

```
private Vector3      m_OriginalColor = new Vector3(0,0,0);
```

The PolyParticleEx constructor calls the Object3d constructor and then initializes class member variables. (See Listing 7-32.)

Listing 7-32. The PolyParticleEx Constructor

```
public PolyParticleEx(Context iContext, Mesh iMesh, MeshEx iMeshEx, Texture[] iTextures,
Material iMaterial, Shader iShader)
{
        super(iContext, iMesh, iMeshEx, iTextures, iMaterial, iShader);
        m_Color.Clear();
        m_TimeStamp             = 0;
        m_TimeDelay             = 1000;
        m_Locked                = false;
        m_Active                = false;
        m_ColorBrightness       = 1;
        m_OriginalColor.Clear();
        m_FadeDelta             = 0.0000f;
}
```

The SetColor() function sets the color of the particle, including the particle's material, in terms of the material's ambient, diffuse, and emissive properties. (See Listing 7-33.)

Listing 7-33. Setting the Particle's Color

```
void SetColor(Vector3 value)
{
        m_Color.x = value.x;
        m_Color.y = value.y;
        m_Color.z = value.z;

        GetMaterial().SetAmbient(value.x, value.y, value.z);
        GetMaterial().SetDiffuse(value.x, value.y, value.z);
        GetMaterial().SetEmissive(value.x, value.y, value.z);
}
```

The SetActiveStatus() function sets the active status of the particle and also resets the particle's color brightness level to 100%. (See Listing 7-34.)

Listing 7-34. Setting the Active Status

```
void SetActiveStatus(boolean value)
{
        m_Active        = value;

        // Reset Brightness Level
        m_ColorBrightness= 1;
}
```

The Destroy() function resets the particle to its initial state. (See Listing 7-35.)

Listing 7-35. Destroying the Particle

```
void Destroy()
{
        GetObjectPhysics().GetVelocity().Clear();

        m_Locked    = false;    // Particle is now free to be used again by the Particle Manager.
        m_Active    = false;    // Do not draw on screen
        m_TimeStamp = 0;

        // Restore Particle to Original Color
        m_Color.x = m_OriginalColor.x;
        m_Color.y = m_OriginalColor.y;
        m_Color.z = m_OriginalColor.z;
}
```

The `Create()` function sets the color of the particle to `Color` and sets the `m_OriginalColor` variable to `Color` as well. (See Listing 7-36.)

Listing 7-36. Creating a New Particle

```
void Create(Vector3 Color)
{
        m_Color.x = Color.x;
        m_Color.y = Color.y;
        m_Color.z = Color.z;

        m_OriginalColor.x = m_Color.x;
        m_OriginalColor.y = m_Color.y;
        m_OriginalColor.z = m_Color.z;
}
```

The `LockParticle()` function can be used to set up the particle for use. (See Listing 7-37.)

The `Lock Particle()` function does the following:

1. Sets up the particle for use by setting `m_Active` to false and `m_Locked` to true

2. Applies a translational force taken from the input parameter `Force` along the direction `DirectionNormalized`

3. Sets the `m_TimeStamp` variable to the `CurrentTime` input parameter

4. Sets the particle color to the original color for the particle at the time of first creation

Listing 7-37. Getting a Particle Ready for Use

```
void LockParticle(float Force, Vector3 DirectionNormalized, long CurrentTime)
{
        // 1. Setup particle for use
        m_Active = false;
        m_Locked = true;
```

```
        // 2. Apply Initial Force
        Vector3 FVector = new Vector3(DirectionNormalized.x, DirectionNormalized.y,
DirectionNormalized.z);
        FVector.Multiply(Force);
        GetObjectPhysics().ApplyTranslationalForce(FVector);

        // 3. Apply Time
        m_TimeStamp = CurrentTime;

        // 4. Calculate Color for Fade
        m_Color.x = m_OriginalColor.x;
        m_Color.y = m_OriginalColor.y;
        m_Color.z = m_OriginalColor.z;
}
```

The FadeColor() function takes as input a reference to a color, ColorIn, and reduces the brightness (m_ColorBrightness) by m_FadeDelta, with a minimum value of 0 for m_ColorBrightness. The color is then scaled by m_ColorBrightness. (See Listing 7-38.)

Listing 7-38. Fading the Color of the Particle

```
void FadeColor(Vector3 ColorIn)
{
        // Fade Color to Black.

        // Adjust Brightness Level Down from full brightness = 1 to no brightness = 0;
        m_ColorBrightness -= m_FadeDelta;

        if (m_ColorBrightness < 0)
        {
                m_ColorBrightness = 0;
        }

        // 1. Adjust Color so that everything is at the same Brightness Level
        ColorIn.x *= m_ColorBrightness;
        ColorIn.y *= m_ColorBrightness;
        ColorIn.z *= m_ColorBrightness;
}
```

The FadeColor() function fades the particle's color by calling FadeColor() and then calls SetColor() to set the color of the particle. (See Listing 7-39.)

Listing 7-39. Fading the Particle's Color

```
void FadeColor(long ElapsedTime)
{
        FadeColor(m_Color);
        SetColor(m_Color);
}
```

The UpdateParticle() function (see Listing 7-40) updates the particle in the following way:

1.   If the particle is active, that is m_Active = true, then it continues with the update; otherwise, it returns.

2.   It applies a rotational force to the particle.

3.   It updates the physics of the particle.

4.   If the time that passes since the particle was created is greater than the particle life span, which is m_TimeDelay, it destroys the particle by calling Destroy(). Otherwise, it calls FadeColor() to fade the color of the particle toward black.

Listing 7-40. Updating the Particle

```
void UpdateParticle(long current_time)
{
        // If particle is Active (on the screen)
        if (m_Active)
        {
                // Update Particle Physics and position
                GetObjectPhysics().ApplyRotationalForce(40, 1);
                GetObjectPhysics().UpdatePhysicsObject(m_Orientation);

                long TimePassed = current_time - m_TimeStamp;
                if (TimePassed > m_TimeDelay)
                {
                        // Destroy Particle
                        Destroy();
                }
                else
                {
                        FadeColor(TimePassed);
                }
        }
}
```

The Render() function draws the particle to the screen by calling DrawObject(), located in the Object3d class. (See Listing 7-41.)

Listing 7-41. Rendering a Particle

```
void Render(Camera Cam, PointLight light)
{
        DrawObject(Cam, light);
}
```

Creating the SphericalPolygonExplosion Class

Next, we have to create the class that represents our explosion, which is the SphericalPolygonExplosion class. What this class does is create a group of PolyParticleEx polygons and uses them to create an explosion.

The MAX_POLYGONS variable holds the maximum number of polygons that our explosion can be composed of.

```
private int MAX_POLYGONS = 1000;
```

The m_Particles variable holds the PolyParticleEx polygons that are used to create the explosion.

```
private PolyParticleEx[] m_Particles = new PolyParticleEx[MAX_POLYGONS];
```

The m_ExplosionDirection variable holds the velocity of each of the particles in m_Particles.

```
private Vector3[] m_ExplosionDirection = new Vector3[MAX_POLYGONS];
```

The m_NumberParticles variable holds the number of particles that make up the explosion.

```
int m_NumberParticles;
```

The m_ParticleColor variable holds the color of the particles.

```
Vector3        m_ParticleColor;
```

The m_ParticleSize variable holds the scale of the particle, with 1 representing the normal scale of the particle mesh.

```
Vector3        m_ParticleSize;
```

The m_ParticleLifeSpan variable holds the time in milliseconds that the particle is to be active and displayed on the screen.

```
long           m_ParticleLifeSpan;
```

The m_ExplosionCenter variable holds the starting position for all the particles in the explosion.

```
Vector3        m_ExplosionCenter;
```

The m_RandomColors variable is true if the colors of the particles are to be random.

```
boolean        m_RandomColors;                   // true if Particles set to have Random colors
```

The m_ParticleColorAnimation variable is true if the particles in the explosion change colors randomly for every rendering.

```
boolean        m_ParticleColorAnimation;         // true if Particles change colors during explosion
```

The m_ExplosionActive variable is true if the explosion is still active and has to be rendered to the screen and updated.

```
boolean        m_ExplosionActive;
```

The m_RandNumber variable is used to generate random numbers.

```
private Random m_RandNumber = new Random();
```

The GenerateRandomColor() function generates and returns a random color using m_RandNumber. The nextFloat() function generates and returns a random number in the range of 0–1. (See Listing 7-42.)

Listing 7-42. Generating a Random Color

```
Vector3 GenerateRandomColor()
{
        Vector3 Color = new Vector3(0,0,0);

        // 1. Generate Random RGB Colors in Range of 0-1;
        Color.x = m_RandNumber.nextFloat();
        Color.y = m_RandNumber.nextFloat();
        Color.z = m_RandNumber.nextFloat();

        return Color;
}
```

The GenerateRandomRotation() function generates and returns the value of a random rotation in the range 0–MaxValue. (See Listing 7-43.)

Listing 7-43. Generating a Random Rotation

```
float GenerateRandomRotation(float MaxValue)
{
        float Rotation;

        // 1. Generate Random Rotation in Range of 0-1 * MaxValue;
        Rotation = MaxValue * m_RandNumber.nextFloat();

        return Rotation;
}
```

The GenerateRandomRotationAxis() function generates and returns a normalized randomly generated rotation axis. (See Listing 7-44.)

Listing 7-44. Generating a Random Rotation Axis

```
Vector3 GenerateRandomRotationAxis()
{
        Vector3 RotationAxis = new Vector3(0,0,0);

        // 1. Generate Random Rotation in Range of 0-1
        RotationAxis.x = m_RandNumber.nextFloat();
        RotationAxis.y = m_RandNumber.nextFloat();
        RotationAxis.z = m_RandNumber.nextFloat();
        RotationAxis.Normalize();

        return RotationAxis;
}
```

The SphericalPolygonExplosion constructor creates and initializes a new explosion. It creates and initializes m_NumberParticles new particles that will have random directions when the explosion is started. (See Listing 7-45.)

Listing 7-45. SphericalPolygonExplosion() Constructor

```
SphericalPolygonExplosion(int NumberParticles, Vector3      Color,long ParticleLifeSpan,boolean
RandomColors, boolean ColorAnimation,float FadeDelta,Vector3 ParticleSize,Context iContext, Mesh
iMesh, MeshEx iMeshEx, Texture[] iTextures, Material iMaterial, Shader iShader )
{
        m_NumberParticles       = NumberParticles;
        m_ParticleColor         = new Vector3(Color.x, Color.y, Color.z);
        m_ParticleLifeSpan      = ParticleLifeSpan;
        m_RandomColors          = RandomColors;      // true if Particles set to have Random colors
        m_ParticleColorAnimation = ColorAnimation;
        m_ExplosionActive       = false;
        m_ParticleSize          = new Vector3(ParticleSize.x, ParticleSize.y, ParticleSize.z);

        if (NumberParticles > MAX_POLYGONS)
        {
                m_NumberParticles = MAX_POLYGONS;
        }

        // For each new Particle
        for (int i = 0; i < m_NumberParticles; i++)
        {
                int signx = 1;
                int signy = 1;
                int signz = 1;

                if (m_RandNumber.nextFloat() > 0.5f)
                {
                        signx = -1;
                }
                if (m_RandNumber.nextFloat() > 0.5f)
                {
                        signy = -1;
                }
                if (m_RandNumber.nextFloat() > 0.5f)
                {
                        signz = -1;
                }

                // Find random direction for particle
                float randomx = (float)signx * m_RandNumber.nextFloat();
                float randomy = (float)signy * m_RandNumber.nextFloat();
                float randomz = (float)signz * m_RandNumber.nextFloat();
```

```
                // Generate random x,y,z coords
                Vector3        direction = new Vector3(0,0,0);
                direction.x = randomx;
                direction.y = randomy;
                direction.z = randomz;
                direction.Normalize();

                // Set Particle Explosion Direction Array
                m_ExplosionDirection[i]        =        direction;

                // Create New Particle
                m_Particles[i] = new PolyParticleEx(iContext, iMesh, iMeshEx, iTextures,
iMaterial, iShader);

                // Set Particle Array Information
                if (RandomColors)
                {
                        m_Particles[i].SetColor(GenerateRandomColor());
                }
                else
                {
                        m_Particles[i].Create(m_ParticleColor);
                }

                m_Particles[i].SetTimeDelay(ParticleLifeSpan);
                m_Particles[i].SetFadeDelta(FadeDelta);

                // Generate Random Rotations
                Vector3 Axis = GenerateRandomRotationAxis();
                m_Particles[i].m_Orientation.SetRotationAxis(Axis);

                float rot = GenerateRandomRotation(360);
                m_Particles[i].m_Orientation.SetRotationAngle(rot);
        }
}
```

The GetRandomParticleVelocity() function (see Listing 7-46) creates and returns a random velocity based on the pdirection that the particle was set when the explosion was created and a random speed.

The function does the following:

1. Gets the normalized direction of particle ParticleNumber from the m_ExplosionDirection array

2. Creates a new vector variable, ParticleVelocity, to hold the final particle velocity and initializes it with the explosion direction found in step 1

3. Generates a random speed for the particle between the MinVelocity and MaxVelocity input parameters

4. Calculates the final new random particle velocity by multiplying the ParticleVelocity variable that holds the direction of the particle by the random speed in the RandomVelocityMagnitude variable

Listing 7-46. Getting a Random Particle Velocity

```
Vector3 GetRandomParticleVelocity(int ParticleNumber, float MaxVelocity, float MinVelocity)
{
        Vector3 ExplosionDirection = m_ExplosionDirection[ParticleNumber];
        Vector3 ParticleVelocity= new Vector3(ExplosionDirection.x, ExplosionDirection.y,
        ExplosionDirection.z);
        float RandomVelocityMagnitude = MinVelocity + (MaxVelocity - MinVelocity)*
        m_RandNumber.nextFloat();
        ParticleVelocity.Multiply(RandomVelocityMagnitude);

        return ParticleVelocity;
}
```

The StartExplosion() function is called to start the actual explosion at Position location with particles with speeds from MinVelocity to MaxVelocity.

The StartExplosion() function (see Listing 7-47) does the following:

1. Sets the m_ExplosionActive variable to true to indicate the explosion is in progress

2. Sets the particles in the explosion to active, which means that they will be rendered and updated

3. Sets the timestamp on all the particles to the current system time, which is the start of the explosion

4. Sets the position of all the particles to the input parameter Position

5. Sets random velocities for all the particles

6. Sets the scale of the particles to m_ParticleSize

7. Sets a random color for all the particles, if random colors for the particles are selected, to m_RandomColors = true; otherwise, sets the color of the particle to m_ParticleColor

8. Sets the particle's life span time to m_ParticleLifeSpan

Listing 7-47. Starting the Explosion

```
void StartExplosion(Vector3        Position,float MaxVelocity, float MinVelocity)
{
        // 1. Set Position of Particles
        m_ExplosionActive = true;
        for (int i = 0; i < m_NumberParticles; i++)
        {
                m_Particles[i].SetActiveStatus(true);
                m_Particles[i].SetTimeStamp(System.currentTimeMillis());

                m_ExplosionCenter = new Vector3(Position.x, Position.y, Position.z);
                m_Particles[i].m_Orientation.SetPosition(m_ExplosionCenter);
```

```
                    m_Particles[i].GetObjectPhysics().SetVelocity(GetRandomParticleVelocity(i,MaxVeloci
ty,MinVelocity));
                    m_Particles[i].m_Orientation.SetScale(m_ParticleSize);

                    if (m_RandomColors)
                    {
                            m_Particles[i].SetColor(GenerateRandomColor());
                    }
                    else
                    {
                            m_Particles[i].SetColor(m_ParticleColor);
                    }
                    m_Particles[i].SetTimeDelay(m_ParticleLifeSpan);
            }
}
```

The RenderExplosion() function draws all the particles that make up the explosion and are active to the screen. (See Listing 7-48.)

Listing 7-48. Rendering the Explosion

```
void RenderExplosion(Camera Cam, PointLight light)
{
        // Render Explosion
        for (int i = 0;  i < m_NumberParticles; i++)
        {
                if (m_Particles[i].GetActiveStatus() == true)
                {
                        m_Particles[i].Render(Cam, light);
                }
        }
}
```

The UpdateExplosion() function updates the explosion. (See Listing 7-49.)

The function does the following:

1. If the explosion is not active, then it returns.

2. For the active particles, it sets the color randomly, if particle color animation is turned on.

3. For the active particles, it updates the particle by calling UpdateParticle().

4. If any particle is active, then the entire explosion is set to active.

Listing 7-49. Updating the Explosion

```
void UpdateExplosion()
{
        if (!m_ExplosionActive)
        {
                return;
        }
```

```
        boolean ExplosionFinished = true;
        for (int i = 0; i < m_NumberParticles; i++)
        {
                // If all Particles are not active then explosion is finished.
                if (m_Particles[i].GetActiveStatus() == true)
                {
                        // If Color Animation is on then set particle to random color
                        if(m_ParticleColorAnimation)
                        {
                                m_Particles[i].SetColor(GenerateRandomColor());
                        }

                        // For each particle update particle
                        m_Particles[i].UpdateParticle(System.currentTimeMillis());
                        ExplosionFinished = false;
                }
        }
        if (ExplosionFinished)
        {
                m_ExplosionActive = false;
        }
}
```

Modifying the Object3d Class

Next, we have to add some code to our Object3d class, so it can use the explosions.

The MAX_EXPLOSIONS variable holds the maximum number of explosions associated with this object.

```
private int MAX_EXPLOSIONS = 3;
```

The m_NumberExplosions variable holds the actual number of explosions associated with this object.

```
private int m_NumberExplosions = 0;
```

The m_Explosions variable holds references to the SphericalPolygonExplosion objects associated with this object, if any.

```
private SphericalPolygonExplosion[] m_Explosions = new SphericalPolygonExplosion[MAX_EXPLOSIONS];
```

We also add the functions RenderExplosions(), to render explosions; UpdateExplosions(), to update explosions; and ExplodeObject(), to start explosions. These functions are fairly simple, so in the interest of saving space, please refer to the Source Code/Download area for this chapter at apress.com for the full details.

The DrawObject() function must be modified to render the explosions, by calling the RenderExplosions() function. (See Listing 7-50.)

Listing 7-50. Modifying the DrawObject() Function to Render Explosions

```
void DrawObject(Camera Cam, PointLight light)
{
        RenderExplosions(Cam,light);
        // Rest of Code
}
```

The UpdateObject3d() function has to be modified to update the explosions, by calling UpdateExplosions(). (See Listing 7-51.)

Listing 7-51. Modifying the UpdateObject3d() Function

```
void UpdateObject3d()
{
        if (m_Visible)
        {
                // Update Object3d Physics
                UpdateObjectPhysics();
        }

        // Update Explosions associated with this object
        UpdateExplosions();
}
```

Creating Game Object Statistics

In order to track the game-related properties, such as health, we have to create a new class that holds these statistics.

Creating the Stats Class

The Stats class holds the game-related statistics for our game objects. The game-related properties we will use for our Drone Grid example are health, kill value, and damage value.

The m_Health variable holds a game object's health and is defaulted to 100 to indicate full health.

```
private int m_Health = 100;
```

The m_KillValue variable indicates the point value of this object when it is destroyed.

```
private int m_KillValue = 50;
```

The m_DamageValue variable is the amount that is subtracted from the health of the player when this object collides with the player's power pyramid.

```
private int m_DamageValue = 25;
```

The SaveStats() function saves the game-related statistics. For your own game, you can modify the stats in this class as needed. For example, you could add hit points and character levels if you are creating a role-playing game. If so, you would have to modify the SaveStats() function to save these new stats. (See Listing 7-52.)

Listing 7-52. Saving the Stats

```
void SaveStats(String Handle)
{
        SharedPreferences settings = m_Context.getSharedPreferences(Handle, 0);
        SharedPreferences.Editor editor = settings.edit();

        // Health
        String HealthHandle = Handle + "Health";
        editor.putInt(HealthHandle, m_Health);

        // Commit the edits!
        editor.commit();
}
```

The LoadStats() function loads in the game-related stats. (See Listing 7-53.)

Listing 7-53. Loading in the Stats

```
void LoadStats(String Handle)
{
        // Restore preferences
        SharedPreferences settings = m_Context.getSharedPreferences(Handle, 0);

        // Health
        String HealthHandle = Handle + "Health";
        m_Health = settings.getInt(HealthHandle, 100);
}
```

The functions to retrieve and set the game-related statistics in this class are in Listing 7-54. They consist of functions relating to damage value, health, and kill value.

Listing 7-54. Getting and Setting the Game-Related Statistics

```
int GetDamageValue(){return m_DamageValue;}
int GetHealth(){return m_Health;}
int GetKillValue(){return m_KillValue;}
void SetDamageValue(int value){m_DamageValue = value;}
void SetHealth(int health){m_Health = health;}
void SetKillValue(int value){m_KillValue = value;}
```

Modifying the Object3d Class

The Object3d class has to be modified to integrate in the Stats class.

The m_ObjectStats variable holds this object's game-related statistics.

```
private Stats m_ObjectStats;
```

The Object3d constructor has to be modified to create a new Stats object.

```
m_ObjectStats = new Stats(iContext);
```

The m_ObjectStats variable can be accessed through the GetObjectStats() function.

```
Stats GetObjectStats(){return m_ObjectStats;}
```

The TakeDamage() function adjusts the object's health stats by the amount of damage done by the DamageObj input object. (See Listing 7-55.)

Listing 7-55. Taking Damage from Another Object

```
void TakeDamage(Object3d DamageObj)
{
        int DamageAmount = DamageObj.GetObjectStats().GetDamageValue();
        int Health = m_ObjectStats.GetHealth();

        Health = Health - DamageAmount;

        // Health can never be negative
        if (Health < 0)
        {
                Health = 0;
        }
        m_ObjectStats.SetHealth(Health);
}
```

The SaveObjectState() function has to be modified by first adding a StatsHandle variable to hold the handle to the game object statistics to be saved.

```
String StatsHandle = Handle + "Stats";
```

Next, we need to add code to call the SaveStats() function with the StatsHandle.

```
m_ObjectStats.SaveStats(StatsHandle);
```

The LoadObjectState() function has to be modified by adding in a StatsHandle variable to hold the handle to our game stats.

```
String StatsHandle = Handle + "Stats";
```

Next, we add code to load in the previously saved game stats.

```
m_ObjectStats.LoadStats(StatsHandle);
```

Hands-on Example: Target Shooting!

Now, it's time to use some of the new classes that I discussed previously in this chapter. This example builds upon previous examples by adding the player's power pyramid, a moveable player's view that you can turn left and right, and a weapon that can fire projectiles by the user touching the screen. A cube is placed in front of the player's pyramid, and another cube is placed behind the pyramid.

We have to make modifications to the MyGLRenderer class to add code that creates the player's pyramid, creates the player's weapon, and processes the collisions between the cube in front of the pyramid and the pyramid and between the player's weapon's projectiles and the cubes.

Creating the Player

In order to create the player's power pyramid, we have to add some new variables and functions.

The player's power pyramid is m_Pyramid.

```
private PowerPyramid m_Pyramid;
```

The textures that are used with the player's pyramid are held in m_TexPyramid1 and m_TexPyramid2.

```
private Texture m_TexPyramid1;
private Texture m_TexPyramid2;
```

The PyramidCreateTexture() function creates the textures for our player's power pyramid and stores these textures in m_TexPyramid1 and m_TexPyramid2. (See Listing 7-56.)

Listing 7-56. Creating the Textures for the Pyramids

```
public void PyramidCreateTexture(Context context)
{
        m_TexPyramid1 = new Texture(context,R.drawable.pyramid1);
        m_TexPyramid2 = new Texture(context,R.drawable.pyramid2);
}
```

The CreatePyramid() function creates the player's pyramid graphic. (See Listing 7-57.)

The function does the following:

1. Creates a shader for use in rendering the pyramid.

2. Creates a new Mesh object using the data from the PyramidVertices array in the Pyramid class.

3. Creates a new Material object that sets the glow animation to true, so that when the material is updated, the emissive color property cycles between the values set with GetEmissiveMin() and GetEmissiveMax().

4. Creates the pyramid's textures by calling PyramidCreateTexture() and setting them up in an array called PyramidTex for use with the player's pyramid.

5. Creates a new power pyramid object.

6. Sets the texture animation properties for the pyramid, so that the textures cycle between one another.

7. Sets the initial position, rotation, and scale of the pyramid.

8. Sets the effect of gravity on the pyramid to none.

9. Sets the pyramid's grid spotlight and spotlight radius.

10. Sets the pyramid's mass to 2000 to indicate that this is a very large structure compared to other enemy objects in the game that will be colliding with it. The collisions that occur with other objects in the game will reflect this.

11. Creates the explosion sound associated with the pyramid and sets the sound effects on for the pyramid.

12. Creates a SphericalPolygonExplosion explosion and adds it to the pyramid using the AddExplosion() function.

Listing 7-57. Creating the Player's Power Pyramid

```
void CreatePyramid(Context iContext)
{
        //Create Cube Shader
        Shader Shader = new Shader(iContext, R.raw.vsonelight, R.raw.fsonelight);        // ok

        // Create Debug Local Axis Shader
        Mesh PyramidMesh = new Mesh(8,0,3,5,Pyramid.PyramidVertices);

        // Create Material for this object
        Material Material1 = new Material();
        Material1.SetEmissive(0.0f, 0.0f, 0.5f);

        Material1.SetGlowAnimation(true);
        Material1.GetEmissiveMax().Set(0.45f, 0.45f, 0.25f);
        Material1.GetEmissiveMin().Set(0, 0, 0);

        // Create Texture
        PyramidCreateTexture(iContext);
        Texture[] PyramidTex = new Texture[2];
        PyramidTex[0] = m_TexPyramid1;
        PyramidTex[1] = m_TexPyramid2;

        m_Pyramid = new PowerPyramid(iContext, PyramidMesh, null, PyramidTex, Material1, Shader );
        m_Pyramid.SetAnimateTextures(true, 0.3f, 0, 1);
```

```
        // Set Initial Position and Orientation
        Vector3 Axis     = new Vector3(0,1,0);
        Vector3 Position = new Vector3(0.0f, 0.0f, 0.0f);
        Vector3 Scale    = new Vector3(0.25f,0.30f,0.25f);

        m_Pyramid.m_Orientation.SetPosition(Position);
        m_Pyramid.m_Orientation.SetRotationAxis(Axis);
        m_Pyramid.m_Orientation.SetScale(Scale);
        m_Pyramid.m_Orientation.AddRotation(45);

        m_Pyramid.GetObjectPhysics().SetGravity(false);

        Vector3 ColorGrid = new Vector3(1.0f, 0.0f, 0.5f);
        m_Pyramid.SetGridSpotLightColor(ColorGrid);
        m_Pyramid.GetObjectPhysics().SetMassEffectiveRadius(7);

        m_Pyramid.GetObjectPhysics().SetMass(2000);

        //SFX
        m_Pyramid.CreateExplosionSFX(m_SoundPool, R.raw.explosion2);
        m_Pyramid.SetSFXOnOff(true);

        // Create Explosion
        int        NumberParticles  = 20;
        Vector3    Color            = new Vector3(1,1,0);
        long       ParticleLifeSpan = 2000;
        boolean    RandomColors     = false;
        boolean    ColorAnimation   = true;
        float      FadeDelta        = 0.001f;
        Vector3    ParticleSize     = new Vector3(0.5f,0.5f,0.5f);

        // No textures
        Mesh PolyParticleMesh = new Mesh(6,0,-1,3,PolyParticleEx.PolyParticleVertices);

        // Create Material for this object
        Material Material2 = new Material();
        Material2.SetSpecular(0, 0, 0);

        //Create Cube Shader
Shader Shader2 = new Shader(iContext, R.raw.vsonelightnotexture, R.raw.fsonelightnotexture);
// ok

        SphericalPolygonExplosion explosion = new SphericalPolygonExplosion(NumberParticle
s, Color, ParticleLifeSpan, RandomColors, ColorAnimation, FadeDelta, ParticleSize, m_Context,
PolyParticleMesh, null, null, Material2, Shader2 );
        m_Pyramid.AddExplosion(explosion);
}
```

Finally, in the onSurfaceCreated() function, the CreatePyramid() function was added and called to actually create the pyramid when the GL surface is created.

```
CreatePyramid(m_Context);
```

Creating the Player's Weapon

Next, the player's weapon and the ammunition it uses need to be created in the MyGLRenderer class.

The player's weapon is m_Weapon.

```
private Weapon  m_Weapon           = null;
```

The sound effects for the player's weapon are held in m_PlayerWeaponSFX.

```
private Sound m_PlayerWeaponSFX = null;
```

The player's weapon is created in the CreateWeapon() function.

The CreateWeapon() function (see Listing 7-58) does the following:

1. Creates a new shader that uses vertex shaders and fragment shaders that require no texture. This is for the ammunition that does not use a texture.

2. Creates a cube mesh for use as the 3D model for the weapon's ammunition.

3. Creates a new Material object and sets its emissive property to green.

4. Creates a new weapon and sets its ammunition range and speed.

5. Creates and loads in new ammunition into the weapon by calling the weapon's LoadAmmunition() function. These pieces of ammunition are green cubes that are created from the cube mesh and Material object created in previous steps.

Listing 7-58. Creating the Weapon

```
 void CreateWeapon(Context iContext)
{
        //Create Cube Shader
        Shader Shader = new Shader(iContext, R.raw.vsonelightnotexture, R.raw.fsonelightnotexture);
// ok

        // Create
        MeshEx CubeMesh = new MeshEx(6,0,-1,3,Cube.CubeDataNoTexture, Cube.CubeDrawOrder);

        // Create Material for this object
        Material Material1 = new Material();
        Material1.SetEmissive(0.0f, 1.0f, 0.0f);

        // Create Weapon
        m_Weapon = new Weapon(iContext, null, null, null, Material1, Shader);
        float AmmunitionRange = 100;
        float AmmunitionSpeed = 0.5f;
```

```
for (int i = 0; i < m_Weapon.GetMaxAmmunition(); i++)
{
        Ammunition Ammo = new Ammunition(iContext, null, CubeMesh, null, Material1, Shader,
        AmmunitionRange,AmmunitionSpeed);

        // Set Intial Position and Orientation
        Vector3 Axis = new Vector3(1,0,1);
        Vector3 Scale = new Vector3(0.3f,0.3f,0.3f);

        Ammo.m_Orientation.SetRotationAxis(Axis);
        Ammo.m_Orientation.SetScale(Scale);

        Ammo.GetObjectPhysics().SetGravity(false);
        Ammo.GetObjectPhysics().SetGravityLevel(0.003f);

        Vector3 GridColor = new Vector3(1,0f,0);
        Ammo.SetGridSpotLightColor(GridColor);
        Ammo.GetObjectPhysics().SetMassEffectiveRadius(10);
        Ammo.GetObjectPhysics().SetMass(100);
        Ammo.GetObjectStats().SetDamageValue(25);

        m_Weapon.LoadAmmunition(Ammo, i);
    }
}
```

The MapWindowCoordsToWorldCoords() function uses the gluUnProject() function to translate screen coordinates generated by user touches to world coordinates, which are returned in a float array.

Using the MapWindowCoordsToWorldCoords() function (see Listing 7-59), the following are accomplished:

1. A new float array, ObjectCoords, is created to return in homogeneous coordinates the 3D world coordinates that correspond to the screen coordinates.

2. The y screen position is translated from screen coordinates to the y coordinate system that OpenGL uses, by subtracting the y location in screen coordinates from the Android's screen height. For example, a point that was input as (0,0) is now sent to the gluUnProject() function as (0, screenheight). A point that was input as (0, screenheight) would be transformed into (0,0).

3. The GLU.gluUnProject() function is then called to convert the screen touch coordinates into 3D world coordinates.

4. The ObjectCoords float array that holds the 3D world coordinates is returned.

Listing 7-59. Mapping Window Coordinates to 3D World Coordinates

```
float[] MapWindowCoordsToWorldCoords(int[] View, float WinX, float WinY, float WinZ)
{
        // Set modelview matrix to just camera view to get world coordinates

        // Map window coordinates to object coordinates. gluUnProject maps the specified
        // window coordinates into object coordinates using model, proj, and view. The result is
        // stored in obj.
        // view          the current view, {x, y, width, height}
        float[] ObjectCoords = new float[4];
        float realy = View[3] - WinY;
        int result = 0;

        //public static int gluUnProject (float winX, float winY, float winZ,
        //                          float[] model, int modelOffset,
        //                          float[] project, int projectOffset,
        //                          int[] view, int viewOffset,
        //                          float[] obj, int objOffset)
        result = GLU.gluUnProject (WinX, realy, WinZ, m_Camera.GetViewMatrix(), 0, m_Camera.
        GetProjectionMatrix(), 0, View, 0, ObjectCoords, 0);

        if (result == GLES20.GL_FALSE)
        {
                Log.e("class Object3d :", "ERROR = GLU.gluUnProject failed!!!");
                Log.e("View = ", View[0] + "," + View[1] + ", " + View[2] + ", " + View[3]);
        }
         return ObjectCoords;
}
```

The CreatePlayerWeaponSound() function creates a new sound effect for the player's weapon. (See Listing 7-60.)

Listing 7-60. Creating the Player's Weapon Sound Effects

```
 void CreatePlayerWeaponSound(Context iContext)
{
        m_PlayerWeaponSFX = new Sound(iContext, m_SoundPool, R.raw.playershoot2);
}
```

The PlayPlayerWeaponSound() function plays the weapon sound effects. (See Listing 7-61.)

Listing 7-61. Playing the Weapon's Sound Effect

```
 void PlayPlayerWeaponSound()
{
        if (m_SFXOn)
        {
                m_PlayerWeaponSFX.PlaySound();
        }
}
```

The CheckTouch() function (see Listing 7-62) is called when the user touches the screen and fires the player's weapon.

The function does the following:

1. Creates an integer array View that holds the current screen view parameters of the Android device.

2. Calls MapWindowCoordsToWorldCoords() function with the view parameters, along with the x and y locations that the user has touched and a z value of 1. The 3D homogeneous world coordinates are returned in the WorldCoords float array.

3. Converts the homogeneous coordinates to Cartesian coordinates by dividing WorldCoords by the w value or fourth element in the WorldCoords array. The result is stored in TargetLocation.

4. Defines the WeaponLocation variable as the location of the camera or viewer.

5. Defines the Direction variable, which is the direction the weapon is to be fired, as a vector going from the WeaponLocation to the TargetLocation.

6. Fires the player's weapon with the projectile starting at the WeaponLocation location and in the direction Direction.

7. Plays the player's weapon sound effect.

Listing 7-62. Checking the User's Touch for Firing the Weapon

```
void CheckTouch()
{
        // Player Weapon Firing
        int[] View = new int[4];

        View[0] = 0;
        View[1] = 0;
        View[2] = m_ViewPortWidth;
        View[3] = m_ViewPortHeight;
        float[] WorldCoords = MapWindowCoordsToWorldCoords(View, m_TouchX, m_TouchY, 1);
// 1 = far clipping plane
        Vector3 TargetLocation = new Vector3(WorldCoords[0]/WorldCoords[3],
WorldCoords[1]/WorldCoords[3], WorldCoords[2]/WorldCoords[3]);
        Vector3 WeaponLocation = m_Camera.GetCameraEye();

        Vector3 Direction = Vector3.Subtract(TargetLocation, WeaponLocation);
        if ((Direction.x == 0) && (Direction.y == 0) && (Direction.z == 0))
        {
            return;
        }
        if (m_Weapon.Fire(Direction, WeaponLocation) == true)
        {
            // WeaponFired
            PlayPlayerWeaponSound();
        }
}
```

Finally, new code has to be added to create the player's weapon in the onSurfaceCreated() function.

```
CreateWeapon(m_Context);
```

New code also has to be added to create the sound effects for the weapon.

```
CreatePlayerWeaponSound(m_Context);
```

Processing Collisions

The ProcessCollisions() function processes collisions between game objects.

The ProcessCollisions() function (see Listing 7-63) does the following:

1. Checks for collision between the player's ammunition and m_Cube2. If there is a collision, then it applies a linear force to the two colliding objects and increases the player's score by the kill value of m_Cube2.

2. Checks for collision between the player's ammunition and m_Cube. If there is a collision, it applies a linear force to the two colliding objects and increases the player's score by the kill value of m_Cube.

3. Checks for a collision between the player's pyramid and m_Cube2, which is the cube in front of the pyramid.

4. If there is a collision, it

    a. Starts the explosion associated with the pyramid

    b. Plays the explosion sound effect associated with the pyramid

    c. Applies the linear impulse to both objects

    d. Resets the pyramid state to eliminate any accelerations, because the pyramid is stationary

    e. Calculates the damage to the pyramid

Listing 7-63. Processing the Game Object Collisions

```
void ProcessCollisions()
{
        Object3d CollisionObj = m_Weapon.CheckAmmoCollision(m_Cube2);
          if (CollisionObj != null)
          {
                CollisionObj.ApplyLinearImpulse(m_Cube2);
                m_Score = m_Score + m_Cube2.GetObjectStats().GetKillValue();
          }
```

```
        CollisionObj = m_Weapon.CheckAmmoCollision(m_Cube);
        if (CollisionObj != null)
        {
                CollisionObj.ApplyLinearImpulse(m_Cube);
                m_Score = m_Score + m_Cube.GetObjectStats().GetKillValue();
        }

        float ExplosionMinVelocity = 0.02f;
        float ExplosionMaxVelocity = 0.4f;

        //Check Collision with Cube2
        Physics.CollisionStatus result = m_Pyramid.CheckCollision(m_Cube2);
        if ((result == Physics.CollisionStatus.COLLISION) ||
                (result == Physics.CollisionStatus.PENETRATING_COLLISION))
        {
                m_Pyramid.ExplodeObject(ExplosionMaxVelocity, ExplosionMinVelocity);
                m_Pyramid.PlayExplosionSFX();
                m_Pyramid.ApplyLinearImpulse(m_Cube2);

                // Set Pyramid Velocity and Acceleration to 0
                m_Pyramid.GetObjectPhysics().ResetState();

                m_Pyramid.TakeDamage(m_Cube2);
        }
}
```

Modifying the `onDrawFrame()` Function

The `onDrawFrame()` function must also be modified in order to render and update the player's view, graphic, and weapons. (See Listing 7-64.)

The following modifications have to be made:

1. The collisions between the player's ammunition and the two cubes and the collision between the cubes and the player's pyramid must all be processed.

2. If the camera has moved, this movement must be processed.

3. The player's power pyramid must be updated and drawn.

4. If the user has touched the screen, you must process the touch and fire the player's weapon.

5. Update and draw the player's weapon and ammunition.

Listing 7-64. Modifying the onDrawFrame() Function

```
@Override
public void onDrawFrame(GL10 unused)
{
        GLES20.glClearColor(0.0f, 0.0f, 0.0f, 1.0f);
        GLES20.glClear( GLES20.GL_DEPTH_BUFFER_BIT | GLES20.GL_COLOR_BUFFER_BIT);
```

```
// Player Update
// Player's Weapon
ProcessCollisions();
if (m_CameraMoved)
{
        ProcessCameraMove();
}
  m_Camera.UpdateCamera();
//////////////////////////// Update Object Physics
// Cube1
m_Cube.UpdateObject3d();

// Cube2
m_Cube2.UpdateObject3d();

// Process Collisions
Physics.CollisionStatus TypeCollision = m_Cube.GetObjectPhysics().CheckForCollisionSphereBounding(m_Cube, m_Cube2);

if ((TypeCollision == Physics.CollisionStatus.COLLISION) ||
(TypeCollision == Physics.CollisionStatus.PENETRATING_COLLISION))
{
        m_Cube.GetObjectPhysics().ApplyLinearImpulse(m_Cube, m_Cube2);

        // SFX
        m_Cube.PlaySound(m_SoundIndex1);
        m_Cube2.PlaySound(m_SoundIndex2);
  }

//////////////////////////// Draw Objects
m_Cube.DrawObject(m_Camera, m_PointLight);
m_Cube2.DrawObject(m_Camera, m_PointLight);

//////////////////////////// Update and Draw Grid
UpdateGravityGrid();
m_Grid.DrawGrid(m_Camera);

// Player's Pyramid
m_Pyramid.UpdateObject3d();
m_Pyramid.DrawObject(m_Camera, m_PointLight);

// Did user touch screen
if (m_ScreenTouched)
{
        // Process Screen Touch
        CheckTouch();
        m_ScreenTouched = false;
}

m_Weapon.UpdateWeapon();
m_Weapon.RenderWeapon(m_Camera, m_PointLight, false);
```

```
/////////////////////// HUD
// Update HUD
UpdateHUD();
m_HUD.UpdateHUD(m_Camera);

// Render HUD
m_HUD.RenderHUD(m_Camera, m_PointLight);
}
```

Now, run the application. You should see something like Figure 7-3.

Figure 7-3. Initial screen

Touch the screen to fire your weapon. Try to knock the cube in front of the pyramid into the pyramid. When the ammunition hits the cube, the score should be increased. When the cube hits the pyramid, the player's health should be decreased, an explosion graphic animation displayed, and a sound effect played. (See Figure 7-4.)

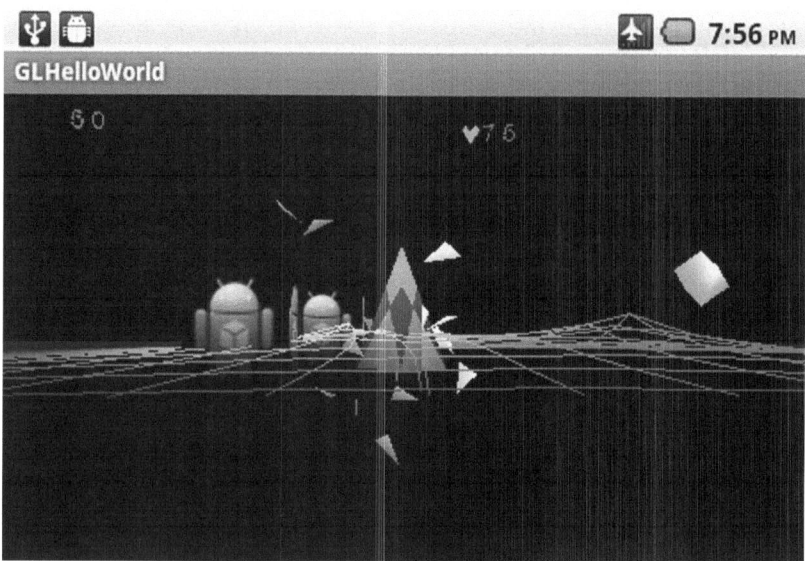

Figure 7-4. Hitting the cube into the player's pyramid

Summary

In this chapter, I covered how to build the player-related code for our Drone Grid case study game. I started with creating the classes needed to build the player's power pyramid. I then covered the code needed to implement the player's view and the player's input. Next, the code to implement the player's weapon and the ammunition that is used in that weapon were covered. This was followed by a discussion of explosions and managing the game-play properties of our game objects. Finally, we went through a hands-on example in which we demonstrated how the player's weapon can be used to hit 3D objects and also how these 3D objects can cause damage to our pyramid.

Chapter 8

Drone Grid Case Study: Creating the Enemies

This chapter will cover the creation of the enemies for the Drone Grid case study. Covered first are arena objects, which have a relatively simple behavior. Next, I cover the tank enemy, which has much more complex behavior. In order to understand this behavior, I cover the basics of a finite state machine. Next, I cover the finite state machine and the related classes specific to our computer controlled tank. I then cover other classes needed to implement the tank in our game. Finally, I discuss a hands-on example that uses the classes I have previously covered in a working example.

Creating Arena Objects

Arena objects are simple enemy objects that move through the gravity grid area. They travel in straight lines inside the gravity grid, until they hit a grid boundary or the player's power pyramid. After hitting a grid boundary or the power pyramid, they reverse direction with the same speed they had before.

The ArenaObject3d class derives from the Object3d class.

```
public class ArenaObject3d extends Object3d
```

The m_ArenaObjectID variable can hold a string identifying an individual object in the game world.

```
private String m_ArenaObjectID = "None";
```

The m_XmaxBoundary and m_XminBoundary variables hold the maximum and minimum boundaries of the game grid or game arena along the x axis.

```
private float m_XMaxBoundary = 1;
private float m_XMinBoundary = 0;
```

The m_ZmaxBoundary and m_ZminBoundary variables hold the maximum and minimum boundaries of the game arena along the z axis.

```
private float m_ZMaxBoundary = 1;
private float m_ZMinBoundary = 0;
```

The m_HitGroundSFXIndex variable holds the handle to the sound effect played, if any, when the arena object hits the ground.

```
private int m_HitGroundSFXIndex = -1;
```

The m_ExplosionSFXIndex variable holds the handle to the sound effect played when the arena object explodes.

```
private int m_ExplosionSFXIndex = -1;
```

The ArenaObject3d constructor calls the constructor of the Object3d class and then sets the game arena boundaries for this arena object. (See Listing 8-1.)

Listing 8-1. ArenaObject3d Constructor

```
ArenaObject3d(Context iContext, Mesh iMesh, MeshEx iMeshEx, Texture[] iTextures, Material iMaterial,
Shader iShader, float XMaxBoundary,float XMinBoundary,float ZMaxBoundary,float ZMinBoundary)
{
        super(iContext, iMesh, iMeshEx, iTextures, iMaterial, iShader);
        m_XMaxBoundary = XMaxBoundary;
        m_XMinBoundary = XMinBoundary;
        m_ZMaxBoundary = ZMaxBoundary;
        m_ZMinBoundary = ZMinBoundary;
}
```

The CreateExplosionSFX() creates a new explosion sound effect by calling the AddSound() function in the Object3d class with the resource id of the sound to create and the sound pool to hold it in. The index to the newly created sound is held in m_ExplosionSFXIndex. (See Listing 8-2.)

Listing 8-2. Creating the Explosion Sound Effect

```
void CreateExplosionSFX(SoundPool Pool, int ResourceID)
{
        m_ExplosionSFXIndex = AddSound(Pool, ResourceID);
}
```

The PlayExplosionSFX() function plays the sound effect for an explosion, if there is one, by calling the PlaySound() function in the Object3d class with the index of the explosion sound effect. (See Listing 8-3.)

Listing 8-3. Playing the Explosion Sound Effect

```
void PlayExplosionSFX()
{
        if (m_ExplosionSFXIndex >= 0)
        {
                PlaySound(m_ExplosionSFXIndex);
        }
}
```

The `CreateHitGroundSFX()` function creates a new sound effect by calling the `AddSound()` function in the Object3d class with the resource id of the sound effect and the sound pool in which the sound will be stored. The index to the newly created sound is returned and put in `m_HitGroundSFXIndex`. (See Listing 8-4.)

Listing 8-4. Creating the Hit Ground Sound Effect

```
void CreateHitGroundSFX(SoundPool Pool, int ResourceID)
{
        m_HitGroundSFXIndex = AddSound(Pool, ResourceID);
}
```

The `PlayHitGoundSFX()` function plays the sound effect for the arena object hitting the ground, if there is one. (See Listing 8-5.)

Listing 8-5. Playing the Hit Ground Sound Effect

```
void PlayHitGoundSFX()
{
        if (m_HitGroundSFXIndex >= 0)
        {
                PlaySound(m_HitGroundSFXIndex);
        }
}
```

The `RenderArenaObject()` function draws the arena object to the screen. It also tests to see if the object has just hit the ground. If it has just hit the ground, the hit ground sound effect is played, and the hit ground status in the Physics class is reset. (See Listing 8-6.)

Listing 8-6. Rendering the Arena Object

```
void RenderArenaObject(Camera Cam, PointLight light)
{
        // Object hits ground
        boolean ShakeCamera = GetObjectPhysics().GetHitGroundStatus();
        if (ShakeCamera)
        {
                GetObjectPhysics().ClearHitGroundStatus();
                PlayHitGoundSFX();
        }
        DrawObject(Cam, light);
}
```

The UpdateArenaObject() function updates the arena object enemy. (See Listing 8-7.)

Listing 8-7. Updating the Arena Object

```
void UpdateArenaObject()
{
      if (IsVisible() == true)
      {
            // Check Bounds for Z
            if (m_Orientation.GetPosition().z >= m_ZMaxBoundary)
            {
                  Vector3 v = GetObjectPhysics().GetVelocity();
                  if (v.z > 0)
                  {
                        v.z = -v.z;
                  }
            }
            else
            if (m_Orientation.GetPosition().z <= m_ZMinBoundary)
            {
                  Vector3 v = GetObjectPhysics().GetVelocity();
                  if (v.z < 0)
                  {
                        v.z = -v.z;
                  }
            }

            // Check bounds for X
            if (m_Orientation.GetPosition().x >= m_XMaxBoundary)
            {
                  Vector3 v = GetObjectPhysics().GetVelocity();
                  if (v.x > 0)
                  {
                        v.x = -v.x;
                  }
            }
            if (m_Orientation.GetPosition().x <= m_XMinBoundary)
            {
                  Vector3 v = GetObjectPhysics().GetVelocity();
                  if (v.x < 0)
                  {
                        v.x = -v.x;
                  }
            }
      }
      // Update Physics for this object
      UpdateObject3d();
}
```

The function does the following:

1. If the object is visible, it continues with the update, otherwise it returns.

2. Checks the max z boundary to see if the object is outside it. If it is, and it is going farther outside the boundary, then it reverses the z component of the object's velocity. Then it checks the minimum z boundary to see if the object is outside it. If it is, and it is going farther outside the boundary, then it reverses the z component of the object's velocity.

3. Repeats step 2 for the x boundary and reverses the x component of the object's velocity instead of the z component.

Finally, an example of an arena object that we use in a later hands-on example in this chapter is shown in Figure 8-1.

Figure 8-1. *An example of an arena object*

Overview of Artificial Intelligence

The way complex artificial intelligence (AI) is generally applied in video games is through the use of the finite state machine. A finite state machine consists of a set of states that represents the behavior of the person, entity, or vehicle you want to simulate. Each state contains code to implement this behavior and checks to see if there is a change in the game conditions that will have to result in a change of state. If there needs to be a change of state, then the finite state machine sets the current executing state to the one specified by the previous state, based on the previous state's transition rules.

For example, let's say you want a player to be able to control a squad of robots and have these robots perform specific tasks according to what command the player sends the robots. The behaviors which the player can select from are

- Retreat
- Patrol
- Attack Enemy

Using a finite state machine, each of these behaviors would be implemented in a separate state. The initial state would process the player's current command and transition to the corresponding state implementing the command. Once each of these commands was completed, the finite state machine would transition back to the state that processes the player's commands. The process would be repeated, and the state machine would transition to the state implementing the player's current command. (See Figure 8-2.)

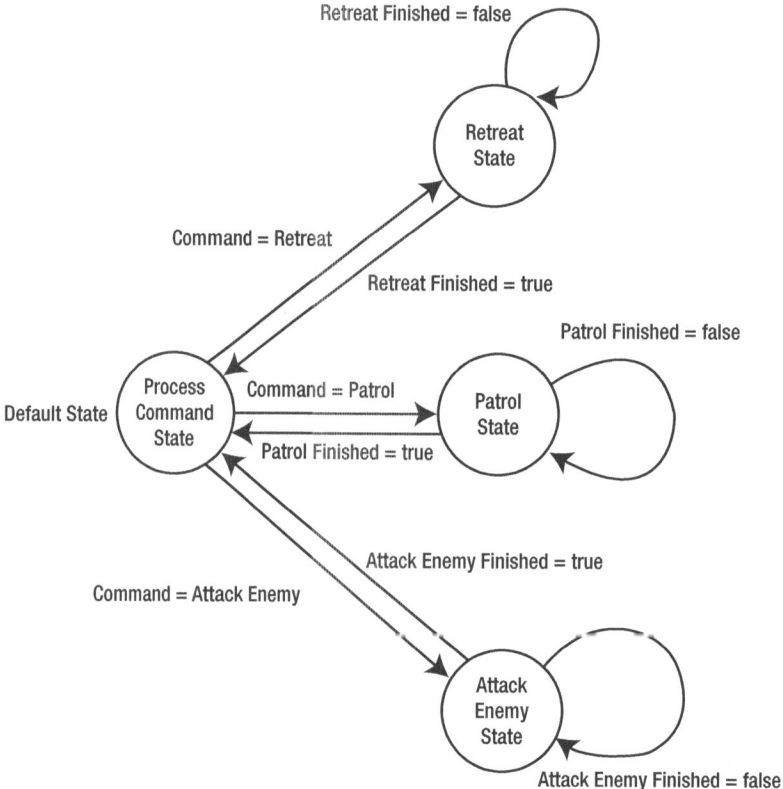

Figure 8-2. *The finite state machine*

In terms of implementing a finite state machine in code, we have a Finite State Machine class for the enemy type we want to control. Each of the behaviors for this type of enemy is represented by a separate class and is loaded into the finite state machine for processing. The UpdateMachine() function in the Finite State Machine class is called to update the state machine. See Figure 8-3 for a diagram illustrating a general class level view of how our state machine in Figure 8-2 would be implemented in code.

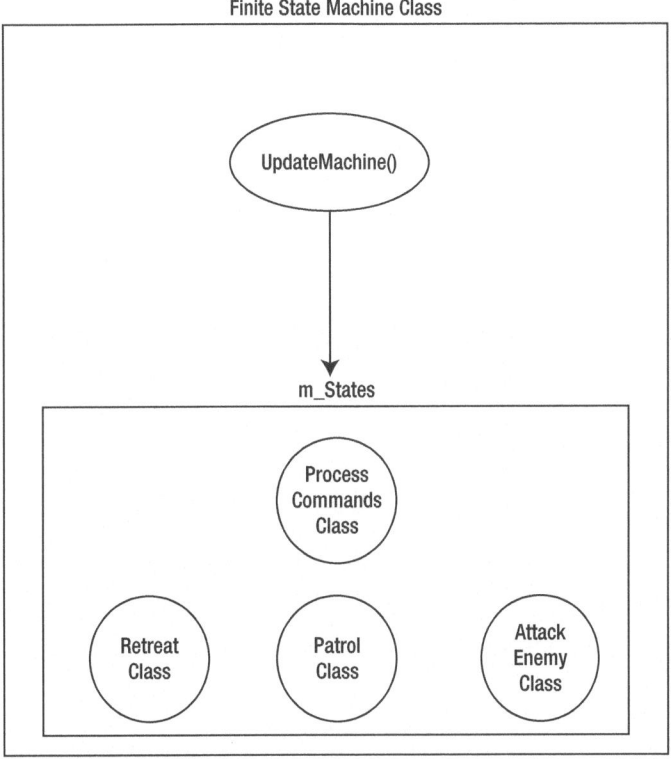

Figure 8-3. Implementing the finite state machine

Creating the Tank Enemy

The tank enemy consists of the tank's graphics as well as the tank's artificial intelligence.

Creating the Tank Graphic

The graphics for the tank are two pyramid-shaped polygons. The lower polygon representing the tank body is identical to the top polygon representing the tank's turret, except for being slightly larger.

The Pyramid2 class holds the data for the tank body and tank turret in the `Pyramid2Vertices` array Listing 8-8.)

Listing 8-8. The Tank Graphics

```
public class Pyramid2  extends Object3d
{
        static float[] Pyramid2Vertices =
        {
                // Triangle Shape
                // Left Side          u   v    nx, ny, nz
```

```
                    -0.5f, -0.5f, -0.5f,    0, 1,  -1, -1, -1,  // v0 = left, bottom, back
                    0.0f, -0.5f,  0.5f,    1, 1,   0,  0,  1,  // v1 = left, bottom, front
                    0.0f,  0.5f, -0.5f, 0.5f, 0,   0,  1,  0,  // v2 = top point

                    // Right Side
                    0.5f, -0.5f, -0.5f,    0, 1,   1, -1, -1,  // v3 = right, bottom, back
                    0.0f, -0.5f,  0.5f,    1, 1,   0,  0,  1,  // v4 = right, bottom, front
                    0.0f,  0.5f, -0.5f, 0.5f, 0,   0,  1,  0,  // v2 = top point

                    // Back
                    -0.5f, -0.5f, -0.5f,    0, 1,  -1, -1, -1,  // v0 = left, bottom, back
                    0.5f, -0.5f, -0.5f,    1, 1,   1, -1, -1,  // v3 = right, bottom, back
                    0.0f,  0.5f, -0.5f, 0.5f, 0,   0,  1,  0,  // v2 = top point

                    // Bottom
                    -0.5f, -0.5f, -0.5f,    0, 1,  -1, -1, -1,  // v0 = left, bottom, back
                    0.5f, -0.5f, -0.5f,    1, 1,   1, -1, -1,  // v3 = right, bottom, back
                    0.0f, -0.5f,  0.5f, 0.5f, 0,   0,  0,  1,  // v4 = right, bottom, front
            };
        Pyramid2(Context iContext, Mesh iMesh, MeshEx iMeshEx, Texture[] iTextures, Material iMaterial,
Shader iShader)
        {
                super(iContext, iMesh, iMeshEx, iTextures, iMaterial, iShader);
        }
}
```

The final composite enemy tank object with both turret and main tank body is shown in Figure 8-4.

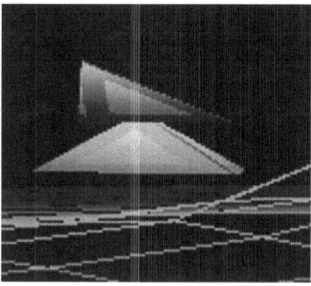

Figure 8-4. The tank 3D object

Creating the Tank State

The tank state machine consists of two states. One is the process command state, which processes the commands sent to the tank and selects the behavior of the tank that implements the command. The other state is the patrol/attack state, which drives the tank around the playfield according to waypoints while firing at the player's power pyramid. The tank's finite state machine is shown in Figure 8-5.

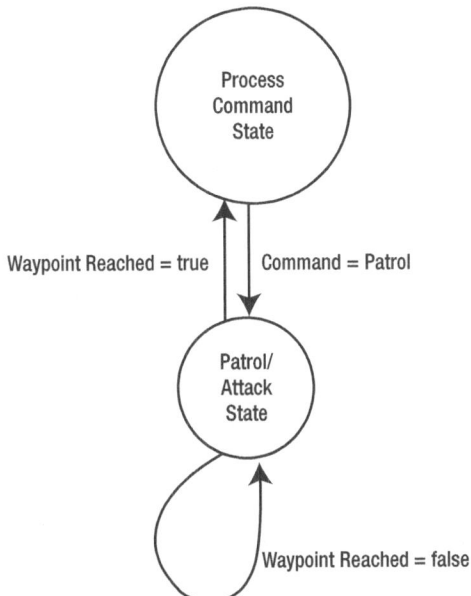

Figure 8-5. *The tank's finite state machine*

The FSM_StatesTank enumeration identifies the different state ids used internally within the tank state machine so that the state machine can identify and transition to the corresponding state when needed.

The FSM_StatesTank enumeration (see Listing 8-9) consists of

> **FSM_STATE_NONE**: This indicates no state.
>
> **FSM_STATE_STEER_WAYPOINT**: This state id corresponds to the patrol/attack state.
>
> **FSM_STATE_PROCESS_COMMAND**: This state id corresponds to the process command state.

Listing 8-9. *The Tank States*

```
enum FSM_StatesTank
{
    FSM_STATE_NONE,
    FSM_STATE_STEER_WAYPOINT,
    FSM_STATE_PROCESS_COMMAND,
};
```

The StateTank class holds the base class from which all other states for the tank are derived.

The m_Parent variable holds a reference to the driver of the tank and provides access to the commands and data given to the AI-controlled vehicle and the tank object itself.

```
private Driver m_Parent;
```

The m_StateID variable holds the id that identifies the states that the tank can be in. See Listing 8-9 for the full list.

```
private FSM_StatesTank m_StateID;
```

The Init() function is called once, when the finite state machine that this class object is part of is first created and reset.

```
void Init()    {}
```

The Enter() function is called when the state is first entered from a different state.

```
void Enter()   {}
```

The Exit() function is called before exiting the state to a different state.

```
void Exit() {}
```

The Update() function is called to update the state.

```
void Update() {}
```

The CheckTransitions() function checks for transitions to another state, based on game conditions, and returns no state by default, unless overridden by a subclass.

The entire StateTank class is shown in Listing 8-10.

Listing 8-10. The Base Class for the Tank States

```
public class StateTank
{
        private Driver m_Parent;
        private FSM_StatesTank m_StateID;

        StateTank(FSM_StatesTank ID, Driver Parent)
        {
                m_StateID = ID;
                m_Parent = Parent;
        }

        void Init()    {}
        void Enter()   {}
        void Exit() {}
        void Update() {}

        FSM_StatesTank  CheckTransitions()
        {
                return FSM_StatesTank.FSM_STATE_NONE;
        }

        Driver GetParent() {return m_Parent;}
        FSM_StatesTank GetStateID() { return m_StateID;}
}
```

Creating Vehicle Commands

Next, we have to create the class that will hold the commands for our tank vehicle to execute. First, we will have to define a couple of enumerations.

The `AIVehicleCommand` enumeration (see Listing 8-11) holds the actual command type for the vehicle to execute.

Listing 8-11. The Vehicle Command

```
enum AIVehicleCommand
{
      None,
      Patrol,
};
```

The vehicle commands can be

> **None**: Specifies that there is no command

> **Patrol**: Tells the tank to move toward the waypoints specified in the VehicleCommand class object and at the same time fire at the player's power pyramid

The `AIVehicleObjectsAffected` enumeration (see Listing 8-12) holds the item that is affected by the command, such as

> **None**: Indicates that there are no objects affected by the command.

> **WayPoints**: Indicates that the waypoints will be affected by the command.

> **PrimaryWeapon**: Indicates that the primary weapon will be used.

> **SecondaryWeapon**: Indicates that the secondary weapon will be used.

Listing 8-12. Objects Affected by the Command

```
enum AIVehicleObjectsAffected
{
      None,
      WayPoints,
      PrimaryWeapon,
      SecondaryWeapon
};
```

The VehicleCommand class represents the command that we will send to the tank.

The `m_Command` variable holds the actual vehicle command in the form of an `AIVehicleCommand` enumeration, mentioned previously.

```
private AIVehicleCommand m_Command;
```

The m_ObjectsAffected variable holds the objects affected by the command, if any, in an AIVehicleObjectsAffected enumeration, mentioned previously.

```
private AIVehicleObjectsAffected m_ObjectsAffected;
```

The m_NumberObjectsAffected variable holds the number of objects affected by the command, if any.

```
private int m_NumberObjectsAffected;
```

The m_DeltaAmount variable holds the number of rounds to fire in one burst.

```
private float  m_DeltaAmount;
```

The m_DeltaIncrement variable holds the firing delay time in milliseconds between bursts of the weapon's fire.

```
private float  m_DeltaIncrement;
```

The m_MaxValue variable holds a maximum value related to the command, if any.

```
private float  m_MaxValue;
```

The m_MinValue variable holds a minimum value related to the command, if any.

```
private float  m_MinValue;
```

The MAX_WAYPOINTS variable holds the maximum number of waypoints that can be held in this command. Waypoints are a series of locations that are reached in sequence one after the other.

```
static int MAX_WAYPOINTS = 50;
```

The m_NumberWayPoints variables hold the number of waypoints that are actually being held in this command.

```
private int m_NumberWayPoints = 0;
```

The m_CurrentWayPointIndex holds the index of the current waypoint that the vehicle is moving to.

```
private int m_CurrentWayPointIndex = 0;
```

The m_WayPoints array holds the values of the locations that the vehicle will move to.

```
private Vector3[] m_WayPoints = new Vector3[MAX_WAYPOINTS];
```

The m_Target variable holds the location that the tank will fire at, if any.

```
private Vector3 m_Target;
```

The m_TargetObject variable holds the object that the tank will fire at, if any.

```
private Object3d m_TargetObject;
```

The VehicleCommand constructor creates the vehicle command by setting the class member variables discussed previously. (See Listing 8-13.)

Listing 8-13. Vehicle Constructor

```
VehicleCommand(Context iContext,AIVehicleCommand Command, AIVehicleObjectsAffected ObjectsAffected,
int NumberObjectsAffected,float DeltaAmount,float DeltaIncrement,float MaxValue,float MinValue, int
NumberWayPoints,Vector3[] WayPoints,Vector3 Target,Object3d  TargetObject)
{
      m_Context = iContext;
      m_Command = Command;
      m_ObjectsAffected       =ObjectsAffected;
      m_NumberObjectsAffected=      NumberObjectsAffected;
      m_DeltaAmount = DeltaAmount;
      m_DeltaIncrement =   DeltaIncrement;
      m_MaxValue = MaxValue;
      m_MinValue = MinValue;

      m_NumberWayPoints =  NumberWayPoints;
      m_WayPoints = WayPoints;
      m_Target =Target;
      m_TargetObject= TargetObject;
}
```

The SaveState() function saves the VehicleCommand class member data using the input Handle string as the main index name under which to store the vehicle command data.

The SaveState() function does the following:

1. Gets a SharedPreferences object associated with the Activity specified in m_Context and the input Handle.

2. Sets up an editor variable from the SharedPreferences object in step 1 that is used to put the class member data into the shared preferences file.

3. Saves the value of each class member variable. The general procedure is to

    a. Create a handle to the class member variable by concatenating the Handle that is an input parameter to the function and a string representing the variable. For example, Handle + "Command".

    b. Next, if needed, convert the variable into a form that can be written into a SharedPreferences object such as a string or integer. For example, an enumeration can be converted to a string by adding it to a null string "". The statement m_Command + "" will convert the enumeration to a string.

    c. Store the variable in a key value pair format using the handle specific to that class member variable as the key and the value of the class member variable as the value.

4. Saves the changes made to all the data to the shared preferences file by calling commit() on the editor variable created in step 2.

See Listing 8-14. In order to save space, the listing has been shortened. For the full code for this chapter, please see the Source Code/Download area located on apress.com.

Listing 8-14. Saving the Vehicle Command

```
void SaveState(String Handle)
{
        SharedPreferences settings = m_Context.getSharedPreferences(Handle, 0);
        SharedPreferences.Editor editor = settings.edit();

        // Command
        String CommandHandle = Handle + "Command";
        String CommandStr = m_Command + "";
        editor.putString(CommandHandle, CommandStr);

        //  Code to save reset of class member variables

        // Commit the edits!
        editor.commit();
}
```

The MatchCommand() function converts a string value to a AIVehicleCommand enumeration and returns it. This function is used in the LoadState() function to load a saved VehicleCommand object. (See Listing 8-15.)

Listing 8-15. Matching a String Command to an Enumeration

```
static AIVehicleCommand MatchCommand(String CommandStr)
{
        AIVehicleCommand Command = AIVehicleCommand.None;

        if (CommandStr.equalsIgnoreCase("None"))
        {
                Command = AIVehicleCommand.None;
        }
        else
        if (CommandStr.equalsIgnoreCase("Patrol"))
        {
                Command = AIVehicleCommand.Patrol;
        }
        return Command;
}
```

The MatchObjectsAffected() function converts a string to a AIVehicleObjectsAffected enumeration. This function is used in the LoadState() function to load a VehicleCommand object. (See Listing 8-16.)

Listing 8-16. Converting a String to an AIVehicleObjectsAffected Enumeration

```
static AIVehicleObjectsAffected MatchObjectsAffected(String ObjectsAffectedStr)
{
        AIVehicleObjectsAffected ObjectsAffected = AIVehicleObjectsAffected.None;

        if (ObjectsAffectedStr.equalsIgnoreCase("None"))
        {
                ObjectsAffected = AIVehicleObjectsAffected.None;
        }
        else
        if (ObjectsAffectedStr.equalsIgnoreCase("WayPoints"))
        {
                ObjectsAffected = AIVehicleObjectsAffected.WayPoints;
        }
        else
        if (ObjectsAffectedStr.equalsIgnoreCase("PrimaryWeapon"))
        {
                ObjectsAffected = AIVehicleObjectsAffected.PrimaryWeapon;
        }
        else
        if (ObjectsAffectedStr.equalsIgnoreCase("SecondaryWeapon"))
        {
                ObjectsAffected = AIVehicleObjectsAffected.SecondaryWeapon;
        }
        return ObjectsAffected;
}
```

The LoadState() function loads data from a shared preferences entry into the class member variables.

The LoadState() function does the following:

1. Gets the shared preferences object associated with the input parameter Handle.

2. Loads data for each of the class member variables. The general format for doing this follows:

    a. Create the handle specific to that class member variable by concatenating the input Handle parameter and the string identifier for that variable.

    b. Read in the data using this new handle created in step 1.

    c. If necessary, convert the data into a data type that can be stored in that class member variable.

See Listing 8-17. In order to save space, the listing has been shortened. For the full code for this chapter, please see the Source Code/Download area located on apress.com.

Listing 8-17. Loading a Vehicle Command

```
void LoadState(String Handle)
{
        SharedPreferences settings = m_Context.getSharedPreferences(Handle, 0);

        // Command
         String CommandHandle = Handle + "Command";
         String CommandStr = settings.getString(CommandHandle, "None");
         m_Command =  MatchCommand(CommandStr);

        // Rest of Code
        ....
}
```

The IncrementWayPointIndex() function increments the m_CurrentWayPointIndex variable, which holds the index of the current waypoint that the vehicle is moving toward. If the last point in the set of waypoints has already been reached, then the next waypoint is the starting waypoint. (See Listing 8-18.)

Listing 8-18. Incrementing the WayPoint Index

```
void IncrementWayPointIndex()
{
        int NextWayPointIndex = m_CurrentWayPointIndex + 1;
        if (NextWayPointIndex < m_NumberWayPoints)
        {
                m_CurrentWayPointIndex = NextWayPointIndex;
        }
        else
        {
                // Loop Waypoints
                m_CurrentWayPointIndex = 0;
        }
}
```

The ClearCommand() function clears the vehicle command and the m_ObjectsAffected variable to None. (See Listing 8-19.)

Listing 8-19. Clearing the Vehicle Command

```
void ClearCommand()
{
        m_Command = AIVehicleCommand.None;
        m_ObjectsAffected  = AIVehicleObjectsAffected.None;
}
```

Creating the Tank State to Process Commands

The state that processes the tank commands is the StateTankProcessCommand class that is derived from the StateTank class.

```
public class StateTankProcessCommand extends StateTank
```

The ProcessAIVehicleCommand() function sets the next tank state based on the command given the tank. (See Listing 8-20.)

Listing 8-20. Processing the AI Vehicle Command

```
void ProcessAIVehicleCommand()
{
        VehicleCommand CurrentOrder = GetParent().GetCurrentOrder();

        if (CurrentOrder == null)
        {
                return;
        }
        if (CurrentOrder.GetCommand() == AIVehicleCommand.None)
        {
                return;
        }

        AIVehicleCommand Command = CurrentOrder.GetCommand();

        // Process Commands
        if (Command == AIVehicleCommand.Patrol)
        {
                m_NextState = FSM_StatesTank.FSM_STATE_STEER_WAYPOINT;
        }
        else
        {
                m_NextState = FSM_StatesTank.FSM_STATE_PROCESS_COMMAND;
        }
}
```

The function does the following:

1. It gets the current order for this vehicle.

2. If the order is nonexistent or the command is none (AIVehicleCommand.None), the function returns.

3. If there is a valid order, the command is retrieved. If the command is to patrol the arena (AIVehicleCommand.Patrol), then set the tank's state to the patrol/attack state, which has the state id of FSM_STATE_STEER_WAYPOINT. Otherwise, set the tank's state to the process command state to wait for a command that can be executed by the tank.

The CheckTransitions() function calls ProcessAIVehicleCommand() to set the next state in the m_NextState variable that is returned. (See Listing 8-21.)

Listing 8-21. Checking for a State Transition

```
FSM_StatesTank  CheckTransitions()
{
        ProcessAIVehicleCommand();
        return m_NextState;
}
```

Creating the Vehicle Steering Class

The Steering class represents the steering controls of a vehicle.

The HorizontalSteeringValues enumeration (see Listing 8-22) holds the values for the horizontal steering of a vehicle and the values are

> **None**: Indicates that there is no steering value

> **Right**: Tells the vehicle to turn to the right

> **Left**: Tells the vehicle to turn to the left

Listing 8-22. Horizontal Steering

```
enum HorizontalSteeringValues
{
        None,
        Right,
        Left
}
```

The VerticalSteeringValues enumeration (see Listing 8-23) holds the values for the vertical steering of a vehicle. The values are

> **None**: Indicates that there is no vertical steering value

> **Up**: Tells the vehicle to move upward

> **Down**: Tells the vehicle to move downward

Listing 8-23. Vertical Steering Values

```
enum VerticalSteeringValues
{
        None,
        Up,
        Down
}
```

The SpeedSteeringValues enumeration (see Listing 8-24) holds the acceleration control values for the vehicle. The values are

None: There is no value for the acceleration. Tells the vehicle to keep the same speed

Accelerate: Tells the vehicle to increase its speed

Decelerate: Tells the vehicle to decrease its speed

Listing 8-24. Acceleration Values

```
enum SpeedSteeringValues
{
      None,
      Accelerate,
      Deccelerate
}
```

The m_HoriontalSteering variable holds the left/right vehicle steering value.

```
private HorizontalSteeringValues m_HoriontalSteering;
```

The m_VerticalSteering variables hold the up/down steering values of a vehicle, if applicable.

```
private VerticalSteeringValues m_VerticalSteering;
```

The m_SpeedSteering variable holds the acceleration steering value of the vehicle.

```
private SpeedSteeringValues    m_SpeedSteering;
```

The m_MaxPitch variable holds the maximum amount of tilt in degrees of the vehicle up or down, if applicable.

```
private float m_MaxPitch = 45; // degrees
```

The m_TurnDelta variable holds the amount in degrees the vehicle turns in one update.

```
private float m_TurnDelta = 1; // degrees
```

The m_MaxSpeed variable holds the maximum speed or change in position per update.

```
private float m_MaxSpeed = 0.1f;
```

The m_MinSpeed variable holds the minimum speed or change in position per update.

```
private float m_MinSpeed = 0.05f;
```

The m_SpeedDelta variable holds the amount the speed will change per vehicle update.

```
private float m_SpeedDelta = 0.01f;
```

The ClearSteering() function clears the horizontal, vertical, and speed vehicle inputs to None. This function is called when the steering object is first constructed. (See Listing 8-25.)

Listing 8-25. Clearing the Steering

```
void ClearSteering()
{
        m_HoriontalSteering = HorizontalSteeringValues.None;
        m_VerticalSteering = VerticalSteeringValues.None;
        m_SpeedSteering = SpeedSteeringValues.None;
}
```

The SetSteeringHorizontal() function sets the horizontal value of the vehicle's steering input and the turn rate or turn delta per vehicle update. (See Listing 8-26.)

Listing 8-26. Setting the Vehicle's Horizontal Steering Value

```
void SetSteeringHorizontal(HorizontalSteeringValues Horizontal, float TurnDelta)
{
        m_HoriontalSteering = Horizontal;
        m_TurnDelta = TurnDelta;
}
```

The SetSteeringVertical() function sets the vertical up/down value of the vehicle's steering input and the maximum pitch up or down the vehicle can tilt. (See Listing 8-27.)

Listing 8-27. Setting the Vertical Steering Value

```
void SetSteeringVertical(VerticalSteeringValues Vertical, float MaxPitch)
{
        m_VerticalSteering = Vertical;
        m_MaxPitch = MaxPitch;
}
```

The SetSteeringSpeed() function sets the acceleration or deceleration input to the vehicle, the maximum speed of the vehicle, the minimum speed of the vehicle, and the rate of change of the speed or speed delta. (See Listing 8-28.)

Listing 8-28. Setting the Speed of the Vehicle

```
void SetSteeringSpeed(SpeedSteeringValues Speed, float MaxSpeed, float MinSpeed, float SpeedDelta)
{
        m_SpeedSteering = Speed;
        m_MaxSpeed = MaxSpeed;
        m_MinSpeed = MinSpeed;
        m_SpeedDelta = SpeedDelta;
}
```

Creating the Tank's Patrol/Attack State

The main tank state is the patrol/attack state, where the tank moves around the playfield according to waypoints, while firing its weapon at the player's power pyramid. More specifically, the bottom part of the tank turns toward the current waypoint and moves toward it, while the top part of the tank turns toward the pyramid and fires at it.

The StateTankSteerWayPoint class implements the patrol/attack state for the tank and is derived from the StateTank class discussed previously.

```
public class StateTankSteerWayPoint extends StateTank
```

The m_WayPoint variable holds the location of the current waypoint that the vehicle is moving toward.

```
private Vector3 m_WayPoint  = new Vector3(0,0,0);
```

The m_WayPointRadius variable holds the radius of the waypoint. If the tank is within the area denoted by the current waypoint and the waypoint's radius, the tank is considered to have reached the waypoint.

```
private float m_WayPointRadius = 0;
```

The m_LastWayPoint variable holds the waypoint that was reached previously, just before the current waypoint.

```
private Vector3 m_LastWayPoint   = new Vector3(5000,5000,5000);
```

If the tank's turret is pointing at the target plus or minus the m_TargetAngleTolerance value, the tank will fire at the target.

```
private float m_TargetAngleTolerance = Physics.PI/16.0f;
```

The m_Target variable holds the location of the target to fire at, if any.

```
private Vector3 m_Target;
```

The m_TargetObj variable holds the target object to fire at, if any.

```
private Object3d m_TargetObj;
```

The m_WeaponType variable holds the type of weapon either primary or secondary to fire at the target.

```
private AIVehicleObjectsAffected m_WeaponType;
```

The m_RoundsToFire variable holds the number of rounds to fire in one burst.

```
private float m_RoundsToFire = 0;
```

The m_NumberRoundsFired variable keeps track of the number of rounds of ammunition fired for each burst.

```
private int m_NumberRoundsFired = 0;
```

The m_TimeIntervalBetweenFiring variable sets the time interval between successive bursts of the tank's weapon.

```
private long m_TimeIntervalBetweenFiring = 0;
```

The m_StartTimeFiring variable holds the last time that the tank has fired its weapon.

```
private long m_StartTimeFiring = 0;
```

The m_FireWeapon variable holds true if the tank's weapon should be fired and false otherwise.

```
private boolean m_FireWeapon = false;
```

The constructor for the StateTankSteerWayPoint class calls the constructor for the StateTank superclass with the FSM_StatesTank id, which identifies which state this is to the finite state machine and the Driver parent object that contains information about this vehicle's commands and the tank object. (See Listing 8-29.)

Listing 8-29. The Constructor

```
StateTankSteerWayPoint(FSM_StatesTank ID, Driver Parent)
{
        super(ID, Parent);
}
```

The Enter() function is called when this state is first entered by the finite state machine. The function initializes key variables. Some data is from the parent Driver class object, such as the current waypoint, waypoint radius, or data from the vehicle command. The Enter() function also lets the parent Driver class object know that the currently executing command is the patrol/attack command. (See Listing 8-30.)

Listing 8-30. Entering the State for the First Time

```
void Enter()
{
        // Weapon is not firing when state is entered initially
        m_NumberRoundsFired = 0;
        m_FireWeapon = false;

        // Get WayPoint Data
        m_WayPoint = GetParent().GetWayPoint();
        m_WayPointRadius = GetParent().GetWayPointRadius();

        // Get Targeting and firing parameters
        m_Target = GetParent().GetCurrentOrder().GetTarget();
        m_TargetObj = GetParent().GetCurrentOrder().GetTargetObject();
```

```
m_WeaponType = GetParent().GetCurrentOrder().GetObjectsAffected();
m_RoundsToFire = GetParent().GetCurrentOrder().GetDeltaAmount();
m_TimeIntervalBetweenFiring = (long)GetParent().GetCurrentOrder().GetDeltaIncrement();

// Tell the Pilot class what command is actually being executed in the FSM
GetParent().SetCommandExecuting(AIVehicleCommand.Patrol);
}
```

The Exit() function is called before this state is exited by the finite state machine. This function increments the current waypoint to the next waypoint, indicating that the current waypoint has been reached and the tank needs to move to the next waypoint in the list. (See Listing 8-31.)

Listing 8-31. Exiting the State

```
void Exit()
{
    // Update Current Waypoint to next WayPoint
    GetParent().IncrementNextWayPoint();
}
```

The TurnTurretTowardTarget() function determines the horizontal or left/right steering direction of the tank's turret, so that the tank's weapon turns to face the target. (See Listing 8-32 for the full source code.)

Listing 8-32. Turning the Tank Turret Toward the Target

```
void TurnTurretTowardTarget(Vector3 Target)
{
    // 1. Find vector from front of vehicle to target
    Vector3 ForwardXZPlane = new Vector3(0,0,0);
    ForwardXZPlane.x = GetParent().GetAIVehicle().GetTurret().m_Orientation.
GetForwardWorldCoords().x;
    ForwardXZPlane.z = GetParent().GetAIVehicle().GetTurret().m_Orientation.
GetForwardWorldCoords().z;

    Vector3 TurretPosition = new Vector3(0,0,0);
    TurretPosition.x = GetParent().GetAIVehicle().GetTurret().m_Orientation.GetPosition().x;
    TurretPosition.z = GetParent().GetAIVehicle().GetTurret().m_Orientation.GetPosition().z;

    Vector3 WayPointXZPlane = new Vector3(Target.x, 0, Target.z);
    Vector3 TurretToTarget = Vector3.Subtract(WayPointXZPlane, TurretPosition);

    // 2. Normalize Vectors for Dot Product operation
    ForwardXZPlane.Normalize();
    TurretToTarget.Normalize();

    // P.Q = P*Q*cos(theta)
    // P.Q/P*Q = cos(theta)
    // acos(P.Q/P*Q) = theta;
```

```
// 3. Get current theta
double Theta = Math.acos(ForwardXZPlane.DotProduct(TurretToTarget));

// 4. Get Theta if boat is turned to left by PI/16
Orientation NewO = new Orientation(GetParent().GetAIVehicle().GetTurret().m_Orientation);
Vector3 Up = NewO.GetUp();
NewO.SetRotationAxis(Up);
NewO.AddRotation(Physics.PI/16);

Vector3 NewForwardXZ = NewO.GetForwardWorldCoords();
NewForwardXZ.y = 0;
NewForwardXZ.Normalize();

double Theta2 = Math.acos(NewForwardXZ.DotProduct(TurretToTarget));

// Check if angle within tolerance for firing
float Diff = Math.abs((float)(Theta));

if (!m_FireWeapon)
{
        if (Diff <= m_TargetAngleTolerance)
        {
                m_FireWeapon = true;
                m_StartTimeFiring = System.currentTimeMillis();
        }
}

// 5. Set Steering
if (Theta2 > Theta)
{
GetParent().GetTurretSteering().SetSteeringHorizontal(HorizontalSteeringValues.Right, 1);
}
else
if (Theta2 < Theta)
{
 GetParent().GetTurretSteering().SetSteeringHorizontal(HorizontalSteeringValues.Left, 1);
}
else
{
 GetParent().GetTurretSteering().SetSteeringHorizontal(HorizontalSteeringValues.None,0);
}
}
```

The TurnTurretTowardTarget() function does the following:

1. Finds the vector representing the firing direction of the tank's weapon in world coordinates, which is ForwardXZPlane (see Figure 8-6).

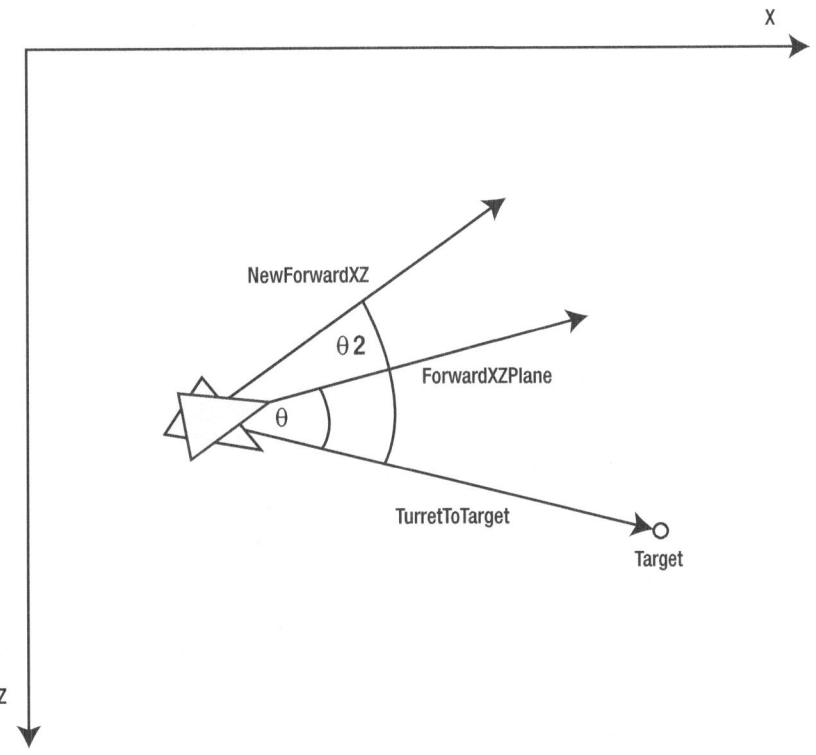

Figure 8-6. Turning the turret toward the target

2. Finds the vector representing the direction from the tank's turret to the target that the tank needs to hit, which is `TurretToTarget`.

3. Normalizes the `ForwardXZPlane` and `TurretToTarget` vectors so that the length of the vectors is 1.

4. Finds the angle Theta between the `ForwardXZPlane` and `TurretToTarget` vectors.

5. Creates a new Orientation class object that is the same as the orientation of the tank's turret but turned by a small angle PI/16, which is NewO.

6. Finds the forward vector of this new tank turret orientation, which is `NewForwardXZ`, and then normalizes it.

7. Finds the angle Theta2 between the `NewForwardXZ` and the `TurretToTarget` vectors.

8. If the tank weapon is not firing and the tank turret is directly facing the target plus or minus the `m_TargetAngleTolerance` value, set the tank weapon to fire and set the `m_StartTimeFiring` variable to the current time.

9. If Theta2 is greater than Theta, turn the tank turret to the right.

10. If Theta2 is less than Theta, turn the tank turret to the left.

The FireTurretWeapon() function fires the tank's primary or secondary weapon, depending on the AIVehicleObjectsAffected value. If the weapon is successfully fired, then the m_NumberRoundsFired variable is incremented by 1. (See Listing 8-33.)

Listing 8-33. Firing the Tank's Turret Weapon

```
void FireTurretWeapon()
{
        Vector3 Direction = GetParent().GetAIVehicle().GetTurret().m_Orientation.
GetForwardWorldCoords();
        boolean IsFired = false;

        if (m_WeaponType == AIVehicleObjectsAffected.PrimaryWeapon)
        {
                IsFired = GetParent().GetAIVehicle().FireWeapon(0, Direction);
        }
        else
        if (m_WeaponType == AIVehicleObjectsAffected.SecondaryWeapon)
        {
                IsFired = GetParent().GetAIVehicle().FireWeapon(1, Direction);
        }
        if (IsFired)
        {
                m_NumberRoundsFired++;
        }
}
```

The SteerVehicleToWayPointHorizontal() function steers the main tank body left or right toward the waypoint. (See Listing 8-34 for the full source code.)

Listing 8-34. Steering the Vehicle to the Current Waypoint

```
void SteerVehicleToWayPointHorizontal(Vector3 WayPoint)
{
        // 1. Find vector from front of vehicle to target
        Vector3 ForwardXZPlane = new Vector3(0,0,0);
        ForwardXZPlane.x = GetParent().GetAIVehicle().GetMainBody().m_Orientation.
GetForwardWorldCoords().x;
        ForwardXZPlane.z = GetParent().GetAIVehicle().GetMainBody().m_Orientation.
GetForwardWorldCoords().z;

        Vector3 VehiclePosition = new Vector3(0,0,0);
        VehiclePosition.x = GetParent().GetAIVehicle().GetMainBody().m_Orientation.GetPosition().x;
        VehiclePosition.z = GetParent().GetAIVehicle().GetMainBody().m_Orientation.GetPosition().z;

        Vector3 WayPointXZPlane = new Vector3(WayPoint.x, 0, WayPoint.z);
        Vector3 VehicleToWayPoint = Vector3.Subtract(WayPointXZPlane, VehiclePosition);

        // 2. Normalize Vectors for Dot Product operation
        ForwardXZPlane.Normalize();
        VehicleToWayPoint.Normalize();
```

```
// P.Q = P*Q*cos(theta)
// P.Q/P*Q = cos(theta)
// acos(P.Q/P*Q) = theta;

// 3. Get current theta
double Theta = Math.acos(ForwardXZPlane.DotProduct(VehicleToWayPoint));

// 4. Get Theta if boat is turned to left by PI/16
Orientation NewO = new Orientation(GetParent().GetAIVehicle().GetMainBody().m_Orientation);
Vector3 Up = NewO.GetUp();
NewO.SetRotationAxis(Up);
NewO.AddRotation(Physics.PI/16);

Vector3 NewForwardXZ = NewO.GetForwardWorldCoords();
NewForwardXZ.y = 0;
NewForwardXZ.Normalize();

double Theta2 = Math.acos(NewForwardXZ.DotProduct(VehicleToWayPoint));

// 5. Set Steering
if (Theta2 > Theta)
{
 GetParent().GetAISteering().SetSteeringHorizontal(HorizontalSteeringValues.Right, 1);
}
else
if (Theta2 < Theta)
{
 GetParent().GetAISteering().SetSteeringHorizontal(HorizontalSteeringValues.Left, 1);
}
else
{
 GetParent().GetAISteering().SetSteeringHorizontal(HorizontalSteeringValues.None,0);
}
}
```

The function does the following:

1. Finds the main tank body's forward vector, which is ForwardXZPlane.

2. Calculates the vector, which is VehicleToWayPoint, from the tank to the waypoint.

3. Normalizes the ForwardXZPlane and VehicleToWayPoint vectors.

4. Calculates the angle Theta between the ForwardXZPlane and VehicleToWayPoint vectors.

5. Creates a new Orientation class object that is the same as the orientation of the tank's main body but turned by a small angle PI/16, which is NewO.

6. Finds the forward vector of this new tank main body orientation, which is NewForwardXZ, and then normalizes it.

7. Finds the angle Theta2 between the NewForwardXZ and the VehicleToWayPoint vectors.

8. If Theta2 is greater than Theta, turn the tank's main body to the right.

9. If Theta2 is less than Theta, turn the tank's main body to the left.

The SteerVehicleWaypointSpeed() function decelerates the tank around the waypoints. If the tank is within TurnArea radius of the last waypoint or the current waypoint, this function decelerates the tank; otherwise, the function accelerates the tank. (See Listing 8-35.)

Listing 8-35. Changing the Speed of the Vehicle Around Waypoints

```
void SteerVehicleWaypointSpeed(Vector3 WayPoint)
{
        // If vehicle is close to waypoint then slow down vehicle
        // else accelerate vehicle
        Tank AIVehicle = GetParent().GetAIVehicle();

        Vector3 VehiclePos = AIVehicle.GetMainBody().m_Orientation.GetPosition();
        Vector3 DistanceVecLastWayPoint = Vector3.Subtract(VehiclePos,m_LastWayPoint);
        Vector3 DistanceVecCurrentWayPoint = Vector3.Subtract(VehiclePos, m_WayPoint);

        float TurnArea = GetParent().GetTurnArea();
        float DLastWayPoint = DistanceVecLastWayPoint.Length();
        float DCurrentWayPoint = DistanceVecCurrentWayPoint.Length();

        if ((DLastWayPoint <= TurnArea) || (DCurrentWayPoint <= TurnArea))
        {
                // Decrease speed
            GetParent().GetAISteering().SetSteeringSpeed(SpeedSteeringValues.Deccelerate, 0.04f,
0.03f, 0.005f);
            GetParent().GetAISteering().SetTurnDelta(3.0f);
        }
        else
        {
            GetParent().GetAISteering().SetSteeringSpeed(SpeedSteeringValues.Accelerate, 0.04f,
0.03f, 0.005f);
        }
}
```

The SteerVehicleToWayPoint() function steers the vehicle left or right toward the waypoint location by calling SteerVehicleToWayPointHorizontal() and adjusts the tank's speed by calling SteerVehicleWaypointSpeed(). (See Listing 8-36.)

Listing 8-36. Steering the Vehicle to the Waypoint

```
void SteerVehicleToWayPoint(Vector3 WayPoint)
{
        SteerVehicleToWayPointHorizontal(WayPoint);
        SteerVehicleWaypointSpeed(WayPoint);
}
```

The Update() function updates the tank's artificial intelligence for the patrol/attack state. (See Listing 8-37.)

Listing 8-37. Updating the Tank Patrol/Attack State

```
void Update()
{
        // Steer Main Tank Body to Waypoint
        SteerVehicleToWayPoint(m_WayPoint);

        // Turn Tank Turret towards target and fire
        if (m_Target != null)
        {
                TurnTurretTowardTarget(m_Target);
        }
        else
        if (m_TargetObj != null)
        {
                TurnTurretTowardTarget(m_TargetObj.m_Orientation.GetPosition());
        }
        else
        {
                Log.e("STATETANKSTEERWAYPOINT" , "NO TARGET FOR TANK TO SHOOT AT!!!!");
        }

        if (m_FireWeapon)
        {
                if (m_NumberRoundsFired >= m_RoundsToFire)
                {
                        m_NumberRoundsFired = 0;
                        m_FireWeapon = false;
                }
                else
                {
                        // Find Time Elapsed Between firing sequences
                        long ElapsedTime = System.currentTimeMillis() - m_StartTimeFiring;
                        if (ElapsedTime > m_TimeIntervalBetweenFiring)
                        {
                                FireTurretWeapon();
                        }
                }
        }
}
```

The Update() function does the following:

1. Sets the steering and acceleration for the tank by calling SteerVehicleToWayPoint() with the target waypoint.

2. Calls TurnTurretTowardTarget() with the target location, which turns the tank's turret toward the target, if a target location is in m_Target.

3. Calls TurnTurretTowardTarget() with m_TargetObj's position, in order to turn the tank's turret toward that target, if there is no location in m_Target and m_TargetObj is not null.

4. If m_FireWeapon is true, then either of the following occurs:

    a. m_FireWeapon resets to false, if the number of rounds to fire for this burst is equal to the number of rounds that are required to be fired.

    b. The weapon is fired by calling FireTurretWeapon(), if the required time delay has passed since the m_FireWeapon was set to true.

The CheckTransitions() function checks for a transition to a different state, based on the game conditions. If the tank is within m_WayPointRadius distance of the target waypoint, the current waypoint is saved in the m_LastWayPoint variable and the state returned by the function is the process command state. If the tank is not within m_WayPointRadius distance of the target waypoint, there is no change in state, and the patrol/attack state is returned. (See Listing 8-38.)

Listing 8-38. Checking the Transitions

```
FSM_StatesTank  CheckTransitions()
{
        Object3d AIVehicle = GetParent().GetAIVehicle().GetMainBody();

        Vector3 VehiclePos = AIVehicle.m_Orientation.GetPosition();
        Vector3 Distance   = Vector3.Subtract(VehiclePos,m_WayPoint);
        float   D = Distance.Length();

        if (D <= m_WayPointRadius)
        {
                m_LastWayPoint.Set(m_WayPoint.x, m_WayPoint.y, m_WayPoint.z);
                return FSM_StatesTank.FSM_STATE_PROCESS_COMMAND;
        }
        else
        {
                return FSM_StatesTank.FSM_STATE_STEER_WAYPOINT;
        }
}
```

Creating the Tank Finite State Machine

The Finite State Machine class for the tank is the FSMDriver class. The finite state machine actually executes the tank AI by calling the appropriate functions in each of the tank states.

The way the FSMDriver class is used is that you

1. Add new tank states with the AddState() function.

2. Set the default state that the machine starts in with the SetDefaultState() command.

3. Initialize the machine by calling `Reset()`.

4. Update the machine by calling `UpdateMachine()`.

The `MAX_STATES` variable holds the maximum number of states that this machine can hold.

```
private int MAX_STATES = 20;
```

The `m_NumberStates` variable holds the number of states in this state machine.

```
private int m_NumberStates = 0;
```

The `m_States` array holds the tank states that make up the finite state machine.

```
protected StateTank[] m_States = new StateTank[MAX_STATES];
```

The `m_CurrentState` variable holds a reference to the currently executing tank state.

```
protected StateTank m_CurrentState = null;
```

The `m_DefaultState` variable holds a reference to the default state that the finite state machine starts in.

```
protected StateTank m_DefaultState = null;
```

The `m_GoalState` is the state that the finite state machine is going to transition to.

```
protected StateTank m_GoalState = null;
```

The `m_GoalID` is the enumeration that identifies the type of tank state to transition to either the process command state or the patrol/attack state.

```
protected FSM_StatesTank m_GoalID;
```

The `Reset()` function initializes the finite state machine and does the following (see Listing 8-39):

1. If there is a current state being executed, it exits that state by calling `Exit()` on the state object.

2. It sets the current state to the default state of the machine.

3. For all the states in the machine, it initializes them by calling each state's `Init()` function.

4. If there is a current state, it then "enters" that state by calling `Enter()` on that state.

Listing 8-39. Resetting the Finite State Machine

```
void Reset()
{
        if(m_CurrentState != null)
        {
                m_CurrentState.Exit();
        }

        m_CurrentState = m_DefaultState;

        for(int i = 0;i < m_NumberStates;i++)
        {
                m_States[i].Init();
        }

        if(m_CurrentState != null)
        {
                m_CurrentState.Enter();
        }
}
```

The AddState() function adds a tank state State to the finite state machine. The function first checks to see if there is room for more states and, if there is, adds them to the m_States array, increases the number of states in the machine, and returns true. False is returned otherwise. (See Listing 8-40.)

Listing 8-40. Adding a State to the Finite State Machine

```
boolean AddState(StateTank State)
{
        boolean result = false;
        if (m_NumberStates < MAX_STATES)
        {
                m_States[m_NumberStates] = State;
                m_NumberStates++;
                result = true;
        }
        return result;
}
```

The TransitionState() function searches through all the states in the machine and tries to match the Goal input state id FSM_StatesTank enumeration with the id from each of the states. If a match is found, then m_GoalState is set to that state and true is returned. Otherwise, false is returned. (See Listing 8-41.)

Listing 8-41. Transitioning Between States

```
boolean TransitionState(FSM_StatesTank Goal)
{
        if(m_NumberStates == 0)
        {
                return false;
        }
```

```
        for(int i = 0; i < m_NumberStates;i++)
        {
                if(m_States[i].GetStateID() == Goal)
                {
                        m_GoalState = m_States[i];
                        return true;
                }
        }
        return false;
}
```

The UpdateMachine() function updates the finite state machine for the tank. (See Listing 8-42.)

Listing 8-42. Updating the Finite State Machine

```
void UpdateMachine()
{
        if(m_NumberStates == 0)
        {
                return;
        }
        if(m_CurrentState == null)
        {
                m_CurrentState = m_DefaultState;
        }
        if(m_CurrentState == null)
        {
                return;
        }
        FSM_StatesTank OldStateID = m_CurrentState.GetStateID();
        m_GoalID = m_CurrentState.CheckTransitions();
        if(m_GoalID != OldStateID)
        {
                if(TransitionState(m_GoalID))
                {
                        m_CurrentState.Exit();
                        m_CurrentState = m_GoalState;
                        m_CurrentState.Enter();
                }
        }
        m_CurrentState.Update();
}
```

The function does the following:

1. Returns, if there are no states in the machine.

2. Sets the current state to the default state, if there is no current state.

3. Returns, if the current state is still nonexistent.

4. Gets the current state's id.

5. Checks the current state for a state transition by calling CheckTransitions().

6. If the goal id that is returned is not the same as the id from the current state, then a transition to a new state must be made. The `TransitionState()` function is then called and, if the goal state has been found, is processed. The current state is exited by calling `Exit()` on the state object. The current state is then set to the goal state in `m_GoalState` that was set by the `TransitionState()` function. The current state is then entered by calling the `Enter()` function.

7. Updates the current state by calling the state's `Update()` function.

Creating the Driver for the Tank

The Driver class holds the finite state machine that serves as the tank's brain. It also holds other key information, such as the vehicle command that is to be executed and other game information.

The `m_CurrentOrder` variable holds a reference to the tank's current `VehicleCommand` order that is to be executed in the tank's finite state machine.

```
private VehicleCommand  m_CurrentOrder      = null; // Order to be executed in the FSM
```

The `m_LastOrder` variable holds a reference to the tank's last order that was executed.

```
private VehicleCommand  m_LastOrder  = null;
```

The `m_CommandExecuting` refers to the actual vehicle command, either None or Patrol, that is currently executing in the finite state machine.

```
private AIVehicleCommand m_CommandExecuting = null; // Command that is currently being executed in
the Finite State Machine
```

The `m_FiniteStateMachine` refers to the finite state machine for the tank that implements the tank's artificial intelligence.

```
private FSMDriver m_FiniteStateMachine = null;
```

The `m_AISteer` variable holds the steering input values for the tank that the finite state machine generates.

```
private Steering m_AISteer = new Steering();
```

The `m_TurretSteering` variable holds the turning input values for the tank's turret that the finite state machine generates.

```
private Steering m_TurretSteering = new Steering();
```

The `m_TurnArea` is the area near the waypoint where the vehicle slows down for turning toward the next waypoint.

```
private float  m_TurnArea = 2.0f;
```

The m_WayPoint variable holds the current waypoint.

```
private Vector3 m_WayPoint = new Vector3(0,0,0);
```

The m_WayPointRadius variable holds the radius of the waypoint.

```
private float m_WayPointRadius = 1.0f;
```

The m_AITank variable holds the tank object that is controlled by the Driver class.

```
private Tank m_AITank = null;
```

The Driver constructor (see Listing 8-43) initializes the class object by

1. Setting the reference to the tank, which is held in m_AITank.

2. Creating the tank's finite state machine by creating a new FSMDriver object.

3. Creating the new tank states StateTankSteerWayPoint and StateTankProcessCommand and adding them to the finite state machine by calling the AddState() function.

4. Setting the default state of the finite state machine to the process command state.

5. Resetting the finite state machine by calling Reset().

Listing 8-43. The Driver Constructor

```
Driver(Tank Vehicle)
{
        // Set Vehicle that is to be controlled
        m_AITank = Vehicle;

        //construct the state machine and add the necessary states
        m_FiniteStateMachine = new FSMDriver();

        StateTankSteerWayPoint SteerWayPoint =
new StateTankSteerWayPoint(FSM_StatesTank.FSM_STATE_STEER_WAYPOINT, this);
        StateTankProcessCommand ProcessCommand =
new StateTankProcessCommand(FSM_StatesTank.FSM_STATE_PROCESS_COMMAND,this);

        m_FiniteStateMachine.AddState(SteerWayPoint);
        m_FiniteStateMachine.AddState(ProcessCommand);

        m_FiniteStateMachine.SetDefaultState(ProcessCommand);
        m_FiniteStateMachine.Reset();
}
```

The SaveDriverState() function saves key class data members from the Driver class. In order to save space, Listing 8-44 has been abbreviated. Please refer to the Source Code/Download area located on apress.com for the full version.

Listing 8-44. Saving the Driver

```
void SaveDriverState(String Handle)
{
        SharedPreferences settings = m_AITank.GetMainBody().GetContext().getSharedPreferences(Handle, 0);
        SharedPreferences.Editor editor = settings.edit();

        // Turn Area
        String TurnAreaKey = Handle + "TurnArea";
        editor.putFloat(TurnAreaKey, m_TurnArea);

        .. Rest of code
}
```

The `LoadDriverState()` function loads key data for the Driver class. The code in Listing 8-45 has been abbreviated. For the full code, please see the Source Code/Download arealocated on apress.com.

Listing 8-45. Loading the Driver State

```
void LoadDriverState(String Handle)
{
        SharedPreferences settings =
m_AITank.GetMainBody().GetContext().getSharedPreferences(Handle, 0);

        // Turn Area
        String TurnAreaKey = Handle + "TurnArea";
        m_TurnArea = settings.getFloat(TurnAreaKey, 4.0f);

        .. Rest of code
}
```

The `IncrementNextWayPoint()` function sets `m_WayPoint` to the next waypoint for the tank, if the current command is the Patrol command. (See Listing 8-46.)

Listing 8-46. Finding the Next Waypoint

```
void IncrementNextWayPoint()
{
        AIVehicleCommand Command = m_CurrentOrder.GetCommand();

        if (Command ==  AIVehicleCommand.Patrol)
        {
                m_CurrentOrder.IncrementWayPointIndex();
                m_WayPoint = m_CurrentOrder.GetCurrentWayPoint();
        }
}
```

The `SetOrder()` function saves the current command in the `m_LastOrder` variable and sets the current order to the vehicle Command input parameter. If the command is the Patrol command, the current waypoint variable `m_WayPoint` is set to the first waypoint. (See Listing 8-47.)

Listing 8-47. Setting a New Order for the Tank

```
void SetOrder(VehicleCommand Command)
{
        m_LastOrder = m_CurrentOrder;
        m_CurrentOrder = Command;

        if (m_CurrentOrder.GetCommand() == AIVehicleCommand.Patrol)
        {
                // Set Inital WayPoint
                Vector3[] WayPoints = m_CurrentOrder.GetWayPoints();
                m_WayPoint = WayPoints[0];
        }
}
```

The Update() function clears the steering input for the tank and updates the finite state machine that controls the tank. (See Listing 8-48.)

Listing 8-48. Updating the Driver

```
void Update()
{
        // Clear AISteering
        m_AISteer.ClearSteering();

        // Update FSM Machine
        m_FiniteStateMachine.UpdateMachine();
}
```

Modifying the Physics Class

The Physics class has to be modified to support our new tank vehicle.

The UpdatePhysicsObjectHeading() function must be added into the Physics class (see Listing 8-49). The UpdatePhysicsObjectHeading() function does the following:

1. Applies gravity to the object, if gravity is on.

2. Updates the linear velocity of the object.

3. Updates the angular velocity of the object.

4. Resets the forces acting on the object to 0 by setting the linear and angular acceleration to 0.

5. Adjusts the velocity so that all the object's speed is redirected along the Heading. If the speed of the object is greater than m_MaxSpeed, set the speed to m_MaxSpeed. If gravity is on, use the y component of the velocity from step 1 in the calculation of the object's new velocity.

6. Updates the object's linear position, taking into account the gravity and ground floor settings in adjusting the object's vertical position.

7. Updates the object's angular position.

Listing 8-49. Updating the Tank's Physics

```
void UpdatePhysicsObjectHeading(Vector3 Heading, Orientation orientation)
{
        // Adjust for Gravity
        if (m_ApplyGravity)
        {
                ApplyGravityToObject();
        }

        // 1. Update Linear Velocity
        ///////////////////////////////////////////////////////////////////////
        m_Acceleration.x = TestSetLimitValue(m_Acceleration.x, m_MaxAcceleration.x);
        m_Acceleration.y = TestSetLimitValue(m_Acceleration.y, m_MaxAcceleration.y);
        m_Acceleration.z = TestSetLimitValue(m_Acceleration.z, m_MaxAcceleration.z);

        m_Velocity.Add(m_Acceleration);
        m_Velocity.x = TestSetLimitValue(m_Velocity.x, m_MaxVelocity.x);
        m_Velocity.y = TestSetLimitValue(m_Velocity.y, m_MaxVelocity.y);
        m_Velocity.z = TestSetLimitValue(m_Velocity.z, m_MaxVelocity.z);

        // 2. Update Angular Velocity
        ///////////////////////////////////////////////////////////////////////
        m_AngularAcceleration = TestSetLimitValue(m_AngularAcceleration, m_MaxAngularAcceleration);

        m_AngularVelocity += m_AngularAcceleration;
        m_AngularVelocity = TestSetLimitValue(m_AngularVelocity, m_MaxAngularVelocity);

        // 3. Reset Forces acting on Object
        //     Rebuild forces acting on object for each update
        ///////////////////////////////////////////////////////////////////////
        m_Acceleration.Clear();
        m_AngularAcceleration = 0;

        // 4. Adjust Velocity so that all the velocity is redirected along the heading.
        ///////////////////////////////////////////////////////////////////////
        float VelocityMagnitude        = m_Velocity.Length();

        if (VelocityMagnitude > m_MaxSpeed)
        {
                VelocityMagnitude = m_MaxSpeed;
        }

        Vector3 NewVelocity = new Vector3(Heading);
        NewVelocity.Normalize();
        NewVelocity.Multiply(VelocityMagnitude);

        Vector3 OldVelocity = new Vector3(m_Velocity);

        if (m_ApplyGravity)
        {
                m_Velocity.Set(NewVelocity.x, OldVelocity.y, NewVelocity.z);
        }
```

```
        else
        {
                m_Velocity.Set(NewVelocity.x, NewVelocity.y, NewVelocity.z);
        }

        //5. Update Object Linear Position
        ///////////////////////////////////////////////////////////////////////////
        Vector3 pos = orientation.GetPosition();
        pos.Add(m_Velocity);
        orientation.SetPosition(pos);

        // Check for object hitting ground if gravity is on.
        if (m_ApplyGravity)
        {
                if ((pos.y < m_GroundLevel)&& (m_Velocity.y < 0))
                {
                        if (Math.abs(m_Velocity.y) > Math.abs(m_Gravity))
                        {
                                m_JustHitGround = true;
                        }
                        pos.y = m_GroundLevel;
                        m_Velocity.y = 0;
                }
        }

        //6. Update Object Angular Position
        ///////////////////////////////////////////////////////////////////////////
        // Add Rotation to Rotation Matrix
        orientation.AddRotation(m_AngularVelocity);
}
```

Modifying the Object3d Class

The UpdateObject3dToHeading() function is added to the Object3d class. The UpdateObject3dToHeading() function updates an Object3d object with the physics model, where the object moves in the direction of the Heading vector. This physics model is for the movement of the tank. (See Listing 8-50.)

Listing 8-50. Updating the Object Along a Heading

```
void UpdateObject3dToHeading(Vector3 Heading)
{
        if (m_Visible)
        {
                // Update Object3d Physics
                m_Physics.UpdatePhysicsObjectHeading(Heading, m_Orientation);
        }

        // Update Explosions associated with this object
        UpdateExplosions();
}
```

Creating the Tank Class

The Tank class represents our tank enemy.

The m_VehicleID can hold a unique id that identifies this specific vehicle or class of vehicles, such as "lasertank03."

```
private String m_VehicleID = "None";
```

The m_Driver variable holds the Driver class object of the tank and contains the finite state machine that implements the tank's artificial intelligence.

```
private Driver m_Driver;
```

The m_MainBody variable holds the lower 3D mesh portion of the tank that turns toward the waypoint and controls the movement of the upper turret portion of the tank.

```
private Object3d m_MainBody;
```

The m_Turret variable holds the upper 3D mesh portion of the tank that turns toward the target and fires projectiles at it.

```
private Object3d m_Turret;
```

The m_Heading holds the direction that the tank is moving.

```
private Vector3 m_Heading = new Vector3(0,0,0);
```

The MAX_WEAPONS variable holds the maximum number of weapons that this tank can have.

```
private int MAX_WEAPONS = 5;
```

The m_NumberWeapons variable holds the number of weapons that this tank currently has.

```
private int m_NumberWeapons = 0;
```

The m_Weapons array holds the tank's weapons.

```
private Weapon[] m_Weapons = new Weapon[MAX_WEAPONS];
```

The m_TurretOffset holds the tank turret's offset position from the center of the tank's main body.

```
private Vector3 m_TurretOffset = new Vector3(0,0,0);
```

The m_HitGroundSFXIndex holds the index to the sound effect to be played when the tank hits the ground.

```
private int m_HitGroundSFXIndex = -1;
```

The m_ExplosionSFXIndex holds the index to the sound effect to be played when the tank explodes or is hit by the player's ammunition.

```
private int m_ExplosionSFXIndex = -1;
```

The Tank constructor initializes the tank by setting the 3D meshes for the tank's main body and the turret, as well as setting the value of the turret offset and creating a new Driver class object for the tank. (See Listing 8-51.)

Listing 8-51. The Tank Constructor

```
Tank(Object3d MainBody, Object3d Turret, Vector3 TurretOffset)
{
        m_MainBody = MainBody;
        m_Turret = Turret;
        m_TurretOffset = TurretOffset;

        // Create new Pilot for this vehicle
        m_Driver = new Driver(this);
}
```

The SaveTankState() function saves the state of the tank. The function saves the state of the tank's main body, turret, and driver. (See Listing 8-52.)

Listing 8-52. Saving the Tank State

```
void SaveTankState(String Handle)
{
        // Main Body
        String MainBodyHandle = Handle + "MainBody";
        m_MainBody.SaveObjectState(MainBodyHandle);

        // Turret
        String TurretHandle = Handle + "Turret";
        m_Turret.SaveObjectState(TurretHandle);

        // Driver
        String DriverHandle = Handle + "Driver";
        m_Driver.SaveDriverState(DriverHandle);
}
```

The LoadTankState() function loads in previously saved data for the tank's main body, turret, and the driver. (See Listing 8-53.)

Listing 8-53. Loading the Tank State

```
void LoadTankState(String Handle)
{
        //Driver
        String DriverHandle = Handle + "Driver";
        m_Driver.LoadDriverState(DriverHandle);
```

```
        // Main Body
        String MainBodyHandle = Handle + "MainBody";
        m_MainBody.LoadObjectState(MainBodyHandle);

        // Turret
        String TurretHandle = Handle + "Turret";
        m_Turret.LoadObjectState(TurretHandle);
}
```

The Reset() function resets the tank's driver and the tank's weapons. (See Listing 8-54.)

Listing 8-54. Resetting the Tank

```
void Reset()
{
        // Reset Driver
        if (m_Driver != null)
        {
                m_Driver.DriverReset();
        }

        // Reset Weapons
        for (int i = 0; i < m_NumberWeapons; i++)
        {
                Weapon TempWeapon = m_Weapons[i];
                TempWeapon.ResetWeapon();
        }
}
```

The AddWeapon() function adds the weapon iWeapon to the tank's set of weapons, if there is more room available. The function returns true if successful and false if not successful. (See Listing 8-55.)

Listing 8-55. Adding a Weapon to the Tank

```
boolean AddWeapon(Weapon iWeapon)
{
        boolean result = false;
        if (m_NumberWeapons < MAX_WEAPONS)
        {
                m_Weapons[m_NumberWeapons] = iWeapon;
                m_NumberWeapons++;
                result = true;
        }
        return result;
}
```

The FireWeapon() function fires the tank's weapon number WeaponNumber in the direction Direction with the weapon's projectile starting at the tank's turret position. The function returns true if the weapon has been fired and false otherwise. (See Listing 8-56.)

Listing 8-56. Firing the Tank's Weapon

```
boolean FireWeapon(int WeaponNumber, Vector3 Direction)
{
      boolean result = false;
      if (WeaponNumber < m_NumberWeapons)
      {
            result = m_Weapons[WeaponNumber].Fire(Direction,
m_Turret.m_Orientation.GetPosition());
      }
      return result;
}
```

The RenderVehicle() function draws the tank's main body, turret, and weapon's projectiles to the screen and also plays a sound effect if the tank has just hit the ground. (See Listing 8-57.)

Listing 8-57. Rendering the Tank

```
void RenderVehicle(Camera Cam, PointLight Light, boolean DebugOn)
{
      // Render Vehicle
      m_MainBody.DrawObject(Cam, Light);
      m_Turret.DrawObject(Cam, Light);

      // Render Vehicles Weapons and Ammunition if any
      for (int i = 0 ; i < m_NumberWeapons; i++)
      {
            m_Weapons[i].RenderWeapon(Cam, Light, DebugOn);
      }

      // Shake Camera if Tank hits ground
      boolean ShakeCamera = m_MainBody.GetObjectPhysics().GetHitGroundStatus();
      if (ShakeCamera)
      {
            m_MainBody.GetObjectPhysics().ClearHitGroundStatus();
            PlayHitGoundSFX();
      }
}
```

The TurnTank() function turns the main body of the tank by TurnDelta degrees. (See Listing 8-58.)

Listing 8-58. Turning the Tank

```
void TurnTank(float TurnDelta)
{
      Vector3 Axis = new Vector3(0,1,0);
      m_MainBody.m_Orientation.SetRotationAxis(Axis);
      m_MainBody.m_Orientation.AddRotation(TurnDelta);
}
```

The ProcessSteering() function (see Listing 8-59) processes the tank's steering input values. The function does the following:

1. Processes the horizontal or left/right steering by calling the TurnTank() to actually turn the tank's main body

2. Processes the tank's acceleration by setting the maximum speed allowed that is associated with the tank's steering input and applying a translation force to the tank's main body

3. Processes the tank's deceleration by calculating a slower speed and setting the maximum speed of the tank to this new slower speed

Listing 8-59. Processing the Tank's Steering

```
void ProcessSteering()
{
        Steering DriverSteering = m_Driver.GetAISteering();

        HorizontalSteeringValues  HorizontalTurn  = DriverSteering.GetHorizontalSteering();
        SpeedSteeringValues  Acceleration  = DriverSteering.GetSpeedSteering();

        float TurnDelta  = DriverSteering.GetTurnDelta();
        float MaxSpeed = DriverSteering.GetMaxSpeed();
        float MinSpeed = DriverSteering.GetMinSpeed();
        float SpeedDelta = DriverSteering.GetSpeedDelta();

        // Process Tank Steering

        // Process Right/Left Turn
        if (HorizontalTurn == HorizontalSteeringValues.Left)
        {
                TurnTank(TurnDelta);
        }
        else if (HorizontalTurn == HorizontalSteeringValues.Right)
        {
                TurnTank(-TurnDelta);
        }

        // Process Acceleration
        if (Acceleration == SpeedSteeringValues.Accelerate)
        {
                m_MainBody.GetObjectPhysics().SetMaxSpeed(MaxSpeed);

                Vector3 Force = new Vector3(0,0,30.0f);
                m_MainBody.GetObjectPhysics().ApplyTranslationalForce(Force);
        }
        else
        if (Acceleration == SpeedSteeringValues.Deccelerate)
        {
                float Speed = m_MainBody.GetObjectPhysics().GetVelocity().Length();
                if (Speed > MinSpeed)
```

```
    {
            float NewSpeed = Speed - SpeedDelta;
            m_MainBody.GetObjectPhysics().SetMaxSpeed(NewSpeed);
    }
  }
}
```

The TurnTurret() function turns the tank's turret based on the TurnDelta input parameter. (See Listing 8-60.)

Listing 8-60. Turning the Tank's Turret

```
void TurnTurret(float TurnDelta)
{
    Vector3 Axis = new Vector3(0,1,0);
    m_Turret.m_Orientation.SetRotationAxis(Axis);
    m_Turret.m_Orientation.AddRotation(TurnDelta);
}
```

The ProcessTurret() function turns the tank's turret left or right, depending on the steering input value for the turret. (See Listing 8-61.)

Listing 8-61. Processing the Tank's Turret Steering

```
void ProcessTurret()
{
    Steering TurretSteering = m_Driver.GetTurretSteering();
    HorizontalSteeringValues HorizontalTurn = TurretSteering.GetHorizontalSteering();

    float TurnDelta  = TurretSteering.GetTurnDelta();

    // Process Right/Left Turn
    if (HorizontalTurn == HorizontalSteeringValues.Left)
    {
            TurnTurret(TurnDelta);
    }
    else if (HorizontalTurn == HorizontalSteeringValues.Right)
    {
            TurnTurret(-TurnDelta);
    }
}
```

The UpdateVehicle() function (see Listing 8-62) updates the Tank class object. The function does the following:

1. If the main tank body is visible, it updates the tank's driver object, processes the tank's steering, and processes the tank's turret movements.

2. Updates the tank's physics by calling UpdateObject3dToHeading(), using the forward vector of the tank's main body in world coordinates as the heading.

3. If the main tank body is visible, it calculates and sets the turret's final position, based on the position of the main tank body and the turret offset.

4. Updates the tank's weapons and the active ammunition that has been fired from those weapons.

Listing 8-62. Updating the Tank

```
void UpdateVehicle()
{
       if (m_MainBody.IsVisible())
       {
              // Update AIPilot
              m_Driver.Update();

              // Update Right/Left and Up/Down Rotation of Vehicle based on AIPilot's Steering
              ProcessSteering();

              // Process Turret Steering
              ProcessTurret();
       }

       // Update Vehicle Physics, Position, Rotation, and attached emitters and explosions
       m_Heading = m_MainBody.m_Orientation.GetForwardWorldCoords();
       m_MainBody.UpdateObject3dToHeading(m_Heading);

       if (m_MainBody.IsVisible())
       {
              // Tie Turret to Main Body
              Vector3 Pos = m_MainBody.m_Orientation.GetPosition();
              Vector3 ZOffset = Vector3.Multiply(m_TurretOffset.z,
m_MainBody.m_Orientation.GetForwardWorldCoords());
              Vector3 XOffset = Vector3.Multiply(m_TurretOffset.x,
m_MainBody.m_Orientation.GetRightWorldCoords());
              Vector3 YOffset = Vector3.Multiply(m_TurretOffset.y,
m_MainBody.m_Orientation.GetUpWorldCoords());

              Vector3 OffsetPos = new Vector3(Pos);
              OffsetPos.Add(XOffset);
              OffsetPos.Add(YOffset);
              OffsetPos.Add(ZOffset);

              m_Turret.m_Orientation.GetPosition().Set(OffsetPos.x, OffsetPos.y,OffsetPos.z);
       }

       // Update Weapons and Ammunition
       for (int i = 0 ; i < m_NumberWeapons; i++)
       {
              m_Weapons[i].UpdateWeapon();
       }
}
```

Hands-on Example: Arena Objects and Tanks

Now, let's use the Arena Object class and the Tank class in a working hands-on example. We must make some modifications to our code from Chapter 7's hands-on example. This hands-on example should also be available from the Source Code/Download area of `apress.com`.

Modifying the MyGLRenderer Class

The type for the `m_Cube` variable has been changed to an ArenaObject class.

```
private ArenaObject3d m_Cube;
```

In the `CreateCube()` function, we create a new ArenaObject instead of a Cube object.

```
 m_Cube = new ArenaObject3d(iContext, null,CubeMesh, CubeTex, Material1, Shader,XMaxBoundary,XMinBoundary,ZMaxBoundary,ZMinBoundary);
```

The `onDrawFrame()` function now updates `m_Cube` with `UpdateArenaObject()` instead of `UpdateObject3d()`.

```
 m_Cube.UpdateArenaObject();
```

Next, we have to add in some code for our new tank. I will discuss the key functions in this section.

The `m_Tank` variable is added in to represent our enemy tank.

```
private Tank m_Tank;
```

The `CreateTankType1()` function (see Listing 8-63) creates the tank object. The `CreateTankType1()` function does the following:

1. Creates the tank's weapon and ammunition

2. Creates the material for the tank's main body

3. Creates the tank's main body texture

4. Creates the mesh for the main tank body with data from `Pyramid2.Pyramid2Vertices`

5. Creates the material for the tank's turret

6. Creates the texture for the tank's turret

7. Creates the tank's turret mesh with data from `Pyramid2.Pyramid2Vertices`

8. Creates the tank's turret offset

9. Creates the tank's shader

10. Initializes the tank properties, such as the tank's position, scale, sound effects, color that it makes on the grid, etc.

11. Creates an explosion that is used when the tank is hit by the player's weapon projectiles

12. Creates the tank, sets the sound effects associated with the tank to true, and returns the tank

Listing 8-63. Creating the Tank

```
Tank CreateTankType1(Context iContext)
{
        //Weapon
        Weapon TankWeapon = CreateTankWeaponType1(iContext);

        // Material
        Material MainBodyMaterial = new Material();
        MainBodyMaterial.SetEmissive(0.0f, 0.4f, 0.0f);

        // Texture
        Texture TexTankMainBody = new Texture(iContext,R.drawable.ship1);
        int NumberMainBodyTextures = 1;
        Texture[] MainBodyTexture = new Texture[NumberMainBodyTextures];
        MainBodyTexture[0] = TexTankMainBody;
        boolean AnimateMainBodyTex = false;
        float MainBodyAnimationDelay = 0;

        // Mesh
        Mesh MainBodyMesh = new Mesh(8,0,3,5,Pyramid2.Pyramid2Vertices);
        MeshEx MainBodyMeshEx= null;

        // Turret
        //Material
        Material TurretMaterial=new Material();
        TurretMaterial.SetEmissive(0.4f, 0.0f, 0.0f);

        // Texture
        Texture TexTankTurret = new Texture(iContext,R.drawable.ship1);
        int NumberTurretTextures = 1;
        Texture[] TurretTexture = new Texture[NumberTurretTextures];
        TurretTexture[0] = TexTankTurret;
        boolean AnimateTurretTex = false;
        float TurretAnimationDelay = 0;

        // Mesh
        Mesh TurretMesh= new Mesh(8,0,3,5,Pyramid2.Pyramid2Vertices);
        MeshEx TurretMeshEx = null;

        // Turret Offset
        Vector3 TurretOffset = new Vector3(0, 0.2f, -0.3f);

        // Shaders
        Shader iShader = new Shader(iContext, R.raw.vsonelight, R.raw.fsonelight); // ok
```

```
    // Initilization
    Vector3 Position = new Vector3(-2.0f, 7.0f, 2.0f);
    Vector3 ScaleTurret = new Vector3(1.5f/2.0f, 0.5f/2.0f, 1.3f/2.0f);
    Vector3 ScaleMainBody = new Vector3(1, 0.5f/2.0f, 1);

    float GroundLevel = 0.0f;
    Vector3 GridColor= new Vector3(0.0f,1.0f,0.0f);
    float MassEffectiveRadius = 7.0f;
    int HitGroundSFX =R.raw.explosion2;
    int ExplosionSFX=R.raw.explosion1;

    // Create Explosion
    int NumberParticles = 20;
    Vector3 Color = new Vector3(0,0,1);
    long ParticleLifeSpan= 3000;
    boolean RandomColors   = false;
    boolean ColorAnimation = true;
    float   FadeDelta = 0.001f;
    Vector3 ParticleSize= new Vector3(0.5f,0.5f,0.5f);

    SphericalPolygonExplosion Explosion = CreateExplosion(iContext,NumberParticles,Color,
ParticleSize, ParticleLifeSpan,RandomColors,ColorAnimation,FadeDelta);
    Tank TankType1 = CreateInitTank(iContext,TankWeapon,MainBodyMaterial,NumberMainBodyTextures,
MainBodyTexture,AnimateMainBodyTex,MainBodyAnimationDelay,MainBodyMesh, MainBodyMeshEx,
TurretMaterial, NumberTurretTextures, TurretTexture, AnimateTurretTex,TurretAnimationDelay,
TurretMesh,TurretMeshEx, TurretOffset, iShader, Explosion,Position,ScaleMainBody,ScaleTurret,
GroundLevel, GridColor, MassEffectiveRadius,HitGroundSFX, ExplosionSFX);
    TankType1.GetMainBody().SetSFXOnOff(true);
    TankType1.GetTurret().SetSFXOnOff(true);
    return TankType1;
}
```

The GenerateTankWayPoints() function creates four waypoints for the tank and returns the number of waypoints created. (See Listing 8-64.)

Listing 8-64. Generating the Tank's Waypoints

```
int GenerateTankWayPoints(Vector3[] WayPoints)
{
    int NumberWayPoints = 4;

    WayPoints[0] = new Vector3( 5, 0, 10);
    WayPoints[1] = new Vector3( 10, 0,-10);
    WayPoints[2] = new Vector3(-10, 0,-10);
    WayPoints[3] = new Vector3(-5, 0, 10);

    return NumberWayPoints;
}
```

The CreatePatrolAttackTankCommand() function creates the patrol/attack vehicle command that is sent to the enemy tank. (See Listing 8-65.)

Listing 8-65. Creating the Patrol/Attack Command

```
VehicleCommand CreatePatrolAttackTankCommand(AIVehicleObjectsAffected ObjectsAffected,int
NumberWayPoints,Vector3[] WayPoints, Vector3 Target, Object3d TargetObj, int NumberRoundToFire,int
FiringDelay)
{
        VehicleCommand TankCommand = null;
        AIVehicleCommand Command = AIVehicleCommand.Patrol;

        Int    NumberObjectsAffected = 0;
        Int    DeltaAmount = NumberRoundToFire;
        Int    DeltaIncrement = FiringDelay;

        Int     MaxValue = 0;
        Int     MinValue = 0;

        TankCommand = new VehicleCommand(m_Context,Command,ObjectsAffected,NumberObjectsAffected,
DeltaAmount,DeltaIncrement,MaxValue,MinValue,NumberWayPoints,WayPoints,Target,TargetObj);
        return TankCommand;
}
```

The CreateTanks() function (see Listing 8-66) creates the enemy tank by

1. Creating the tank object and assigning it to m_Tank.

2. Setting the tank main body material to emit a green glow.

3. Setting the tank turret material to emit a red glow.

4. Setting the tank id to "tank1".

5. Creating the patrol/attack tank command by calling the
    CreatePatrolAttackTankCommand() function. The tank is ordered to fire three
    rounds of projectiles at the player's pyramid, then pause for five seconds,
    then repeat the firing sequence.

6. Sending the Patrol/Attack command to the Driver of the tank by calling the
    SetOrder() function.

Listing 8-66. Creating the Tank

```
void CreateTanks()
 {
        m_Tank= CreateTankType1(m_Context);

        // Set Material
        m_Tank.GetMainBody().GetMaterial().SetEmissive(0.0f, 0.5f, 0f);
        m_Tank.GetTurret().GetMaterial().SetEmissive(0.5f, 0, 0.0f);

        // Tank ID
        m_Tank.SetVehicleID("tank1");
```

```
        // Set Patrol Order
        int MAX_WAYPOINTS = 10;
        Vector3[] WayPoints = new Vector3[MAX_WAYPOINTS];
        int NumberWayPoints = GenerateTankWayPoints(WayPoints);
        AIVehicleObjectsAffected ObjectsAffected = AIVehicleObjectsAffected.PrimaryWeapon;
        Vector3 Target = new Vector3(0,0,0);
        Object3d TargetObj = null;
        int NumberRoundToFire = 3;
        int FiringDelay = 5000;
        VehicleCommand Command = CreatePatrolAttackTankCommand(ObjectsAffected,NumberWayPoints,
WayPoints, Target,TargetObj,NumberRoundToFire,FiringDelay);

        m_Tank.GetDriver().SetOrder(Command);
 }
```

The `ProcessTankCollisions()` function processes collisions between the tank and the player's ammunition and the tank weapon's ammunition and the player's power pyramid. (See Listing 8-67.)

Listing 8-67. Processing the Tank Collisions

```
void ProcessTankCollisions()
{
        float ExplosionMinVelocity = 0.02f;
        float ExplosionMaxVelocity = 0.4f;

        if (!m_Tank.GetMainBody().IsVisible())
        {
                return;
        }

        // Check Collisions between Tank and Player's Ammunition
        Object3d CollisionObj = m_Weapon.CheckAmmoCollision(m_Tank.GetMainBody());
        if (CollisionObj != null)
        {
                CollisionObj.ApplyLinearImpulse(m_Tank.GetMainBody());

                m_Tank.GetMainBody().ExplodeObject(ExplosionMaxVelocity, ExplosionMinVelocity);
                m_Tank.PlayExplosionSFX();

                // Process Damage
                m_Tank.GetMainBody().TakeDamage(CollisionObj);
                int Health = m_Tank.GetMainBody().GetObjectStats().GetHealth();
                if (Health <= 0)
                {
                        int KillValue = m_Tank.GetMainBody().GetObjectStats().GetKillValue();
                        m_Score = m_Score + KillValue;

                        m_Tank.GetMainBody().SetVisibility(false);
                        m_Tank.GetTurret().SetVisibility(false);
                }
        }
```

```
// Tank Weapons and Pyramid
int NumberWeapons = m_Tank.GetNumberWeapons();
for (int j=0; j < NumberWeapons; j++)
{
        CollisionObj = m_Tank.GetWeapon(j).CheckAmmoCollision(m_Pyramid);
        if (CollisionObj != null)
        {
                CollisionObj.ApplyLinearImpulse(m_Pyramid);

                //Process Damage
                m_Pyramid.TakeDamage(CollisionObj);

                // Obj Explosion
                m_Pyramid.ExplodeObject(ExplosionMaxVelocity, ExplosionMinVelocity);
                m_Pyramid.PlayExplosionSFX();

                // Set Pyramid Velocity and Acceleration to 0
                m_Pyramid.GetObjectPhysics().ResetState();
        }
    }
}
}
```

The ProcessTankCollisions() function does the following:

1. Returns, if the tank's main body is not visible.

2. Checks for a collision with the player's ammunition and the enemy tank's main body.

3. If there is a collision, it

   a. Applies forces to the player's ammunition and the tank

   b. Starts an explosion sequence for the tank

   c. Plays an explosion sound effect

   d. Processes the damage caused by the player's ammunition on the tank's main body

   e. Adds the kill value of the tank to the player's score, if the health of the tank's main body is less than or equal to 0. It also sets the visibility of the tank's main body and the tank's turret to false.

4. Checks for a collision between the weapon's ammunition and the player's power pyramid for each of the tank's weapons.

5. If there is a collision, it

   a. Applies forces to the ammunition and the pyramid

   b. Processes the damage to the pyramid

   c. Starts the pyramid's explosion effect

   d. Plays the pyramid's explosion sound effect

   e. Resest the pyramid's physics so that it does not move from the collision with the tank's ammunition.

Next, we must make modifications to the UpdateGravityGrid() function. If the tank's main body is visible, we have to add the tank's main body and the tank's active ammunition to the gravity grid. (See Listing 8-68.)

Listing 8-68. Modifying the UpdateGravityGrid() Function

```
if (m_Tank.GetMainBody().IsVisible())
{
        m_Grid.AddMass(m_Tank.GetMainBody());
        NumberMasses = m_Tank.GetWeapon(0).GetActiveAmmo(0, Masses);
        m_Grid.AddMasses(NumberMasses, Masses);
}
```

The onDrawFrame() function must also be modified with additional code.

The tank has to be updated by calling the UpdateVehicle() function.

```
m_Tank.UpdateVehicle();
```

The tank is rendered using the RenderVehicle() function.

```
m_Tank.RenderVehicle(m_Camera, m_PointLight, false);
```

Next, run the project, and you should see something similar to Figure 8-7 and Figure 8-8.

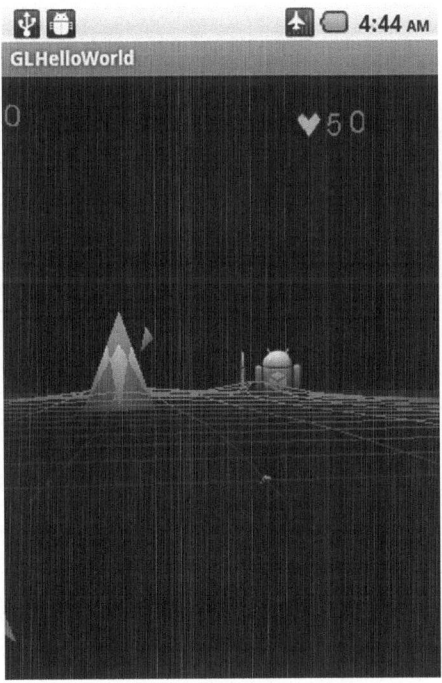

Figure 8-7. The arena object

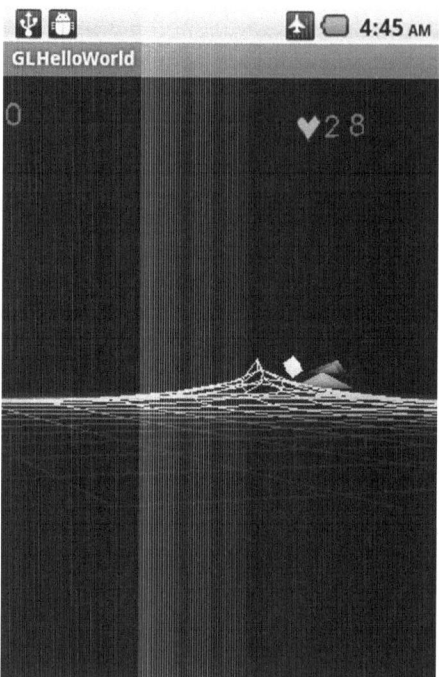

Figure 8-8. *The tank*

Summary

In this chapter, I covered the enemies that will be in our Drone Grid case study. First, I discussed arena objects. Next, I went through an overview of finite state machines. I then discussed specifically what the finite state machine for our tank will be. The classes that implemented the finite state machine and the states for our tank were then examined. Required modifications to other classes were also presented. Finally, a hands-on example was given to create and implement these enemies in a working game demo.

Chapter 9

Drone Grid Case Study: The User Interface

This chapter will cover the user interface for our Drone Grid case study game. The Main Menu System is covered first, which allows the player to choose between a new game, continuing an old game, or viewing the high score table. Next, the creation of the high score table is covered, including the class that is used to implement it. The high score entry menu is then discussed with the class that is used to implement that menu system. Finally, a hands-on example is given to demonstrate these user interfaces.

Creating the Main Menu System

The menu system for our game will consist of a main menu represented by a MainMenu class. Each item in the main menu is represented by a MenuItem class.

The MenuItem Class

The MenuItem class holds an item for a menu that also contains the actual 3D graphic.

The MenuItem class is derived from the BillBoard class.

```
public class MenuItem extends BillBoard
```

The MenuItem constructor calls the constructor for its parent BillBoard class. (See Listing 9-1.)

Listing 9-1. The MenuItem Constructor

```
MenuItem(Context iContext, Mesh iMesh, MeshEx iMeshEx, Texture[] iTextures, Material iMaterial,
Shader iShader)
{
        super(iContext, iMesh, iMeshEx, iTextures, iMaterial, iShader);
}
```

The GetObject3dWindowCoords() function gets the window coordinates of a point on a menu item object that can be offset by ObjOffset and that is displayed within the viewport defined by ViewPortWidth and ViewPortHeight input parameters. (See Listing 9-2.)

Listing 9-2. Getting the Window Coordinates for the MenuItem Object

```
float[] GetObject3dWindowCoords(int ViewPortWidth,int ViewPortHeight,Vector3 ObjOffset)
{
        float[] WindowCoords;
        int[] View = new int[4];

        View[0] = 0;
        View[1] = 0;
        View[2] = ViewPortWidth;
        View[3] = ViewPortHeight;

        WindowCoords = MapObjectCoordsToWindowCoords(View, 0, ObjOffset);

        // Flip Y starting point so that 0 is at top of window
        WindowCoords[1] = ViewPortHeight - WindowCoords[1];

        return WindowCoords;
}
```

The function does the following:

1. Creates the View variable, which is a viewport window defined by the coordinates (0,0) and (ViewPortWidth, ViewPortHeight)

2. Gets the window coordinates of the menu item object at ObjOffset offset position by calling the MapObjectCoordsToWindowCoords() function

3. Converts the y component of the window coordinates to screen space from OpenGL space

4. Returns the window coordinates

The Touched() function returns true if the input screen touch coordinates TouchX and TouchY map to within this menu item. (See Listing 9-3.)

Listing 9-3. Testing for the User's Touch Input

```
boolean Touched(float TouchX, float TouchY,int ViewPortWidth,int ViewPortHeight)
{
        boolean result = false;

        float Radius = GetRadius();
        Vector3 ObjCoordsUpperLeft   = new Vector3(-Radius,  Radius, 0);
        Vector3 ObjCoordsUpperRight  = new Vector3( Radius,  Radius, 0);
        Vector3 ObjCoordsLowerLeft   = new Vector3(-Radius, -Radius, 0);
```

```
float[] UpperLeft  = GetObject3dWindowCoords(ViewPortWidth, ViewPortHeight, ObjCoordsUpperLeft);
float[] UpperRight = GetObject3dWindowCoords(ViewPortWidth, ViewPortHeight, ObjCoordsUpperRight);
float[] LowerLeft  = GetObject3dWindowCoords(ViewPortWidth, ViewPortHeight, ObjCoordsLowerLeft);

if ((TouchX >= UpperLeft[0]) && (TouchX <= UpperRight[0]) &&
    (TouchY >= UpperLeft[1]) && (TouchY <= LowerLeft[1]))
{
        result = true;
 }

 return result;
}
```

The following are attributes of the Touched() function:

1. For our menu items, we will use a Cube class object that has equal lengths on all sides. We get the radius of this cube by calling the GetRadius() function and store it in the Radius variable.

2. Using the Radius value, we create the object's upper left, upper right, and lower left object coordinates. We store these values in the variables ObjCoordsUpperLeft, ObjCoordsUpperRight, and ObjCoordsLowerLeft.

3. The function retrieves the window coordinates of the upper left, upper right, and lower left corners of the menu item object by calling GetObject3dWindowCoords() using the object's upper left, upper right, and lower left object coordinates created in step 2.

4. The function tests to see if the screen coordinates (TouchX, TouchY) are within the bounds of the screen coordinates of the menu item. If they are, then return true; otherwise, return false.

The MainMenu Class

The MainMenu class represents the actual main menu interface for our Drone Grid case study game.

The MenuStates enumeration represents the available menu selections (see Listing 9-4) for the main menu, such as

None: Which means that no menu item is selected

NewGame: Which means that the player wants to begin playing a completely new game

ContinueCurrentGame: Which means that the player wants to continue the last saved game

HighScoreTable: Which means that the player wants to view the high score table

Copyright: Used in debug mode to help test the high score entry system

Listing 9-4. The Menu States

```
enum MenuStates
{
        None,
        NewGame,
        ContinueCurrentGame,
        HighScoreTable,
        Copyright
}
```

The `m_NewGameItem` variable holds a reference to a menu item that the user touches, if he or she wants to start a new game.

```
MenuItem m_NewGameItem;
```

The `m_ContinueGameItem` variable holds a reference to a menu item that the user touches if he or she wants to continue the last saved game.

```
MenuItem m_ContinueGameItem;
```

The `m_HighScoresItem` variable holds a reference to a menu item that the user touches if he or she wants to view the high score table.

```
MenuItem m_HighScoresItem;
```

The `m_CopyRightItem` variable holds a reference to a menu item that is used in debugging the high score entry system.

```
MenuItem m_CopyRightItem;
```

The MainMenu constructor initializes the MainMenu class by assigning objects to each of the menu items in the main menu. (See Listing 9-5.)

Listing 9-5. The MainMenu Constructor

```
MainMenu(MenuItem NewGameItem,MenuItem ContinueGameItem,MenuItem HighScoresItem,MenuItem
CopyRightItem)
{
        m_NewGameItem       = NewGameItem;
        m_ContinueGameItem = ContinueGameItem;
        m_HighScoresItem    = HighScoresItem;
        m_CopyRightItem     = CopyRightItem;
}
```

The `GetMainMenuStatus()` function tests the main menu items to see if any has been touched at the screen location (TouchX, TouchY). Each menu item button is tested to see if it has been touched by calling the `Touched()` function on that button. The button type that is touched is returned or None is returned if no menu buttons have been touched. (See Listing 9-6.)

Listing 9-6. Getting the Main Menu Status

```
MenuStates GetMainMenuStatus(float TouchX, float TouchY,int ViewPortWidth,int ViewPortHeight)
{
        MenuStates Selection = MenuStates.None;

        boolean Touched = false;

        // New Game Menu Item
        Touched = m_NewGameItem.Touched(TouchX, TouchY, ViewPortWidth, ViewPortHeight);
        if (Touched)
        {
                Selection = MenuStates.NewGame;
        }

        // New ContinueGame Menu Item
        Touched = m_ContinueGameItem.Touched(TouchX, TouchY, ViewPortWidth, ViewPortHeight);
        if (Touched)
        {
                Selection = MenuStates.ContinueCurrentGame;
        }

        // New HighScoreTable Menu Item
        Touched = m_HighScoresItem.Touched(TouchX, TouchY, ViewPortWidth, ViewPortHeight);
        if (Touched)
        {
                Selection = MenuStates.HighScoreTable;
        }

        // CopyRight Menu Item
        Touched = m_CopyRightItem.Touched(TouchX, TouchY, ViewPortWidth, ViewPortHeight);
        if (Touched)
        {
                Selection = MenuStates.Copyright;
        }
        return Selection;
}
```

The RenderMenu() function draws the main menu items to the screen. (See Listing 9-7.)

Listing 9-7. Rendering the Main Menu

```
void RenderMenu(Camera Cam, PointLight Light, boolean DebugOn)
{
        m_NewGameItem.DrawObject(Cam, Light);
        m_ContinueGameItem.DrawObject(Cam, Light);
        m_HighScoresItem.DrawObject(Cam, Light);
        m_CopyRightItem.DrawObject(Cam, Light);
}
```

The UpdateMenu() function updates each main menu item and also turns each of the items (which are billboards) toward the camera by calling the UpdateObject3d() function. All the main menu items, which are the new game button, the continue game button, the display high scores button, and the copyright graphic, are processed here. (See Listing 9-8.)

Listing 9-8. Updating the Main Menu

```
void UpdateMenu(Camera Cam)
{
        m_NewGameItem.UpdateObject3d(Cam);
        m_ContinueGameItem.UpdateObject3d(Cam);
        m_HighScoresItem.UpdateObject3d(Cam);
        m_CopyRightItem.UpdateObject3d(Cam);
}
```

Creating the High Score Table

The high score table that holds the player's highest scores for the game consists of two classes: the HighScoreEntry class and the HighScoreTable class.

The HighScoreEntry Class

The HighScoreEntry class holds the data for a high score entry. This class implements the Comparable public interface and defines a function that allows the high score entries to be compared and sorted.

```
public class HighScoreEntry implements Comparable<HighScoreEntry>
```

The m_ItemValid variable is set to true, if this high score entry is valid, and should be displayed in the high score table.

```
private boolean m_ItemValid;
```

The m_Initials variable holds the player's initials.

```
private String  m_Initials;
```

The m_Score variable holds the player's score.

```
private int     m_Score;
```

The HighScoreEntry constructor initializes the entry by setting the initials and score. (See Listing 9-9.)

Listing 9-9. HighScoreEntry Constructor

```
HighScoreEntry(String Initials,int Score)
{
        m_Initials = Initials;
        m_Score    = Score;
}
```

The compareTo() function is used in combination with the Collections.sort() function to sort high score table entries in descending order, with high scores ordered first (see Listing 9-10). Normally, the compareTo() function sorts the entries in ascending order. Because we are sorting in descending order, we need to make some key changes.

- The function returns a negative integer if the score in this instance of the entry is greater than the score in the input parameter variable Another's score.

- The function returns a positive integer if the score in this instance of the entry is less than the score in the input parameter variable Another's score.

Listing 9-10. Comparing and Sorting Entries

```
public int compareTo(HighScoreEntry Another)
{
        /*
        Normally ascending sorting - Returns
        a negative integer if this instance is less than another; a positive integer if this
        instance is greater than another; 0 if this instance has the same order as another.
        */
        int result = 0;
        if (m_Score > Another.m_Score)
        {
                result = -1;
        }
        else
        if (m_Score < Another.m_Score)
        {
                result = 1;
        }
        return result;
}
```

The HighScoreTable Class

The HighScoreTable class represents the high score table containing all the player's high score entries.

The HIGH_SCORES string variable holds the handle for loading and saving the high score table.

```
private String HIGH_SCORES = "HighScores";
```

The MAX_RANK variable holds the maximum number of high scores to display.

```
private int MAX_RANK = 10;
```

The MAX_SCORES variable holds the maximum number of scores that are stored internally for processing and calculation purposes.

```
private int MAX_SCORES = 11;
```

The m_HighScoreTable array holds the actual high score entries.

```
private HighScoreEntry[] m_HighScoreTable = new HighScoreEntry[MAX_SCORES];
```

The m_Text variable holds the character set that will be used to print out the text for the high score table graphics.

```
private BillBoardCharacterSet m_Text;
```

The m_FontWidth variable holds the width of each character in the character set that is used to print text for the high score table.

```
private int m_FontWidth;
```

The m_FontHeight variable holds the height of each character in the character set that is used to print text for the high score table.

```
private int m_FontHeight;
```

The m_BackGroundTexture variable holds the texture that is used to clear the high score table texture.

```
private Texture m_BackGroundTexture;
```

The m_HighScoreTableImage variable holds a reference to the BillBoard object that contains the texture with the player's high scores on it.

```
private BillBoard m_HighScoreTableImage;
```

The m_Dirty variable is true if the high score table has been altered since the last update and false otherwise.

```
private boolean m_Dirty = false;
```

The HighScoreTable() constructor (see Listing 9-11) creates a new high score table by

1. Creating a new background texture for the high score table
2. Initializing the high score table by creating blank high score entries in the m_HighScoreTable array
3. Initializing other class member variables
4. Loading in the player's previously saved high scores, if any

Listing 9-11. The HighScoreTable Constructor

```
HighScoreTable(Context iContext,BillBoardCharacterSet CharacterSet,BillBoard HighScoreTableImage)
{
        m_Context = iContext;
        m_BackGroundTexture = new Texture(iContext, R.drawable.background);
```

```
        String  Initials        = "AAA";
        int     Score   = 0;

        // Initialize High Score Entries
        for (int i = 0; i < MAX_SCORES; i++)
        {
                m_HighScoreTable[i] = new HighScoreEntry(Initials,Score);
                m_HighScoreTable[i].SetItemValidState(false);
        }

        m_Text = CharacterSet;
        m_FontWidth = m_Text.GetFontWidth();
        m_FontHeight = m_Text.GetFontHeight();

        m_HighScoreTableImage = HighScoreTableImage;

        // Load In Saved high Scores
        LoadHighScoreTable(HIGH_SCORES);
        m_Dirty = true;
}
```

The SaveHighScoreTable() function saves the player's top MAX_RANK number high score entries consisting of the player's initials and the player's score. (See Listing 9-12.)

Listing 9-12. Saving the High Score Table

```
void SaveHighScoreTable(String Handle)
{
        // We need an Editor object to make preference changes.
        // All objects are from android.context.Context
        SharedPreferences settings = m_Context.getSharedPreferences(Handle, 0);
        SharedPreferences.Editor editor = settings.edit();
        for (int i = 0; i < MAX_RANK; i++)
        {
                editor.putString("Name" + i, m_HighScoreTable[i].GetInitials());
                editor.putInt("Score" + i, m_HighScoreTable[i].GetScore());
        }
        // Commit the edits!
        editor.commit();
}
```

The LoadHighScoreTable() function loads the player's high score data consisting of the player's name or initials and the player's score. If the player's score is greater than 0, the entry is valid. (See Listing 9-13.)

Listing 9-13. Loading the High Score Table

```
void LoadHighScoreTable(String Handle)
{
        // Restore preferences
        SharedPreferences settings = m_Context.getSharedPreferences(Handle, 0);
```

```
        for (int i = 0; i < MAX_RANK; i++)
        {
                String Name = settings.getString("Name" + i, "...");
                int Score = settings.getInt("Score" + i, 0);

                m_HighScoreTable[i].SetName(Name);
                m_HighScoreTable[i].SetScore(Score);

                if (Score > 0)
                {
                        m_HighScoreTable[i].SetItemValidState(true);
                }
        }
}
```

The NumberValidHighScores() function finds the number of valid high score entries in the high score table. (See Listing 9-14.)

Listing 9-14. Finding the Number of Valid High Scores

```
int NumberValidHighScores()
{
        int NumberValidScores = 0;
        for (int i = 0; i < MAX_RANK; i++)
        {
                if (m_HighScoreTable[i].IsValid())
                {
                        NumberValidScores++;
                }
        }
        return NumberValidScores;
}
```

The GetLowestScore() function retrieves the lowest valid player score from the high score table m_HighScoreTable. (See Listing 9-15.)

Listing 9-15. Getting the Lowest Score

```
int GetLowestScore()
{
        // Get Lowest valid score
        int LowestScore = 0;
        int ValidScores = 0;

        for (int i = 0; i < MAX_RANK; i++)
        {
                if (m_HighScoreTable[i].IsValid())
                {
                        ValidScores++;
                }
        }
```

```
        if (ValidScores > 0)
        {
                LowestScore = m_HighScoreTable[ValidScores-1].GetScore();
        }
        return LowestScore;
}
```

The FindEmptySlot() function retrieves the index of an empty (meaning not valid) high score entry slot in m_HighScoreTable. (See Listing 9-16.)

Listing 9-16. Finding an Empty Slot for the High Score Entry

```
int FindEmptySlot()
{
        int EmptySlot = -1;
        for (int i = 0; i < MAX_SCORES; i++)
        {
                if (m_HighScoreTable[i].IsValid() == false)
                {
                        return i;
                }
        }
        return EmptySlot;
}
```

The AddItem() function adds a high score entry to an empty slot in the high score table if an empty slot is found. More specifically, if there is an empty slot, then assign the high score entry to that slot in the m_HighScoreTable array, set the valid state of the entry to true, and set the m_Dirty value to true to indicate that the high score table needs to be sorted and rendered. (See Listing 9-17.)

Listing 9-17. Adding an Item to the High Score Table

```
boolean AddItem(HighScoreEntry Item)
{
        boolean result = false;

        int EmptySlot =  FindEmptySlot();
        if (EmptySlot >= 0)
        {
                m_HighScoreTable[EmptySlot] = Item;
                m_HighScoreTable[EmptySlot].SetItemValidState(true);
                result = true;
                m_Dirty = true;
        }
        return result;
}
```

The SortHighScoreTable() function sorts the high score table in descending order. After sorting in descending order the top 10 entries, the function takes the 11th high score entry and sets the status to invalid, so that new entries to the high score table can be put at the end, if needed. (See Listing 9-18.)

Listing 9-18. Sorting the High Score Table

```
void SortHighScoreTable()
{
        Collections.sort(Arrays.asList(m_HighScoreTable));

        // Only keep top 10 and make room for another to be added to end of array
        m_HighScoreTable[MAX_SCORES-1].SetItemValidState(false);
}
```

The ClearHighScoreTable() function clears the high score table texture image held in
m_HighScoreTableImage by copying the blank texture m_BackGroundTexture over it, using the
CopySubTextureToTexture() function. (See Listing 9-19.)

Listing 9-19. Clearing the High Score Table

```
void ClearHighScoreTable()
{
        Texture HighScoreTableTexture = m_HighScoreTableImage.GetTexture(0);

        // Clear Composite Texture;
        Bitmap BackGroundBitmap = m_BackGroundTexture.GetTextureBitMap();
        HighScoreTableTexture.CopySubTextureToTexture(0, 0, 0, BackGroundBitmap);
}
```

The RenderTitle() function renders the text "High Scores" to m_HighScoreTableImage, which is the
billboard that holds the final composite texture for the high score table. (See Listing 9-20.)

Listing 9-20. Rendering the Title of the High Score Table

```
void RenderTitle()
{
        m_Text.SetText("High".toCharArray());
        m_Text.RenderToBillBoard(m_HighScoreTableImage, 0 , 0);

        m_Text.SetText("Scores".toCharArray());
        m_Text.RenderToBillBoard(m_HighScoreTableImage, 5*m_FontWidth , 0);
}
```

The CopyHighScoreEntryToHighScoreTable() function copies a high score entry to the final
composite m_HighScoreTableImage high score billboard that is used to display the final player's high
score. (See Listing 9-21.)

Listing 9-21. Copying the High Score Entry to the Final Composite Billboard Texture Object

```
void CopyHighScoreEntryToHighScoreTable(int Rank, Camera Cam, HighScoreEntry Item)
{
        // Put HighScore Entry onto Final Composite Bitmap

        // CharacterPosition
        int HeightOffset = 10;
        int CharPosX = 0;
        int CharPosY = m_FontHeight + (Rank * (m_FontHeight + HeightOffset));
```

```
    // Render Rank
    String RankStr = Rank + ".";
    m_Text.SetText(RankStr.toCharArray());
    m_Text.RenderToBillBoard(m_HighScoreTableImage, CharPosX, CharPosY);

    // Render Player Name/Initials and render to composite billboard
    String Name = Item.GetInitials();
    m_Text.SetText(Name.toCharArray());

    CharPosX = CharPosX + m_FontWidth * 3;
    m_Text.RenderToBillBoard(m_HighScoreTableImage, CharPosX, CharPosY);

    // Render Numerical Value and render to composite billboard
    String Score = String.valueOf(Item.GetScore());
    m_Text.SetText(Score.toCharArray());

    int BlankSpace = 4 * m_FontWidth;
    CharPosX = CharPosX + Name.length() + BlankSpace;
    m_Text.RenderToBillBoard(m_HighScoreTableImage, CharPosX, CharPosY);
}
```

The function does the following:

1. Calculates the beginning x,y position for the high score entry, based on the height of the character font to be used, the rank of the score (1st, 2nd, 3rd, etc.), and the HeightOffset variable

2. Renders the high score entry's rank to the m_HighScoreTableImage billboard

3. Renders the high score entry's player's initials to the m_HighScoreTableImage billboard

4. Renders the high score entry's player's score to the m_HighScoreTableImage billboard

The UpdateHighScoreTable() function updates the high score table. (See Listing 9-22.)

Listing 9-22. Updating the High Score Table

```
void UpdateHighScoreTable(Camera Cam)
{
    if (m_Dirty)
    {
        // Sort High Score Table in descending order for score
        SortHighScoreTable();

        // Clear High Score Table and set background texture
        ClearHighScoreTable();

        // Render Title
        RenderTitle();
```

```
            // For the Top Ranked entries copy these to the HighScore Table BillBoard
            for (int i = 0; i < MAX_RANK; i++)
            {
                    if (m_HighScoreTable[i].IsValid())
                    {
                            CopyHighScoreEntryToHighScoreTable(i+1, Cam, m_HighScoreTable[i]);
                    }
            }

            // Save High Scores
            SaveHighScoreTable(HIGH_SCORES);

            m_Dirty = false;
        }

    // Update BillBoard orientation for Score
    m_HighScoreTableImage.UpdateObject3d(Cam);
}
```

The function does the following:

1. If m_Dirty is true, it adds a new high score entry and processes and renders it to the high score table billboard.

2. If m_Dirty is true, it

    a. Sorts the high score table in descending order, with the highest score listed first, by calling the SortHighScoreTable() function

    b. Clears the high score table by calling the ClearHighScoreTable() function

    c. Renders the high score table's heading by calling the RenderTitle() function

    d. Renders the top 10 high scores to the high score table by calling the CopyHighScoreEntryToHighScoreTable() function

    e. Saves the high score table by calling the SaveHighScoreTable() function

    f. Sets m_Dirty to false to indicate that the high score table has been updated

3. Updates the high score table billboard by calling the UpdateObject3d() function to turn the billboard toward the camera

The RenderHighScoreTable() function draws the high score table billboard and the texture containing the player's high scores to the screen. (See Listing 9-23.)

Listing 9-23. Rendering the High Score Table

```
void RenderHighScoreTable(Camera Cam, PointLight Light, boolean DebugOn)
{
        // Render Final High Score Table Composite Image
        m_HighScoreTableImage.DrawObject(Cam, Light);
}
```

Creating the High Score Entry System

The HighScoreEntryMenu class controls the menu for entering a player's high score.

The EntryMenuStates enumeration holds the type of menu button that is clicked (see Listing 9-24), which include

> **None**: There is no button clicked.
>
> **NextCharacterPressed**: The next character button is pressed to change the character in the selection to the next character in the character set for user selection.
>
> **PreviousCharacterPressed**: The previous character button is pressed to change the character in the selection to the previous character in the character set for user selection.
>
> **Enter**: The current character that is displayed is entered as the character for this position.

Listing 9-24. The High Score Entry Menu Buttons

```
enum EntryMenuStates
{
        None,
        NextCharacterPressed,
        PreviousCharacterPressed,
        Enter
}
```

The MAX_ENTRY_CHARACTERS variables hold the maximum number of characters to enter for a player's name or initials.

```
private int MAX_ENTRY_CHARACTERS = 3;
```

The m_EntryIndex variable holds the current character position in the player's name/initial selection input. For example, 0 would indicate that the player is selecting the first initial to enter.

```
private int m_EntryIndex = 0;
```

The m_Entry variable holds the player's name/initials.

```
private char[] m_Entry = new char[MAX_ENTRY_CHARACTERS];
```

The m_NextCharacterButton button cycles forward through the character set that is available for entering the player's name/initials.

```
private MenuItem m_NextCharacterButton;
```

The m_PreviousCharacterButton button cycles backward through the character set that is available for entering the player's name/initials.

```
private MenuItem m_PreviousCharacterButton;
```

The m_EnterButton button is pressed to select the character for the current player's name/initial entry position.

```
private MenuItem m_EnterButton;
```

The m_Text variable holds the character set that will be used for the player's name/initials.

```
private BillBoardCharacterSet m_Text;
```

The m_NumberCharactersInSet variable holds the number of characters in the m_Text character set.

```
private int m_NumberCharactersInSet = 0;
```

The m_CharacterSetIndex variable is used to keep track of the current character that the user has selected for input for his or her name/initials.

```
private int m_CharacterSetIndex = 0;
```

The m_FontWidth is the width of the character font used in the character set m_Text.

```
private int m_FontWidth;
```

The m_FontHeight is the height of the character font used in the character set m_Text.

```
private int m_FontHeight;
```

The m_HighScoreEntryMenuImage variable holds the billboard for the high score entry menu.

```
private BillBoard m_HighScoreEntryMenuImage;
```

The m_Dirty variable is true if m_HighScoreEntryMenuImage has to be updated.

```
private boolean m_Dirty = true;
```

The m_StartingEntryPositionX variable holds the starting x position of the player's name/initials input field.

```
private int m_StartingEntryPositionX;
```

The m_StartingEntryPositionY variable holds the starting y position of the player's name/initials input field.

```
private int m_StartingEntryPositionY;
```

The m_CurrentEntryPositionX variable holds the current x position of the player's name/initials entry location.

```
private int m_CurrentEntryPositionX;
```

The m_CurrentEntryPositionY variable holds the current y position of the player's name/initials entry location.

```
private int m_CurrentEntryPositionY;
```

The m_EntryFinished variable is true if the user has finished entering his or her name/initials.

```
private boolean m_EntryFinished = false;
```

The HighScoreEntryMenu constructor initializes and resets the high score entry menu for use. (See Listing 9-25.)

Listing 9-25. The HighScoreEntryMenu Constructor

```
HighScoreEntryMenu(MenuItem NextCharacterButton,MenuItem PreviousCharacterButton,MenuItem Ente
rButton,BillBoardCharacterSet Text,BillBoard HighScoreEntryMenuImage,int StartingEntryXPos,int
StartingEntryYPos)
{
        m_NextCharacterButton = NextCharacterButton;
        m_PreviousCharacterButton = PreviousCharacterButton;
        m_EnterButton = EnterButton;
        m_Text = Text;
        m_HighScoreEntryMenuImage = HighScoreEntryMenuImage;

        m_FontWidth = m_Text.GetFontWidth();
        m_FontHeight = m_Text.GetFontHeight();

        m_NumberCharactersInSet = m_Text.GetNumberCharactersInSet();

        m_CurrentEntryPositionX = StartingEntryXPos;
        m_CurrentEntryPositionY = StartingEntryYPos;

        m_StartingEntryPositionX = StartingEntryXPos;
        m_StartingEntryPositionY = StartingEntryYPos;

        ResetMenu();
}
```

The ResetMenu() function resets the menu to its initial state, where the user can begin entering his or her initials, starting at the first initial. The default initial is reset to "...". (See Listing 9-26.)

Listing 9-26. Resetting the High Score Entry Menu

```
void ResetMenu()
{
        m_CharacterSetIndex = 10;

        m_EntryIndex = 0;

        m_CurrentEntryPositionX = m_StartingEntryPositionX;
        m_CurrentEntryPositionY = m_StartingEntryPositionY;

        m_Text.SetText("...".toCharArray());
        m_Text.RenderToBillBoard(m_HighScoreEntryMenuImage, m_CurrentEntryPositionX,
m_CurrentEntryPositionY);

        m_EntryFinished = false;
```

The FindCurrentCharacter() function finds the character matching the user's current initial entry selection and returns it. (See Listing 9-27.)

Listing 9-27. Finding the Currently Selected Initial

```
char FindCurrentCharacter()
{
        BillBoardFont Font = m_Text.GetCharacter(m_CharacterSetIndex);
        return Font.GetCharacter();
}
```

The ProcessEnterMenuSelection() function enters the currently selected initial as the entry for the current player's name/initial position and increments the entry point to the next initial entry position. The m_Dirty variable is also set to true, indicating the need to update the menu's billboard. (See Listing 9-28.)

Listing 9-28. Processing the Menu Selection

```
void ProcessEnterMenuSelection()
{
        char EnteredChar = FindCurrentCharacter();
        m_Entry[m_EntryIndex] = EnteredChar;

        m_EntryIndex++;
        if (m_EntryIndex >= MAX_ENTRY_CHARACTERS)
        {
                m_EntryFinished = true;
        }
        m_CurrentEntryPositionX = m_CurrentEntryPositionX + m_FontWidth;
        m_Dirty = true;
}
```

The ProcessPreviousMenuSelection() function processes the Previous character menu button by decrementing the index used to retrieve the current character selection from the character set held in m_Text. If the index is less than 0, it wraps around to point to the last character in the character set. The m_Dirty variable is also set to true. (See Listing 9-29.)

Listing 9-29. Processing the Previous Character Menu Selection Button

```
void ProcessPreviousMenuSelection()
{
        // Go to next character
        m_CharacterSetIndex--;

        if (m_CharacterSetIndex < 0)
        {
                m_CharacterSetIndex = m_NumberCharactersInSet-1;
        }
        m_Dirty = true;
}
```

The ProcessNextMenuSelection() function processes the Next character menu button by incrementing the index used to retrieve the current character selection from the character set held in m_Text. If the index is greater than the last element in the array, it wraps around to point to the first character in the character set. The m_Dirty variable is also set to true. (See Listing 9-30.)

Listing 9-30. Processing the Next Menu Selection Button

```
void ProcessNextMenuSelection()
{
        // Go to next character
        m_CharacterSetIndex++;

        if (m_CharacterSetIndex >= m_NumberCharactersInSet)
        {
                m_CharacterSetIndex = 0;
        }
        m_Dirty = true;
}
```

The RenderTextToMenu() function renders the input character Character at screen position (XPos, YPos) on the high score entry menu, which is the billboard m_HighScoreEntryMenuImage. (See Listing 9-31.)

Listing 9-31. Rendering Text to the Entry Menu

```
void RenderTextToMenu(String Character, int XPos, int YPos)
{
        m_Text.SetText(Character.toCharArray());
        m_Text.RenderToBillBoard(m_HighScoreEntryMenuImage, XPos , YPos);
}
```

The RenderEntryToMenu() function renders the currently selected character for the player's initials to the entry menu. (See Listing 9-32.)

Listing 9-32. Rendering the Current Entry Selection

```
void RenderEntryToMenu()
{
        char CurrentCharacter = FindCurrentCharacter();
        String StringCharacter = CurrentCharacter + "";

        RenderTextToMenu(StringCharacter, m_CurrentEntryPositionX, m_CurrentEntryPositionY);
}
```

The GetEntryMenuStatus() function tests the input screen coordinates TouchX and TouchY to determine if the user has pressed a high score entry menu button. If he or she has, the button type is returned. (See Listing 9-33.)

Listing 9-33. Getting the Entry Menu Status

```
EntryMenuStates GetEntryMenuStatus(float TouchX, float TouchY,int ViewPortWidth,int ViewPortHeight)
{
        EntryMenuStates Selection = EntryMenuStates.None;

        boolean Touched = false;

        // Next character Menu Item
        Touched = m_NextCharacterButton.Touched(TouchX, TouchY, ViewPortWidth, ViewPortHeight);
        if (Touched)
        {
                Selection = EntryMenuStates.NextCharacterPressed;
        }

        // Previous character Menu Item
        Touched = m_PreviousCharacterButton.Touched(TouchX, TouchY, ViewPortWidth, ViewPortHeight);
        if (Touched)
        {
                Selection = EntryMenuStates.PreviousCharacterPressed;
        }

        // Enter Menu Item
        Touched = m_EnterButton.Touched(TouchX, TouchY, ViewPortWidth, ViewPortHeight);
        if (Touched)
        {
                Selection = EntryMenuStates.Enter;
        }
        return Selection;
}
```

The UpdateHighScoreEntryMenu() function updates all the components that make up the high score entry menu, including the input character that the user has selected, the next character button, the previous character button, the enter button, and the high score entry menu billboard. (See Listing 9-34.)

Listing 9-34. Updating the High Score Entry Menu

```
void UpdateHighScoreEntryMenu(Camera Cam)
{
        //Update Menu Texture if changed
        if (m_Dirty)
        {
                // If need to alter Menu texture then render new texture data
                RenderEntryToMenu();
                m_Dirty = false;
        }

        // Update Buttons
        m_NextCharacterButton.UpdateObject3d(Cam);
        m_PreviousCharacterButton.UpdateObject3d(Cam);
        m_EnterButton.UpdateObject3d(Cam);

        // Update Initial Entry Area
        m_HighScoreEntryMenuImage.UpdateObject3d(Cam);
}
```

The RenderHighScoreEntryMenu() function (see Listing 9-35) renders all the components of the high score entry menu to the screen in the following order:

1. Renders the cycle to the Next Character button

2. Renders the cycle to the Previous Character button

3. Renders the Enter player initial selection button

4. Renders the m_HighScoreEntryMenuImage billboard, which contains the player's input and the rest of the high score entry menu graphics data

Listing 9-35. Rendering the High Score Entry Menu

```
void RenderHighScoreEntryMenu(Camera Cam, PointLight Light, boolean DebugOn)
{
        // Render Buttons
        m_NextCharacterButton.DrawObject(Cam, Light);
        m_PreviousCharacterButton.DrawObject(Cam, Light);
        m_EnterButton.DrawObject(Cam, Light);

        // Render Billboard with Entry Menu info
        m_HighScoreEntryMenuImage.DrawObject(Cam, Light);
}
```

Hands-on Example: Demonstrating the User Interface

In this hands-on example, we will hook the user interface to the hands-on example from Chapter 8. We will construct a working main menu from which the user can select a new game or continue a previously saved game. For the high score table, we will use a test button to input some high scores manually, before making this feature fully functional in the final game in Chapter 10.

Modifying the MyGLRenderer Class

The MyGLRenderer class must be modified in order to integrate the main menu, high score table, and high score entry menu.

The GameState enumeration (see Listing 9-36) holds the general state of the game, which can be one of the following:

> **MainMenu**: The state where the main menu is being displayed

> **ActiveGamePlay**: The state where the game is active

> **HighScoreTable**: The state where the high score table is being displayed

> **HighScoreEntry**: The state where a player is entering his or her initials after making a high score

Listing 9-36. The Game State

```
enum GameState
{
        MainMenu,
        ActiveGamePlay,
        HighScoreTable,
        HighScoreEntry
}
```

The m_GameState variable holds the state of the game.

```
private GameState m_GameState = GameState.MainMenu;
```

The m_MainMenu variable holds a reference to the MainMenu class object that implements the main menu.

```
private MainMenu m_MainMenu;
```

The m_HighScoreEntryMenu variable holds a reference to the HighScoreEntryMenu class object that implements the high score entry menu system.

```
private HighScoreEntryMenu m_HighScoreEntryMenu;
```

The m_HighScoreTable variable holds a reference to the class object that implements the high score table.

```
private HighScoreTable m_HighScoreTable;
```

The m_CanContinue variable is true if there is a previously saved game that the user can continue from.

```
private boolean m_CanContinue = false;
```

The CreateInitBillBoard() function creates and returns a new BillBoard object according to the input Texture resource, object position, and object scale values. (See Listing 9-37.)

Listing 9-37. Creating a BillBoard Object

```
BillBoard CreateInitBillBoard(Context iContext,int TextureResourceID,Vector3 Position,Vector3 Scale)
{
        BillBoard NewBillBoard = null;
        Texture BillBoardTexture = new Texture(iContext, TextureResourceID);

        //Create Shader
        Shader Shader = new Shader(iContext, R.raw.vsonelight, R.raw.fsonelight);        // ok
        MeshEx Mesh = new MeshEx(8,0,3,5,Cube.CubeData, Cube.CubeDrawOrder);

        // Create Material for this object
        Material Material1 = new Material();

        // Create Texture for BillBoard
        Texture[] Tex = new Texture[1];
        Tex[0] = BillBoardTexture;

        // Create new BillBoard
        NewBillBoard = new BillBoard(iContext, null, Mesh, Tex, Material1, Shader );

        // Set Initial Position and Orientation
        NewBillBoard.m_Orientation.SetPosition(Position);
        NewBillBoard.m_Orientation.SetScale(Scale);

        NewBillBoard.GetObjectPhysics().SetGravity(false);

        return NewBillBoard;
}
```

The CreateHighScoreTable() function creates the high score table that is assigned to the m_HighScoreTable variable. (See Listing 9-38.)

Listing 9-38. Creating the High Score Table

```
void CreateHighScoreTable(Context iContext)
{
        int TextureResourceID = R.drawable.background;
        Vector3 Position = new Vector3(0.5f, 1, 4);
        Vector3 Scale = new Vector3(4.5f,5,1);

        BillBoard HighScoreTableImage = CreateInitBillBoard(iContext,TextureResourceID,Position,Scale);
        m_HighScoreTable = new HighScoreTable(iContext,m_CharacterSet,HighScoreTableImage);
}
```

The CreateHighScoreEntryMenu() function creates the high score entry menu. (See Listing 9-39.)

Listing 9-39. Creating the High Score Entry Menu

```
void CreateHighScoreEntryMenu(Context iContext)
{
    // Create High Score Entry Menu Billboard
    int TextureResourceID = R.drawable.backgroundentrymenu;
    Vector3 Position = new Vector3(0.0f, 1, 4);
    Vector3 Scale = new Vector3(4.5f,5,1);

    BillBoard HighScoreEntryMenuImage = CreateInitBillBoard(iContext,TextureResourceID,
Position,Scale);

    // Create Menu Buttons
    Shader ObjectShader = new Shader(iContext, R.raw.vsonelight, R.raw.fsonelight);      // ok

    MeshEx MenuItemMeshEx = new MeshEx(8,0,3,5,Cube.CubeData, Cube.CubeDrawOrder);
    Mesh MenuItemMesh = null;

    // Create Material for this object
    Material Material1 = new Material();
    Material1.SetEmissive(0.3f, 0.3f, 0.3f);

    // Create Texture
    int NumberTextures = 1;
    Texture TexNextButton = new Texture(iContext,R.drawable.nextbutton);

    Texture[] Tex = new Texture[NumberTextures];
    Tex[0] = TexNextButton;

    boolean AnimateTextures = false;
    float TimeDelay = 0.0f;

    Position = new Vector3(-1.0f, 1.3f, 4.25f);
    Scale = new Vector3(1.4f,1.0f,1.0f);

    // Next Character Button
    MenuItem NextCharacterButton = CreateMenuItem(iContext, MenuItemMesh,MenuItemMeshEx,
Material1,NumberTextures,Tex,AnimateTextures,TimeDelay,Position,Scale,ObjectShader );

    // Previous Character Button
    Position = new Vector3(0.5f, 1.3f, 4.25f);
    Texture TexPreviousGameButton = new Texture(iContext,R.drawable.previousbutton);
    Tex = new Texture[NumberTextures];
    Tex[0] = TexPreviousGameButton;
    MenuItem PreviousCharacterButton = CreateMenuItem(iContext, MenuItemMesh,MenuItemMeshEx,
Material1, NumberTextures,Tex,AnimateTextures,TimeDelay,Position,Scale,ObjectShader);

    // Enter Button
    Position = new Vector3(0.0f, 0.0f, 4.25f);
    Texture TexEnterButton = new Texture(iContext,R.drawable.enterbutton);
```

```
Tex = new Texture[NumberTextures];
Tex[0] = TexEnterButton;
Scale = new Vector3(3.0f,1.0f,1.0f);
MenuItem EnterButton = CreateMenuItem(iContext, MenuItemMesh, MenuItemMeshEx, Material1,
NumberTextures, Tex, AnimateTextures, TimeDelay, Position, Scale, ObjectShader);

int StartingEntryXPos = 168;
int StartingEntryYPos = 100;
m_HighScoreEntryMenu = new HighScoreEntryMenu(NextCharacterButton, PreviousCharacterButton,
EnterButton, m_CharacterSet, HighScoreEntryMenuImage, StartingEntryXPos, StartingEntryYPos);
}
```

The CreateMenuItem() function creates a new MenuItem object. (See Listing 9-40.)

Listing 9-40. Creating a Menu Item

```
MenuItem CreateMenuItem(Context iContext, Mesh MenuItemMesh, eshEx MenuItemMeshEx, Material
Material1, int NumberTextures, Texture[] Tex, boolean AnimateTextures, float TimeDelay, Vector3
Position, Vector3 Scale, Shader ObjectShader)
{
        MenuItem NewMenuItem = null;
        NewMenuItem = new MenuItem(iContext, MenuItemMesh, MenuItemMeshEx, Tex, Material1,
ObjectShader);
        NewMenuItem.SetAnimateTextures(AnimateTextures, TimeDelay, 0, NumberTextures-1);

        NewMenuItem.m_Orientation.SetPosition(Position);
        NewMenuItem.m_Orientation.SetScale(Scale);
        NewMenuItem.GetObjectPhysics().SetGravity(false);

        return NewMenuItem;
}
```

The CreateMainMenu() function creates the main menu for our game. The function also creates the individual menu items within the main menu, such as the new game menu item, the continue game menu item, the display high score table menu item, and the copyright menu item. (See Listing 9-41.)

Listing 9-41. Creating the Main Menu

```
void CreateMainMenu(Context iContext)
{
        // Create New Game Button
        Shader ObjectShader = new Shader(iContext, R.raw.vsonelight, R.raw.fsonelight);        // ok

        MeshEx MenuItemMeshEx = new MeshEx(8,0,3,5,Cube.CubeData, Cube.CubeDrawOrder);
        Mesh MenuItemMesh = null;

        // Create Material for this object
        Material Material1 = new Material();

        // Create Texture
        int NumberTextures = 1;
        Texture TexNewGameButton = new Texture(iContext,R.drawable.newgamebutton);
```

```
        Texture[] Tex = new Texture[NumberTextures];
        Tex[0] = TexNewGameButton;

        boolean AnimateTextures = false;
        float TimeDelay = 0.0f;

        Vector3 Position = new Vector3(0.0f, 2.5f, 4.25f);
        Vector3 Scale = new Vector3(3.0f,1.0f,1.0f);

        MenuItem NewGameMenuItem = CreateMenuItem(iContext, MenuItemMesh,MenuItemMeshEx,
Material1, NumberTextures, Tex, AnimateTextures, TimeDelay, Position, Scale, ObjectShader);

        // Continue Game
        Position = new Vector3(0.0f, 1.3f, 4.25f);
        Texture TexContinueGameButton = new Texture(iContext, R.drawable.continuegamebutton);
        Tex = new Texture[NumberTextures];
        Tex[0] = TexContinueGameButton;

        MenuItem ContinueGameMenuItem = CreateMenuItem(iContext, MenuItemMesh, MenuItemMeshEx,
Material1, NumberTextures, Tex, AnimateTextures, TimeDelay, Position, Scale, ObjectShader);

        // View High Scores
        Position = new Vector3(0.0f, 0.0f, 4.25f);
        Texture TexHighScoresButton = new Texture(iContext, R.drawable.highscoresbutton);
        Tex = new Texture[NumberTextures];
        Tex[0] = TexHighScoresButton;

        MenuItem HighScoreMenuItem = CreateMenuItem(iContext, MenuItemMesh, MenuItemMeshEx,
Material1, NumberTextures, Tex, AnimateTextures, TimeDelay, Position, Scale, ObjectShader);

        // CopyRight Notice
        Position = new Vector3(0.0f, -1.3f, 4.25f);
        Texture TexCopyrightButton = new Texture(iContext,R.drawable.copyright);
        Tex = new Texture[NumberTextures];
        Tex[0] = TexCopyrightButton;
        Material Material2 = new Material();
        Material2.SetEmissive(0.3f, 0.3f, 0.3f);

        MenuItem CopyrightMenuItem = CreateMenuItem(iContext, MenuItemMesh, MenuItemMeshEx,
Material2, NumberTextures, Tex, AnimateTextures, TimeDelay, Position, Scale, ObjectShader);
        m_MainMenu = new MainMenu(NewGameMenuItem, ContinueGameMenuItem,
HighScoreMenuItem, CopyrightMenuItem);
}
```

The CheckTouch() function has to be modified (see Listing 9-42). The CheckTouch() function is called when the user touches the screen and is used to process user touches.

Listing 9-42. Modifying the CheckTouch() Function

```
if (m_GameState == GameState.MainMenu)
{
        // Reset camera to face main menu
        MenuStates result = m_MainMenu.GetMainMenuStatus(m_TouchX, m_TouchY, m_ViewPortWidth,
m_ViewPortHeight);

        if (result == MenuStates.NewGame)
        {
                ResetGame();
                m_GameState = GameState.ActiveGamePlay;
        }
        else
        if (result == MenuStates.ContinueCurrentGame)
        {
                LoadContinueStatus(MainActivity.SAVE_GAME_HANDLE);
                if (m_CanContinue)
                {
                        LoadGameState(MainActivity.SAVE_GAME_HANDLE);
                }
                else
                {
                        ResetGame();
                }
                m_GameState = GameState.ActiveGamePlay;
        }
        else
        if (result == MenuStates.HighScoreTable)
        {
                m_GameState = GameState.HighScoreTable;
        }
        else
        if (result == MenuStates.Copyright)
                {
                m_GameState = GameState.HighScoreEntry;
                }
        return;
}
else
if (m_GameState == GameState.HighScoreTable)
{
        m_GameState = GameState.MainMenu;
        return;
}
else
if (m_GameState == GameState.HighScoreEntry)
{
        // If User presses finished button from High Score Entry Menu
        EntryMenuStates result = m_HighScoreEntryMenu.GetEntryMenuStatus(m_TouchX, m_TouchY,
m_ViewPortWidth, m_ViewPortHeight);
```

```
        if (result == EntryMenuStates.NextCharacterPressed)
        {
                m_HighScoreEntryMenu.ProcessNextMenuSelection();
        }
        else
        if (result == EntryMenuStates.PreviousCharacterPressed)
                {
                        m_HighScoreEntryMenu.ProcessPreviousMenuSelection();
                }
        else
        if (result == EntryMenuStates.Enter)
        {
                m_HighScoreEntryMenu.ProcessEnterMenuSelection();

                if (m_HighScoreEntryMenu.IsEntryFinished())
                {
                        char[] Initials = m_HighScoreEntryMenu.GetEntry();
                        String StrInitials = new String(Initials);

                        CreateHighScoreEntry(StrInitials, m_Score);

                        m_GameState = GameState.HighScoreTable;
                        m_HighScoreEntryMenu.ResetMenu();
                }
        }
        return;
}
```

For the CheckTouch() function modifications

1. If the main menu is being displayed, get the main menu's status in terms of finding out what menu item has been pressed.

    a. If the new game menu item has been pressed, reset the game and set the game state to ActiveGamePlay.

    b. If the continue current game menu item has been pressed, load the m_CanContinue status. If the status is true, load the previously saved game; otherwise, reset the game. Set the game state to ActiveGamePlay.

    c. If the high score table menu item has been selected, set the game state to HighScoreTable.

    d. If the copyright menu item has been selected, set the game state to HighScoreEntry to activate the high score entry menu. This is a debug button that will be commented out in the final game. In this chapter, we use it to test the high score menu entry system.

2. If the high score table is being displayed, set the game state to the main menu.

3. If the high score entry menu is being displayed, process the menu item being clicked.

    a. If the next character button was clicked, call `ProcessNextMenuSelection()`.

    b. If the previous character button was clicked, call `ProcessPreviousMenuSelection()`.

    c. If the enter button was clicked, call `ProcessEnterMenuSelection()` to process it. If the entry was complete, add the new high score to the high score table and set the game state to display the high score table.

The `onDrawFrame()` function has to be modified to update and render the main menu, the high score table, and the high score entry menu, based on the `m_GameState` variable. (See Listing 9-43.)

Listing 9-43. Modifying the `onDrawFrame()` Function

```
 if (m_GameState == GameState.MainMenu)
{
        m_MainMenu.UpdateMenu(m_Camera);
        m_MainMenu.RenderMenu(m_Camera, m_PointLight, false);
        return;
}
// High Score Table
if (m_GameState == GameState.HighScoreTable)
{
        m_HighScoreTable.UpdateHighScoreTable(m_Camera);
        m_HighScoreTable.RenderHighScoreTable(m_Camera, m_PointLight, false);
         return;
}

// High Score Entry
if (m_GameState == GameState.HighScoreEntry)
{
        m_HighScoreEntryMenu.UpdateHighScoreEntryMenu(m_Camera);
        m_HighScoreEntryMenu.RenderHighScoreEntryMenu(m_Camera, m_PointLight, false);
        return;
}
```

Now, run the program. You should see the following main menu appear. (See Figure 9-1.)

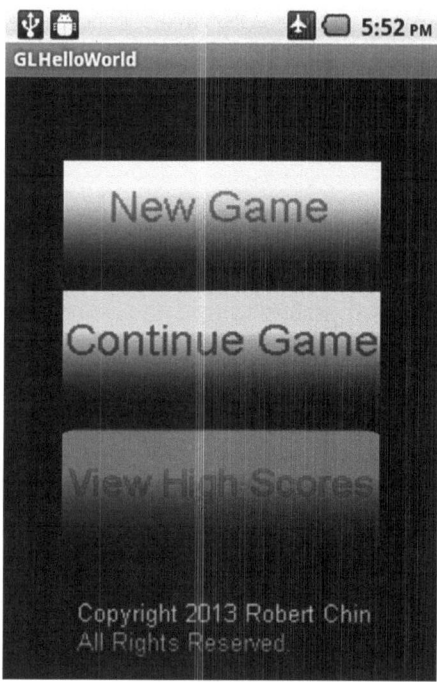

Figure 9-1. *The main menu*

If you click the Copyright button, the high score entry menu should pop up. For this hands-on example, try to enter some high scores manually, with entries containing different values for the score. You can set the score (m_Score) at 93, for example, in the actual source code and then save the file, recompile, run the program, and create a new entry in the high score table. The new entry in the table should be the score you assigned to the m_Score value in the source code. (See Figure 9-2.)

```
private int m_Score = 93;
```

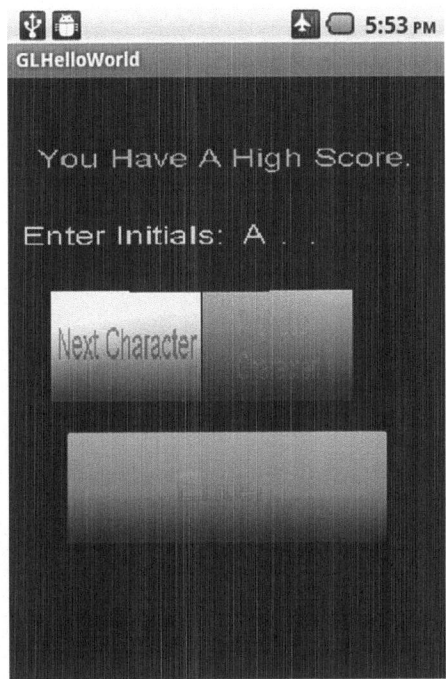

Figure 9-2. Entering a new high score

Enter some initials, and when you are done, you should be taken to the high score table. (See Figure 9-3.)

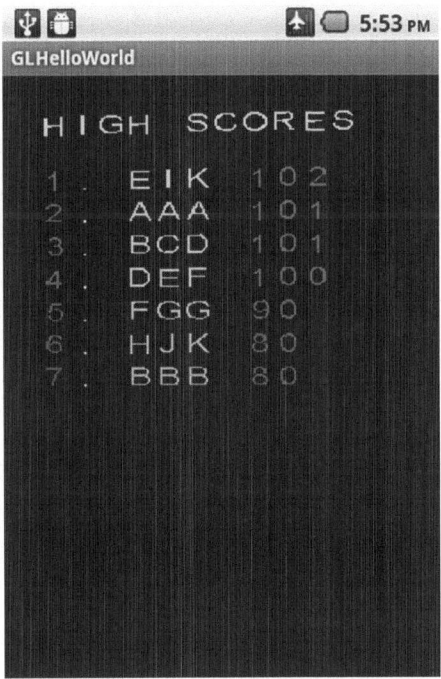

Figure 9-3. The high score table

Summary

In this chapter, I covered the user interfaces for our case study game. I started with a discussion of the main menu, including the Android Java classes and code needed to implement the main menu. Next, I covered the high score table and how to implement it in code. Then, I went over the high score entry menu. Finally, we implemented the main menu, high score table, and high score entry menu in a working demo.

Chapter 10

The Final Drone Grid Game

This chapter covers the final Drone Grid game. I start off by covering classes that are needed to help manage enemy objects in the game. The GamePlayController class, which controls elements such as what types of enemies are added to the playfield, is then discussed. Next, code modifications and new functions that save and load the state of the game are covered. This is followed by a discussion on implementing a game over graphic and game over state code into the Drone Grid game. Finally, a hands-on example is covered that demonstrates the concepts and classes discussed previously in the chapter.

Organizing and Controlling Groups of Enemies

For our final game, we will require some support classes that make the manipulation of our enemy objects easier. The two classes we will need are the ArenaObjectSet class, which controls our arena object enemies, and the TankFleet class, which controls our tank object enemies.

The ArenaObjectSet Class

The ArenaObjectSet class holds a group of arena objects and has functions that manage these objects.

The MAX_ARENA_OBJECTS variable holds the maximum number of arena objects that this set will hold.

```
private int MAX_ARENA_OBJECTS = 25;
```

The m_NumberArenaObjects variable holds the actual number of arena objects in the set.

```
private int m_NumberArenaObjects = 0;
```

The m_ArenaObjectSet variable array holds the set of arena objects that will be used in this class.

```
private ArenaObject3d[] m_ArenaObjectSet = new ArenaObject3d[MAX_ARENA_OBJECTS];
```

The m_Active boolean array holds true for an element if the corresponding arena object in the m_ArenaObjectSet is active and has to be rendered and updated.

```
private boolean[] m_Active = new boolean[MAX_ARENA_OBJECTS];
```

The m_ExplosionMinVelocity variable holds the minimum velocity for particles in an explosion associated with this set of arena objects.

```
private float m_ExplosionMinVelocity = 0.02f;
```

The m_ExplosionMaxVelocity variable holds the maximum velocity for particles in an explosion associated with this set of arena objects.

```
private float m_ExplosionMaxVelocity = 0.4f;
```

The Init() function initializes all the arena objects in this class to null, and the m_Active status is set to false, which means it is inactive and not to be rendered, updated, or processed. (See Listing 10-1.)

Listing 10-1. Initializing the Arena Objects

```
void Init()
{
        for (int i = 0; i < MAX_ARENA_OBJECTS; i++)
        {
                m_ArenaObjectSet[i] = null;
                m_Active[i] = false;
        }
}
```

The ArenaObjectSet constructor initializes the ArenaObject set by calling the Init() function. (See Listing 10-2.)

Listing 10-2. The ArenaObjectSet Constructor

```
ArenaObjectSet(Context iContext)
{
        m_Context = iContext;
        Init();
}
```

The SaveSet() function saves the data for the ArenaObject set. (See Listing 10-3.)

The SaveSet() function does the following:

1. Retrieves a SharedPreferences object for the input parameter Handle by calling the getSharedPreferences() function

2. Retrieves an Editor object by calling edit() on the SharedPreferences object from Step 1

3. Saves the values of the m_Active array to the shared preferences by creating a unique handle consisting of the input parameter Handle, the "Active" keyword, and the index i

4. Saves the arena objects by calling the SaveObjectState() function using a handle composed of the input parameter Handle, the "ArenaObject" keyword, and the index i for all the elements in the m_ArenaObjectSet array, if there is a valid element for that array index slot

5. Saves and commits the changes to the shared preferences

Listing 10-3. Saving the ArenaObject Set

```
void SaveSet(String Handle)
{
        SharedPreferences settings = m_Context.getSharedPreferences(Handle, 0);
        SharedPreferences.Editor editor = settings.edit();

        for (int i = 0; i < MAX_ARENA_OBJECTS; i++)
        {
                // Active Status
                String ActiveHandle = Handle + "Active" + i;
                editor.putBoolean(ActiveHandle, m_Active[i]);

                if (m_ArenaObjectSet[i] != null)
                {
                        String ArenaObjectHandle = Handle + "ArenaObject" + i;
                        m_ArenaObjectSet[i].SaveObjectState(ArenaObjectHandle);
                }
        }
        // Commit the edits!
        editor.commit();
}
```

The LoadSet() function loads in the ArenaObject set data. (See Listing 10-4.)

The LoadSet() function does the following:

1. Retrieves a SharedPreferences object based on the input parameter Handle

2. Reads in all the elements of the m_Active boolean array from the shared preferences using a handle based on the input parameter Handle, "Active" keyword, and the slot index

3. Reads in the saved data by calling the LoadObjectState() function for each element in the m_ArenaObjectSet array that has a valid object

Listing 10-4. Loading the ArenaObject Set

```
void LoadSet(String Handle)
{
        // Restore preferences
        SharedPreferences settings = m_Context.getSharedPreferences(Handle, 0);
```

```
        for (int i = 0; i < MAX_ARENA_OBJECTS; i++)
        {
                // Active Status
                String ActiveHandle = Handle + "Active" + i;
                m_Active[i] = settings.getBoolean(ActiveHandle, false);

                if (m_ArenaObjectSet[i] != null)
                {
                        String ArenaObjectHandle = Handle + "ArenaObject" + i;
                        m_ArenaObjectSet[i].LoadObjectState(ArenaObjectHandle);
                }
        }
}
```

The ResetSet() function resets the entire set of arena objects by changing each object's active state to false and setting the object's visibility to false. (See Listing 10-5.)

Listing 10-5. Resetting the Set

```
void ResetSet()
{
        // Sets all objects to inactive and invisible
        for (int i = 0; i < MAX_ARENA_OBJECTS; i++)
        {
                if (m_ArenaObjectSet[i] != null)
                {
                        m_Active[i] = false;
                        m_ArenaObjectSet[i].SetVisibility(false);
                }
        }
}
```

The NumberActiveArenaObjects() function returns the number of active arena objects in the set. (See Listing 10-6.)

Listing 10-6. Getting the Number of Active Arena Objects

```
int NumberActiveArenaObjects()
{
        int NumberActiveVehicles = 0;
        for (int i = 0; i < MAX_ARENA_OBJECTS; i++)
        {
                if (m_Active[i] == true)
                {
                        NumberActiveVehicles++;
                }
        }
        return NumberActiveVehicles;
}
```

The GetAvailableArenaObject() function (see Listing 10-7) returns an available arena object by

1. Searching through the entire set of arena objects in the m_ArenaObjectSet array and trying to find an ArenaObject3d object that is not null and not active

2. Processing a non-null object by setting the object's visibility to true, setting the object to active by setting the corresponding slot in the m_Active array, and returning the object

3. Returning a null value if no available object is found

Listing 10-7. Getting an Available Arena Object

```
ArenaObject3d GetAvailableArenaObject()
{
        ArenaObject3d temp = null;
        for (int i = 0; i < MAX_ARENA_OBJECTS; i++)
        {
                if (m_ArenaObjectSet[i] != null)
                {
                        if (m_Active[i] == false)
                        {
                                m_ArenaObjectSet[i].SetVisibility(true);
                                m_Active[i] = true;
                                return m_ArenaObjectSet[i];
                        }
                }
        }
        return temp;
}
```

The GetRandomAvailableArenaObject() function (see Listing 10-8) gets a random arena object from the set of arena objects by

1. Creating a random number generator called RandomNumber

2. Creating an array that will hold the indices of available arena objects to choose from for our random selection

3. Building a list of available arena objects and putting the indices into the AvailableObjects array

4. Finding a random arena object based on the output of the random number generator and the list of available arena objects

5. Returning an available object after setting the visibility and active status to true

Listing 10-8. Getting a Random Arena Object

```
ArenaObject3d GetRandomAvailableArenaObject()
{
        ArenaObject3d Obj = null;

        Random RandomNumber = new Random();
        int RandomIndex = 0;

        int AvailableObjectsIndex = 0;
        int[] AvailableObjects = new int[MAX_ARENA_OBJECTS];

        // Build list of available objects
        for (int i = 0; i < MAX_ARENA_OBJECTS; i++)
        {
                if (m_ArenaObjectSet[i] != null)
                {
                        if (m_Active[i] == false)
                        {
                                AvailableObjects[AvailableObjectsIndex] = i;
                                AvailableObjectsIndex++;
                        }
                }
        }

        // If there are Available Objects then choose one at random from the list of available objects
        if (AvailableObjectsIndex > 0)
        {
                // Find Random Object from array of available objects
                RandomIndex = RandomNumber.nextInt(AvailableObjectsIndex);

                int ObjIndex = AvailableObjects[RandomIndex];
                Obj = GetArenaObject(ObjIndex);

                if (Obj != null)
                {
                        Obj.SetVisibility(true);
                        m_Active[ObjIndex] = true;
                }
                else
                {
                        Log.e("ARENAOBJECTSSET", "Random Arena OBJECT = NULL ERROR!!!! ");
                }
        }
        return Obj;
}
```

The AddNewArenaObject() function (see Listing 10-9) adds a new arena object into the set of arena objects by:

1. Finding an empty slot in the m_ArenaObjectSet array

2. If an empty slot is found, setting the new input arena object to invisible; setting the empty slot to point to the new arena object; increasing the number of arena objects in the set by incrementing the m_NumberArenaObjects variable; and returning true

3. Returning false if an empty slot is not found

Listing 10-9. Adding a New Arena Object to the Set

```
boolean AddNewArenaObject(ArenaObject3d ArenaObj)
{
        boolean result = false;
        for (int i = 0; i < MAX_ARENA_OBJECTS; i++)
        {
                if (m_ArenaObjectSet[i] == null)
                {
                        ArenaObj.SetVisibility(false);
                        m_ArenaObjectSet[i] = ArenaObj;
                        m_NumberArenaObjects++;
                        return true;
                }
        }
        return result;
}
```

The SetSoundOnOff() function sets the sound effects for the set of arena objects to on or off, based on the input parameter Value. For each valid arena object, the SetSFXOnOff() function is called. (See Listing 10-10.)

Listing 10-10. Setting the Sound Effects for the ArenaObject Set

```
void SetSoundOnOff(boolean Value)
{
        for (int i = 0; i < MAX_ARENA_OBJECTS; i++)
        {
                if (m_ArenaObjectSet[i] != null)
                {
                        m_ArenaObjectSet[i].SetSFXOnOff(Value);
                }
        }
}
```

The ProcessCollisionsWeapon() function (see Listing 10-11) processes the collision between a weapon's ammunition and arena objects in the set by

1. Determining the active status for each element in the ArenaObject set

2. If the element is active, checking for collisions between that arena object and the ammunition from the input iWeapon weapon

3.   If there is a collision, processing the collision by calling `ApplyLinearImpulse()` to apply action reaction forces to the arena object and the weapon's ammunition

4.   Starting the explosion graphic and sound associated with the arena object involved in the collision

5.   Processing the damage to the arena object by calling the `TakeDamage()` function

6.   Checking the health of the arena object and, if it is 0 or less, updating the total kill value, which is the total value of all the objects destroyed by the weapon's ammunition, and destroying the arena object by setting its active status to false and its visibility to false

7.   Returning the total kill value

Listing 10-11. Collision Detection Between a Weapon's Ammunition and the ArenaObject Set

```
int ProcessCollisionsWeapon(Weapon iWeapon)
{
        int TotalKillValue = 0;

        for (int i = 0; i < MAX_ARENA_OBJECTS; i++)
        {
                if ((m_ArenaObjectSet[i] != null) && (m_Active[i] == true))
                {
                        Object3d CollisionObj = iWeapon.CheckAmmoCollision(m_ArenaObjectSet[i]);
                        if (CollisionObj != null)
                        {
                                CollisionObj.ApplyLinearImpulse(m_ArenaObjectSet[i]);
                                SphericalPolygonExplosion Exp = m_ArenaObjectSet[i].GetExplosion(0);
                                Exp.StartExplosion(m_ArenaObjectSet[i].m_Orientation.GetPosition(),
m_ExplosionMaxVelocity, m_ExplosionMinVelocity);
                                m_ArenaObjectSet[i].PlayExplosionSFX();

                                // Process Damage
                                m_ArenaObjectSet[i].TakeDamage(CollisionObj);
                                int Health = m_ArenaObjectSet[i].GetObjectStats().GetHealth();
                                if (Health <= 0)
                                {
                                        int KillValue =
m_ArenaObjectSet[i].GetObjectStats().GetKillValue();
                                        TotalKillValue = TotalKillValue + KillValue;
                                        m_Active[i] = false;
                                        m_ArenaObjectSet[i].SetVisibility(false);
                                }
                        }
                }
        }
        return TotalKillValue;
}
```

The AddArenaObjectsToGravityGrid() function adds all the arena objects in the set that are active to the gravity grid iGrid that is the input parameter. (See Listing 10-12.)

Listing 10-12. Adding the Arena Objects in the Set to the Gravity Grid

```
void AddArenaObjectsToGravityGrid(GravityGridEx iGrid)
{
        for (int i = 0; i < MAX_ARENA_OBJECTS; i++)
        {
                if ((m_ArenaObjectSet[i] != null) && (m_Active[i] == true))
                {
                        // Add Mass of AirVehicle to grid
                        iGrid.AddMass(m_ArenaObjectSet[i]);
                }
        }
}
```

The GetArenaObject() function returns the arena object from the set of arena objects in m_ArenaObjectSet that contains the input ID object id that is currently active. (See Listing 10-13.)

Listing 10-13. Getting an Arena Object Based on Its ID

```
ArenaObject3d GetArenaObject(String ID)
{
        ArenaObject3d temp = null;
        for (int i = 0; i < MAX_ARENA_OBJECTS; i++)
        {
                if (m_ArenaObjectSet[i] != null)
                {
                        if ((m_Active[i]== true) && (m_ArenaObjectSet[i].GetArenaObjectID() == ID))
                        {
                                return m_ArenaObjectSet[i];
                        }
                }
        }
        return temp;
}
```

The ProcessCollisionWithObject() function (see Listing 10-14) processes the collisions between an input object Obj and the ArenaObject set by doing the following:

1. For each arena object in the set that is active, it checks for a collision with the input Obj object.

2. If there is a collision, then it processes the collision by calling ApplyLinearImpulse() to apply the action and reaction forces on both colliding objects. It starts the explosion graphics and plays the explosion sound effect for the arena object that is colliding. It plays the explosion graphic for input object Obj that is involved in the collision. It processes the damage to the arena object and the object Obj from the collision. If the health

of the arena object is less than or equal to 0, it adds the kill value of the arena object to the total kill value. The arena object is destroyed by setting the active status to false and the visible status to false.

3. The total kill value that equals the total of all the kill values of the destroyed arena objects that have just been destroyed by a collision with the input object Obj is returned.

Listing 10-14. Processing Collisions Between an Object and the ArenaObject Set

```
int ProcessCollisionWithObject(Object3d Obj)
{
        int TotalKillValue = 0;
        for (int i = 0; i < MAX_ARENA_OBJECTS; i++)
        {
                if ((m_ArenaObjectSet[i] != null) && (m_Active[i] == true))
                {
                        Physics.CollisionStatus result = Obj.CheckCollision(m_ArenaObjectSet[i]);
                        if ((result == Physics.CollisionStatus.COLLISION) ||
                        (result == Physics.CollisionStatus.PENETRATING_COLLISION))
                        {
                                // Process Collision
                                Obj.ApplyLinearImpulse(m_ArenaObjectSet[i]);
                                // Arena Object Explosion
                                SphericalPolygonExplosion Exp = m_ArenaObjectSet[i].GetExplosion(0);
                                if (Exp != null)
                                {
                                        Exp.StartExplosion(m_ArenaObjectSet[i].m_Orientation.
GetPosition(), m_ExplosionMaxVelocity,  m_ExplosionMinVelocity);
                                        m_ArenaObjectSet[i].PlayExplosionSFX();
                                }

                                // Pyramid Explosion
                                Exp = Obj.GetExplosion(0);
                                if (Exp != null)
                                {
                                        Exp.StartExplosion(Obj.m_Orientation.GetPosition(),
m_ExplosionMaxVelocity,   m_ExplosionMinVelocity);
                                }

                                // Process Damage
                                Obj.TakeDamage(m_ArenaObjectSet[i]);
                                m_ArenaObjectSet[i].TakeDamage(Obj);
                                int Health = m_ArenaObjectSet[i].GetObjectStats().GetHealth();
                                if (Health <= 0)
                                {
                                        int KillValue = m_ArenaObjectSet[i].GetObjectStats().
GetKillValue();

                                        TotalKillValue = TotalKillValue + KillValue;
                                        m_Active[i] = false;
                                        m_ArenaObjectSet[i].SetVisibility(false);
                                }
```

```
                    }
                }
            }
        }
        return TotalKillValue;
}
```

The RenderArenaObjects() function renders all the arena objects in the set m_ArenaObjectSet array that are valid objects that are not null. (See Listing 10-15.)

Listing 10-15. Rendering the ArenaObject Set

```
void RenderArenaObjects(Camera Cam, PointLight Light, boolean DebugOn)
{
        for (int i = 0; i < MAX_ARENA_OBJECTS; i++)
        {
                if (m_ArenaObjectSet[i] != null)
                {
                        m_ArenaObjectSet[i].RenderArenaObject(Cam, Light);
                }
        }
}
```

The UpdateArenaObjects() function updates all the valid arena objects in the set. (See Listing 10-16.)

Listing 10-16. Updating the ArenaObject Set

```
void UpdateArenaObjects()
{
        for (int i = 0; i < MAX_ARENA_OBJECTS; i++)
        {
                if (m_ArenaObjectSet[i] != null)
                {
                        m_ArenaObjectSet[i].UpdateArenaObject();
                }
        }

}
```

The TankFleet Class

The TankFleet class holds a group of tank objects and contains functions to help manage and manipulate the fleet of tanks. There are many similarities between the TankFleet class and the ArenaObjectSet class, so I will only cover functions in the TankFleet class here that are substantially different or important.

The MAX_TANKS variable holds the maximum number of tanks that can be held in this set.

```
private int MAX_TANKS = 5;
```

The m_TankFleet array holds the tank objects for this set.

```
private Tank[] m_TankFleet = new Tank[MAX_TANKS];
```

The m_InService variable array holds a value of true for an element if the corresponding tank in the m_TankFleet array is active and needs to be updated and rendered.

```
private boolean[] m_InService = new boolean[MAX_TANKS];
```

The TankFleet( ) constructor initializes the tank fleet and is similar to the way the ArenaObjectSet class was initialized. (See Listing 10-17.)

Listing 10-17. The TankFleet Constructor

```
TankFleet(Context iContext)
{
        m_Context = iContext;
        Init();
}
```

The ResetSet() function (see Listing 10-18) resets the tank fleet by

1. Setting existing tanks to inactive, m_InService = false

2. Setting the main body and the turret to invisible

3. Resetting the tank's finite state machine and weapons

Listing 10-18. Resetting the Tank Set

```
void ResetSet()
{
        // Sets all objects to inactive and invisible
        for (int i = 0; i < MAX_TANKS; i++)
        {
                if (m_TankFleet[i] != null)
                {
                        m_InService[i] = false;
                        m_TankFleet[i].GetMainBody().SetVisibility(false);
                        m_TankFleet[i].GetTurret().SetVisibility(false);
                        m_TankFleet[i].Reset();
                }
        }
}
```

The SetSoundOnOff() function turns on/off the sound effects for the tank fleet. (See Listing 10-19.) The sound effects are turned on or off for the components of the tank, which are

1. The tank's main body

2. The tank's turret

3. The tank's weapons

Listing 10-19. Setting the Sound Effects for the Tanks

```
void SetSoundOnOff(boolean Value)
{
      for (int i = 0; i < MAX_TANKS; i++)
      {
            if (m_TankFleet[i] != null)
            {
                  m_TankFleet[i].GetMainBody().SetSFXOnOff(Value);
                  m_TankFleet[i].GetTurret().SetSFXOnOff(Value);

                  int NumberWeapons = m_TankFleet[i].GetNumberWeapons();
                  for (int j = 0; j < NumberWeapons; j++)
                  {
                        m_TankFleet[i].GetWeapon(j).TurnOnOffSFX(Value);
                  }
            }
      }
}
```

The AddTankFleetToGravityGrid() function adds all the active tanks in the fleet to the input parameter gravity grid iGrid. More specifically, the tank's main body and the tank's active ammunition from its weapons are all added to the gravity grid. (See Listing 10-20.)

Listing 10-20. Adding the Tank Fleet to the Gravity Grid

```
// Add in all the Air vehicles in the fleet to the gravity grid
void AddTankFleetToGravityGrid(GravityGridEx iGrid)
{
      Object3d[] Masses = new Object3d[50];
      int NumberMasses = 0;

      for (int i = 0; i < MAX_TANKS; i++)
      {
            if ((m_TankFleet[i] != null) && (m_InService[i] == true))
            {
                  // Add Mass of AirVehicle to grid
                  iGrid.AddMass(m_TankFleet[i].GetMainBody());

                  // Add Weapons Fire from AirVehicle to grid
                  int NumberWeapons = m_TankFleet[i].GetNumberWeapons();
                  for (int j = 0; j < NumberWeapons; j++)
                  {
                        NumberMasses = m_TankFleet[i].GetWeapon(j).GetActiveAmmo(0, Masses);
                        iGrid.AddMasses(NumberMasses, Masses);
                  }
            }
      }
}
```

The ProcessWeaponAmmoCollisionObject() function processes the collision between the live ammunition from the tanks in the tank fleet and the input object Obj. (See Listing 10-21.)

For each tank in the tank fleet that is active

1. Check for a collision between the ammunition from the tank's weapons and the object Obj.

2. If there is a collision, apply a linear force to the ammunition and the object by calling the `ApplyLinearImpulse()` function

3. Process the damage to the object by calling `TakeDamage()`

4. Start the explosion graphic for the object by calling `StartExplosion()`

Listing 10-21. Collision Processing Between an Object and a Tank's Live Ammunition

```
boolean ProcessWeaponAmmoCollisionObject(Object3d Obj)
{
        Object3d CollisionObj = null;
        boolean hitresult = false;

        for (int i = 0; i < MAX_TANKS; i++)
        {
                if ((m_TankFleet[i] != null) && (m_InService[i] == true))
                {
                        int NumberWeapons = m_TankFleet[i].GetNumberWeapons();

                        for (int j=0; j < NumberWeapons; j++)
                        {
                                CollisionObj = m_TankFleet[i].GetWeapon(j).CheckAmmoCollision(Obj);
                                if (CollisionObj != null)
                                {
                                        hitresult = true;
                                        CollisionObj.ApplyLinearImpulse(Obj);

                                        //Process Damage
                                        Obj.TakeDamage(CollisionObj);

                                        // Obj Explosion
                                        SphericalPolygonExplosion Exp = Obj.GetExplosion(0);
                                        if (Exp != null)
                                        {
                                          Exp.StartExplosion(Obj.m_Orientation.GetPosition(),
m_VehicleExplosionMaxVelocity, m_VehicleExplosionMinVelocity);
                                        }
                                }
                        }
                }
        }
        return hitresult;
}
```

The GamePlayController Class

The GamePlayController class is used to control the game play in terms of how many enemies of each type are allowed on the gravity grid, the location they first appear at, and at what rate they appear.

The m_RandNumber variable holds the random number generator.

```
private Random m_RandNumber = new Random();
```

The m_ArenaObjectsSet variable holds the set of arena objects that will be used on the playfield.

```
private ArenaObjectSet m_ArenaObjectsSet;
```

The m_TankFleet variable holds the set of tanks to be used on the playfield.

```
private TankFleet m_TankFleet;
```

The m_Grid variable holds the gravity grid that marks the boundaries of the playfield.

```
private GravityGridEx  m_Grid;
```

The DROP_HEIGHT variable indicates the height at which the arena objects will be dropped onto the playfield.

```
private float DROP_HEIGHT = 13;
```

The m_TimeDeltaAddArenaObject variable holds the number of milliseconds between adding new arena objects.

```
private long m_TimeDeltaAddArenaObject = 1000 * 15;
```

The m_TimeLastArenaObjectAdded variable holds the time that the last arena object was added to the playfield.

```
private long m_TimeLastArenaObjectAdded = 0;
```

The m_MinArenaObjectsOnPlayField variable holds the minimum number of arena objects that must be on the playfield.

```
private int m_MinArenaObjectsOnPlayField = 1;
```

The m_MaxSpeedArenaObjects variable holds the maximum speed for the arena objects.

```
private float m_MaxSpeedArenaObjects = 0.1f;
```

The m_TimeDeltaAddTank variable holds the number of milliseconds between adding new tanks onto the playfield.

```
private long m_TimeDeltaAddTank = 1000 * 25;
```

The m_TimeLastTankOnGrid variable holds the time that a tank enemy was last added to the playfield.

```
private long m_TimeLastTankOnGrid = 0;
```

The m_MaxTanksOnPlayField variable holds the maximum number of tanks allowed on the playfield at one time.

```
private int m_MaxTanksOnPlayField = 2;
```

The m_NumberTankRoutes variable holds the total number of tank routes available to select from.

```
private int m_NumberTankRoutes = 0;
```

The m_TankRouteIndex variable holds the current index into the available tank routes, which is the m_TankRoutes array variable.

```
private int m_TankRouteIndex = 0;
```

The m_TankRoutes array variable holds the tank routes consisting of groups of waypoints.

```
private Route[] m_TankRoutes = null;
```

The GamePlayController() constructor initializes the GamePlayController class by setting key class member variables, such as the arena objects set, the tank fleet, the gravity grid, and the available tank routes. (See Listing 10-22.)

Listing 10-22. GamePlayController Constructor

```
GamePlayController(Context iContext,ArenaObjectSet  ArenaObjectsSet,TankFleet TankFleet,
GravityGridEx   Grid,int NumberTankRoutes,Route[] TankRoutes)
{
        m_Context = iContext;

        m_ArenaObjectsSet = ArenaObjectsSet;
        m_TankFleet = TankFleet;
        m_Grid = Grid;

        m_NumberTankRoutes = NumberTankRoutes;
        m_TankRoutes = TankRoutes;
}
```

The GenerateRandomGridLocation() generates and returns a random location on the gravity grid. (See Listing 10-23.)

The function does the following:

1. Gets the minimum x boundary for the gravity grid

2. Gets the maximum x boundary for the gravity grid

3. Finds the difference between the maximum x boundary and the minimum x boundary

4. Finds a random offset position based on the difference found in step 3 multiplied by a random number that ranges from 0 to 1

5. Calculates the final x position based on the minimum x value added with the randomly generated x offset value

6. Repeats steps 1 through 5 for the z axis, in order to find a random z coordinate

7. Returns a random location vector consisting of the random x and z values calculated in the previous steps and the DROP_HEIGHT value as the y value

Listing 10-23. Generating a Random Grid Location

```
Vector3 GenerateRandomGridLocation()
{
        Vector3 Location = new Vector3(0,0,0);

        // Get Random X
        float MinX = m_Grid.GetXMinBoundary();
        float MaxX = m_Grid.GetXMaxBoundary();
        float DiffX = MaxX - MinX;
        float RandomXOffset = DiffX * m_RandNumber.nextFloat(); // DiffX * (Number from 0-1);
        float PosX = MinX + RandomXOffset;

        // Get Random Z
        float MinZ = m_Grid.GetZMinBoundary();
        float MaxZ = m_Grid.GetZMaxBoundary();
        float DiffZ = MaxZ - MinZ;
        float RandomZOffset = DiffZ * m_RandNumber.nextFloat(); // DiffX * (Number from 0-1);
        float PosZ = MinZ + RandomZOffset;

        // Y is 0 for Ground Level for Playfield
        float PosY = DROP_HEIGHT;

        // Set Random Location
        Location.Set(PosX, PosY, PosZ);

        return Location;
}
```

The GenerateGridLocationRestricted() function generates a random grid location for dropping an enemy within the boundaries of Min and Max. (See Listing 10-24.)

The function does the following:

1. The GenerateRandomGridLocation() function is called to create a random location within the grid.

2. Then, the location is limited or clamped to the maximum location value by taking the lesser value between Max and the randomly generated location from step 1.

3.   Then the location is clamped to the minimum location value in Min by taking the greater value between Min and the clamped location from step 2.

4.   The final clamped location from step 3 is returned.

Listing 10-24. Generating a Random Grid Location Within Boundaries

```
Vector3 GenerateGridLocationRestricted(Vector3 Max, Vector3 Min)
{
        Vector3 ClampedLocation = new Vector3(0,0,0);
        Vector3 OriginalLocation = null;

        OriginalLocation = GenerateRandomGridLocation();

        ClampedLocation.x = Math.min(OriginalLocation.x, Max.x);
        ClampedLocation.y = Math.min(OriginalLocation.y, Max.y);
        ClampedLocation.z = Math.min(OriginalLocation.z, Max.z);

        ClampedLocation.x = Math.max(ClampedLocation.x, Min.x);
        ClampedLocation.y = Math.max(ClampedLocation.y, Min.y);
        ClampedLocation.z = Math.max(ClampedLocation.z, Min.z);

        return ClampedLocation;
}
```

The GenerateRandomVelocityArenaObjects() function generates a random velocity for an arena object on the xz plane. (See Listing 10-25.)

The function does the following:

1.   Generates a random speed along the x axis by multiplying the maximum speed for an arena object by a randomly generated number within the range of 0 to 1

2.   Generates a random speed along the z axis by multiplying the maximum speed for an arena object by a randomly generated number within the range of 0 to 1

3.   Creates the final velocity by using the x and z values from steps 1 and 2 with a y value of 0

4.   Returns the final random velocity

Listing 10-25. Generating a Random Velocity

```
Vector3 GenerateRandomVelocityArenaObjects()
{
        Vector3 Velocity = new Vector3(0,0,0);

        float VelX = m_MaxSpeedArenaObjects * m_RandNumber.nextFloat();
        float VelZ = m_MaxSpeedArenaObjects * m_RandNumber.nextFloat();
```

```
        Velocity.Set(VelX, 0, VelZ);
        return Velocity;
}
```

The AddNewArenaObject() function adds an arena object to the playfield. (See Listing 10-26.)

The function does the following:

1. Tries to retrieve a new available arena object by calling the GetRandomAvailableArenaObject() function

2. If an arena object is available, it

    a. Sets its visibility to true and its health to 100

    b. Creates a location vector called Max that holds the maximum location of the arena object along the x and z axes

    c. Creates a location vector called Min that holds the minimum location of the arena object along the x and z axes

    d. Calls the GenerateGridLocationRestricted() function with the Max and Min locations to retrieve a random location within the bounds of Max and Min

    e. Sets the position of the new arena object to the location generated from the previous step

3. Returns true if a new arena object has been added to the game and false otherwise

Listing 10-26. Adding a New Arena Object

```
boolean AddNewArenaObject()
{
        boolean result = false;
        ArenaObject3d AO = m_ArenaObjectsSet.GetRandomAvailableArenaObject();
        if (AO != null)
        {
                // Respawn
                AO.SetVisibility(true);
                AO.GetObjectStats().SetHealth(100);

                Vector3 Max = new Vector3(m_Grid.GetXMaxBoundary(), DROP_HEIGHT, -5.0f);
                Vector3 Min = new Vector3(m_Grid.GetXMinBoundary(), DROP_HEIGHT, m_Grid.
GetZMinBoundary());

                Vector3 Position = GenerateGridLocationRestricted(Max, Min);
                AO.m_Orientation.GetPosition().Set(Position.x, Position.y, Position.z);
                result = true;
        }
        return result;
}
```

The CreatePatrolAttackTankCommand() function creates a new patrol/attack tank command and returns it. (See Listing 10-27.)

The key data fields for this command are

1. The Command variable, which is set to AIVehicleCommand.Patrol

2. The DeltaAmount variable, which is set to the number of rounds for the tank to fire in one burst

3. The DeltaIncrement variable, which is set to the time delay in milliseconds between the tank's bursts of fire

Listing 10-27. Creating a Patrol/Attack Tank Command

```
VehicleCommand CreatePatrolAttackTankCommand(AIVehicleObjectsAffected ObjectsAffected, int
NumberWayPoints, Vector3[] WayPoints, Vector3 Target, Object3d TargetObj, int NumberRoundToFire,int
FiringDelay)
{
        VehicleCommand TankCommand = null;
        AIVehicleCommand Command = AIVehicleCommand.Patrol;

        int     NumberObjectsAffected = 0;
        int     DeltaAmount = NumberRoundToFire;
        int     DeltaIncrement = FiringDelay;

        int     MaxValue = 0;
        int     MinValue = 0;

        TankCommand = new VehicleCommand(m_Context,Command, ObjectsAffected, NumberObjectsAffected,
DeltaAmount, DeltaIncrement,MaxValue, MinValue, NumberWayPoints,WayPoints, Target, TargetObj);
        return TankCommand;
}
```

The SetTankOrder() function creates a new tank patrol/attack order and gives the order to the tank. (See Listing 10-28.)

The SetTankOrder() function does the following:

1. Sets the tank route index to cycle through all the available routes

2. Retrieves the selected tank route

3. Retrieves the tank's waypoints from the route

4. Retrieves the number of waypoints from the route

5. Sets up the tank's patrol/attack command to fire three-round bursts of ammunition at the target located at the origin every five seconds

6. Creates the tank's patrol/attack command and sets this command for the input TankVehicle

Listing 10-28. Setting the Patrol/Attack Tank Order

```
void SetTankOrder(Tank TankVehicle)
{
        // Set Tank Route Index to cycle through all available routes
        m_TankRouteIndex++;
        if (m_TankRouteIndex >= m_NumberTankRoutes)
        {
                m_TankRouteIndex = 0;
        }

        // Set Patrol Order
        Route SelectedRoute = m_TankRoutes[m_TankRouteIndex];
        Vector3[] WayPoints = SelectedRoute.GetWayPoints();
        int NumberWayPoints = SelectedRoute.GetNumberWayPoints();

        AIVehicleObjectsAffected ObjectsAffected = AIVehicleObjectsAffected.PrimaryWeapon;
        Vector3 Target = new Vector3(0,0,0);
        Object3d TargetObj = null;
        int NumberRoundToFire = 3;
        int FiringDelay = 5000;

        VehicleCommand Command = CreatePatrolAttackTankCommand(ObjectsAffected, NumberWayPoints,
WayPoints, Target, TargetObj,NumberRoundToFire,FiringDelay);
        TankVehicle.GetDriver().SetOrder(Command);
}
```

The AddNewTank() function adds a tank object to the playfield. (See Listing 10-29.)

The function does the following:

1. Tries to retrieve a new available tank object by calling the GetAvailableTank() function.

2. If a tank object was available, it

    a. Resets the tank and its health to 100

    b. Creates a location vector called Max that holds the maximum location of the tank object along the x and z axes

    c. Creates a location vector called Min that holds the minimum location of the tank object along the x and z axes

    d. Calls GenerateGridLocationRestricted() with the Max and Min locations to retrieve a random location within the bounds of Max and Min

    e. Sets the position of the new tank object to the location generated from the previous step

    f. Creates and sets the tank object's order by calling SetTankOrder()

3. Returns true if a new tank object has been added to the game and false otherwise

Listing 10-29. Adding a New Tank to the Playfield

```
boolean AddNewTank()
{
        boolean result = false;
        Tank TankVehicle = m_TankFleet.GetAvailableTank();
        if (TankVehicle != null)
        {
                TankVehicle.Reset();
                TankVehicle.GetMainBody().GetObjectStats().SetHealth(100);

                // Set Position
                Vector3 Max = new Vector3(m_Grid.GetXMaxBoundary(), DROP_HEIGHT, -5.0f);
                Vector3 Min = new Vector3(m_Grid.GetXMinBoundary(), DROP_HEIGHT, m_Grid.
GetZMinBoundary());

                Vector3 Position = GenerateGridLocationRestricted(Max, Min);
                TankVehicle.GetMainBody().m_Orientation.GetPosition().Set( Position.x,
Position.y, Position.z);

                // Set Command
                SetTankOrder(TankVehicle);
                result = true;
        }
        return result;
}
```

The UpdateArenaObjects() function adds more arena objects to the playfield, if needed. (See Listing 10-30.)

The UpdateArenaObjects() function does the following:

1. If there are fewer arena objects on the playfield than the minimum number, it creates a new arena object by calling AddNewArenaObject().

2. If there are enough arena objects on the playfield, it checks to see if the elapsed time since adding the last arena object is greater or equal to m_TimeDeltaAddArenaObject. If it is, it adds another object by calling AddNewArenaObject().

Listing 10-30. Updating the Arena Objects

```
void UpdateArenaObjects(long CurrentTime)
{
        // Check to see if need to add in more Arena Objects
        int NumberObjects = m_ArenaObjectsSet.NumberActiveArenaObjects();

        if (NumberObjects < m_MinArenaObjectsOnPlayField)
        {
                // Add another object to meet minimum
                boolean result = AddNewArenaObject();
```

```
                if (result == true)
                {
                        m_TimeLastArenaObjectAdded = System.currentTimeMillis();
                }
        }
        else
        {
                // Check to see if enough time has elapsed to add in another object.
                long ElapsedTime = CurrentTime - m_TimeLastArenaObjectAdded;
                if (ElapsedTime >= m_TimeDeltaAddArenaObject)
                {
                        // Add New Arena Object
                        boolean result = AddNewArenaObject();
                        if (result == true)
                        {
                                m_TimeLastArenaObjectAdded = System.currentTimeMillis();
                        }
                }
        }
}
```

The UpdateTanks() function adds a new tank to the playfield by calling AddNewTank() if the current number of tanks is less than m_MaxTanksOnPlayField and the elapsed time since the last tank was added is greater than m_TimeDeltaAddTank. (See Listing 10-31.)

Listing 10-31. Updating the Tanks

```
void UpdateTanks(long CurrentTime)
{
        int NumberTanks = m_TankFleet.NumberActiveVehicles();
        long ElapsedTime = CurrentTime - m_TimeLastTankOnGrid;

        if ((NumberTanks < m_MaxTanksOnPlayField)&&
                (ElapsedTime > m_TimeDeltaAddTank))
        {
                // Add New Tank
                boolean result = AddNewTank();
                if (result == true)
                {
                        m_TimeLastTankOnGrid = System.currentTimeMillis();
                }
        }
}
```

The UpdateController() function updates the number of arena objects on the playfield, if needed, by calling UpdateArenaObjects(), and updates the number of tank objects on the playfield, if needed, by calling UpdateTanks(). (See Listing 10-32.)

Listing 10-32. Updating the GamePlay Controller

```
void UpdateController(long CurrentTime)
{
        UpdateArenaObjects(CurrentTime);
        UpdateTanks(CurrentTime);
}
```

Saving and Loading the Game State

In order to save the state of the game and restore this state, new code has to be added to the MainActivity class and the MyGLRenderer class.

Modifying the MainActivity Class

For the MainActivity class, new code is added to the onPause() function that calls the SaveGameState() function in the MyGLRenderer class when the Android game is paused. (See Listing 10-33.)

Listing 10-33. Modifying the onPause() Function

```
@Override
protected void onPause()
{
        super.onPause();
        m_GLView.onPause();

        // Save State
        m_GLView.CustomGLRenderer.SaveGameState(SAVE_GAME_HANDLE);
}
```

Modifying the MyGLRenderer Class

The MyGLRenderer class has to be modified by adding functions that save and load in the game state.

The SaveGameState() function saves the state by saving key game variables if the game play is currently active (GameState.ActiveGamePlay). (See Listing 10-34.)

The key elements of the game that are saved are the

1. Player's score

2. Player's health

3. The m_CanContinue variable, which is true if there is a previously saved game to load and then to continue from

4. Camera's state

5. Arena objects

6. Tank objects

Listing 10-34. Saving the Game State

```
void SaveGameState(String Handle)
{
        // Only save game state when game is active and being played not at
        // menu or high score table etc.
        if (m_GameState != GameState.ActiveGamePlay)
        {
                return;
        }

        // Save Player's Score
        SharedPreferences settings = m_Context.getSharedPreferences(Handle, 0);
        SharedPreferences.Editor editor = settings.edit();

        // Player's Score
        editor.putInt("Score", m_Score);

        // Player's Health
        editor.putInt("Health", m_Pyramid.GetObjectStats().GetHealth());

        // Can Continue Game
        editor.putBoolean("CanContinue", m_CanContinue);

        // Commit the edits!
        editor.commit();

        // Camera
        m_Camera.SaveCameraState("Camera");

        // Arena Objects Set
        m_ArenaObjectsSet.SaveSet(ARENA_OBJECTS_HANDLE);

        // Tank Fleet
        m_TankFleet.SaveSet(TANK_FLEET_HANDLE);
}
```

The LoadGameState() function loads the data that was saved from the SaveGameState() function. (See Listing 10-35.)

Listing 10-35. Loading the Game State

```
void LoadGameState(String Handle)
{
        // Load game state of last game that was interrupted during play
        // Restore preferences
        SharedPreferences settings = m_Context.getSharedPreferences(Handle, 0);

        // Load In Player Score
            m_Score = settings.getInt("Score", 0);
```

```
        // Load in Player's Health
        int Health = settings.getInt("Health", 100);
        m_Pyramid.GetObjectStats().SetHealth(Health);

        // Can Continue
        m_CanContinue = settings.getBoolean("CanContinue", false);

        // Camera
        m_Camera.LoadCameraState("Camera");

        // Arena Objects Set
        m_ArenaObjectsSet.LoadSet(ARENA_OBJECTS_HANDLE);

        // Tank Fleet
        m_TankFleet.LoadSet(TANK_FLEET_HANDLE);
}
```

Adding in the Game Over Game State

One of the final features we must add to the final game is a game over message. We will have to modify the MyGLRenderer class to add new code to handle the game over graphics and game logic.

The m_GameOverBillBoard variable holds the graphic that tells the player that the game is over.

```
private BillBoard m_GameOverBillBoard;
```

The m_GameOverPauseTime variable holds the minimum time for displaying the game over graphic before user input is to be processed for continuing the game.

```
private long m_GameOverPauseTime = 1000;
```

The m_GameOverStartTime variable holds the time the game ended.

```
private long m_GameOverStartTime;
```

The CreateGameOverBillBoard() function creates the game over billboard that contains the game over graphic that is displayed when the player's game ends. The billboard is actually created by calling the function CreateInitBillBoard(). (See Listing 10-36.)

Listing 10-36. Creating the Game Over BillBoard

```
void CreateGameOverBillBoard(Context iContext)
{
// Put Game over Billboard in front of camera
        int TextureResourceID = R.drawable.gameover;

        Vector3 Position= new Vector3(0,0,0);
        Vector3 Scale   = new Vector3(1 , 0.5f, 0.5f);

        m_GameOverBillBoard = CreateInitBillBoard(iContext,TextureResourceID, Position, Scale);
}
```

The `UpdateGameOverBillBoard()` function calculates and positions the game over billboard in front of the camera at a `DistanceToBillBoard` distance to the camera. (See Listing 10-37.)

Listing 10-37. Updating the Game Over BillBoard

```
void UpdateGameOverBillBoard()
{
        Vector3 TempVec = new Vector3(0,0,0);
        float DistanceToBillBoard = 5;
        TempVec.Set(m_Camera.GetOrientation().GetForwardWorldCoords().x, m_Camera.GetOrientation().
GetForwardWorldCoords().y, m_Camera.GetOrientation().GetForwardWorldCoords().z);
        TempVec.Multiply(DistanceToBillBoard);
        Vector3 Position = Vector3.Add(m_Camera.GetOrientation().GetPosition(), TempVec);
        m_GameOverBillBoard.m_Orientation.SetPosition(Position);
}
```

The `IsNewHighScore()` function returns true, which means there will be a new entry in the high score table if the player's score is greater than the lowest score in the high score table or if the player's score is greater than zero and there is at least one blank slot in the top ten scores in the high score table. The latter case handles the situation where the player's score is equal to or less than the lowest score currently in the table but there are blank entries left in the top ten scores in the high score table. (See Listing 10-38.)

Listing 10-38. Testing If the Player Has a New High Score

```
boolean IsNewHighScore()
{
        boolean result = false;
        int LowestScore = m_HighScoreTable.GetLowestScore();
        int MaxScores = m_HighScoreTable.MaxNumberHighScores();
        int NumberValidScores = m_HighScoreTable.NumberValidHighScores();

        boolean SlotAvailable = false;
        if (NumberValidScores < MaxScores)
        {
                SlotAvailable = true;
        }
        if ((m_Score > LowestScore) ||
            ((m_Score > 0) && SlotAvailable))
        {
                result = true;
        }
        return result;
}
```

The `SaveContinueStatus()` function saves the `m_CanContinue` variable that is true if there is a previously saved game that can be loaded and then continued. (See Listing 10-39.)

Listing 10-39. Saving the Continue Status

```
void SaveContinueStatus(String Handle)
{
        SharedPreferences settings = m_Context.getSharedPreferences(Handle, 0);
        SharedPreferences.Editor editor = settings.edit();

        editor.putBoolean("CanContinue", m_CanContinue);

        // Commit the edits!
        editor.commit();
}
```

The CheckTouch() function must be modified to integrate the game over function. (See Listing 10-40.)

If the user touches the screen, and the state is the game over screen, and the time that has passed since the game ended is not yet greater or equal to m_GameOverPauseTime, the program execution returns from the function without processing the user's touch. This ensures that the game over message will be displayed for at least m_GameOverPauseTime milliseconds.

If the required amount of time has passed, the IsNewHighScore() function is called to check to see if the player has made a high score that has to be entered into the high score table. If there is a new high score, the game state is set to GameState.HighScoreEntry, to indicate that the high score entry menu has to be displayed. If there is no new high score, the main menu must be displayed.

The camera is then reset to its initial position and rotation. The m_CanContinue variable is set to false, to indicate that this game is now over and cannot be continued later, and SaveContinueStatus() is called to save the variable.

Listing 10-40. Modifying the CheckTouch() Function

```
if (m_GameState == GameState.GameOverScreen)
{
        long CurTime = System.currentTimeMillis();
        long Delay = CurTime - m_GameOverStartTime;

        if (Delay < m_GameOverPauseTime)
        {
                return;
        }

        // Test for High Score
        if (IsNewHighScore())
        {
                // Go to High Score Entry Screen
                m_GameState = GameState.HighScoreEntry;
        }
        else
        {
                m_GameState = GameState.MainMenu;
        }
        ResetCamera();
```

```
        // Cannot continue further since game is now over
        m_CanContinue = false;
        SaveContinueStatus(MainActivity.SAVE_GAME_HANDLE);

        return;
}
```

The UpdateScene() function has to be modified to process the game over state. (See Listing 10-41.)

If the game state is the game over screen state, the UpdateScene() function updates the position of the game over billboard to make sure it is in front of and facing the camera. If the health of the player's power pyramid is less than or equal to 0, the UpdateScene() function sets the game state to the game over screen state and sets the can continue status to false to indicate that this game is over and cannot be continued later.

Listing 10-41. Modifying the UpdateScene() Function

```
if (m_GameState == GameState.GameOverScreen)
{
        // Update Game Over Screen Here
        UpdateGameOverBillBoard();
        m_GameOverBillBoard.UpdateObject3d(m_Camera);
        return;
}
if (m_Pyramid.GetObjectStats().GetHealth() <= 0)
{
        m_GameState = GameState.GameOverScreen;
        m_GameOverStartTime = System.currentTimeMillis();

        // Game is over cannnot continue current game.
        m_CanContinue = false;
}
```

The RenderScene() function must be modified so that when the game state is in the game over screen state, the game over billboard is rendered to the screen. (See Listing 10-42.)

Listing 10-42. Modifying the RenderScene() Function

```
if (m_GameState == GameState.GameOverScreen)
{
        // Update Game Over Screen Here
        m_GameOverBillBoard.DrawObject(m_Camera, m_PointLight);
}
```

Hands-on Example: The Drone Grid Game

This hands-on example will demonstrate the final Drone Grid game with a fully working menu system and using the classes discussed previously in this chapter for creating and managing groups of arena objects and tanks. In addition, I cover code that controls the frame rate of the game, so that it runs at a smooth constant frame rate. All these additions and changes are made in the MyGLRenderer class.

Modifying the MyGLRenderer Class

The m_GamePlayController variable holds a reference to the GamePlay Controller for this game.

```
private GamePlayController m_GamePlayController;
```

The ARENA_OBJECTS_HANDLE string holds the name of the handle that the set of arena objects is saved under.

```
private String ARENA_OBJECTS_HANDLE = "ArenaObjectsSet";
```

The TANK_FLEET_HANDLE string holds the name of the handle that the fleet of tanks is saved under.

```
private String TANK_FLEET_HANDLE = "TankFleet";
```

The m_ArenaObjectsSet variable holds a reference to the set of arena objects that will be used in this game.

```
private ArenaObjectSet m_ArenaObjectsSet;
```

The m_TankFleet variable holds a reference to the fleet of tanks that will be used in this game.

```
private TankFleet m_TankFleet;
```

The k_SecondsPerTick variable holds the time in milliseconds for each tick or update to the game. This variable is used to help update the game at a constant rate.

```
private float   k_SecondsPerTick   =    0.05f  * 1000.0f/1.0f; // milliseconds 20 frames /sec
```

The m_ElapsedTime variable holds the time that has passed since the last update to the game.

```
private long   m_ElapsedTime       =   0;
```

The m_CurrentTime variable holds the current time in milliseconds.

```
private long   m_CurrentTime       =   0;
```

The m_UpdateTimeCount variable is used in keeping track of the number of updates that are needed in the game, based on the elapsed time since the last update.

```
private long   m_UpdateTimeCount   =   0;
```

The m_TimeInit variable is true if the frame update timing control–related variables have been initialized and false otherwise.

```
private boolean m_TimeInit         =   false;
```

The CreateArenaObjectsSet() function creates the ArenaObject set for the game and fills it with two arena objects. The new arena objects are added by calling the AddNewArenaObject() function. (See Listing 10-43.)

Listing 10-43. Creating the ArenaObject Set for the Game

```
void CreateArenaObjectsSet(Context iContext)
{
      m_ArenaObjectsSet = new ArenaObjectSet(iContext);

      // Cube 1
      float MaxVelocity = 0.1f;
      ArenaObject3d Obj = CreateArenaObjectCube1(iContext);
      Obj.SetArenaObjectID("cube1");
      Obj.GetObjectStats().SetDamageValue(10);
      Obj.GetObjectPhysics().GetMaxVelocity().Set(MaxVelocity, 1, MaxVelocity);
      boolean result = m_ArenaObjectsSet.AddNewArenaObject(Obj);

      // Cube 2
      Obj = CreateArenaObjectCube2(iContext);
      Obj.SetArenaObjectID("cube2");
      Obj.GetObjectStats().SetDamageValue(10);
      Obj.GetObjectPhysics().GetMaxVelocity().Set(MaxVelocity, 1, MaxVelocity);
      result = m_ArenaObjectsSet.AddNewArenaObject(Obj);
}
```

The CreateTankFleet() function creates the fleet of tanks by generating two different types of tanks and adding them to the m_TankFleet array by calling the AddNewTankVehicle() function. (See Listing 10-44.)

Listing 10-44. Creating the Tank Fleet

```
void CreateTankFleet(Context iContext)
{
      m_TankFleet = new TankFleet(iContext);

      // Tank1
      Tank TankVehicle = CreateTankType1(iContext);

      // Set Material
      TankVehicle.GetMainBody().GetMaterial().SetEmissive(0.0f, 0.5f, 0f);
      TankVehicle.GetTurret().GetMaterial().SetEmissive(0.5f, 0, 0.0f);

      // Tank ID
      TankVehicle.SetVehicleID("tank1");

      // Set Patrol Order
      int MAX_WAYPOINTS = 10;
      Vector3[] WayPoints = new Vector3[MAX_WAYPOINTS];
      int NumberWayPoints = GenerateTankWayPoints(WayPoints);
      AIVehicleObjectsAffected ObjectsAffected = AIVehicleObjectsAffected.PrimaryWeapon;
      Vector3 Target = new Vector3(0,0,0);
      Object3d TargetObj = null;
      int NumberRoundToFire = 2;
      int FiringDelay = 5000;
```

```
        VehicleCommand Command = CreatePatrolAttackTankCommand(ObjectsAffected, NumberWayPoints,
WayPoints, Target,TargetObj, NumberRoundToFire,FiringDelay);
        TankVehicle.GetDriver().SetOrder(Command);
        boolean result = m_TankFleet.AddNewTankVehicle(TankVehicle);

        // Tank 2
        TankVehicle = CreateTankType2(iContext);

        // Set Material
        TankVehicle.GetMainBody().GetMaterial().SetEmissive(0, 0.5f, 0.5f);
        TankVehicle.GetTurret().GetMaterial().SetEmissive(0.5f, 0, 0.5f);

        // Tank ID
        TankVehicle.SetVehicleID("tank2");

        // Set Patrol Order
        WayPoints = new Vector3[MAX_WAYPOINTS];
        NumberWayPoints = GenerateTankWayPoints2(WayPoints);
        Target = new Vector3(0,0,0);
        TargetObj = null;
        NumberRoundToFire = 3;
        FiringDelay = 3000;

        Command = CreatePatrolAttackTankCommand(ObjectsAffected, NumberWayPoints, WayPoints, Target,
TargetObj, NumberRoundToFire, FiringDelay);
        TankVehicle.GetDriver().SetOrder(Command);
        result = m_TankFleet.AddNewTankVehicle(TankVehicle);
}
```

The CreateTankRoute1() function creates a route object that consists of waypoints that the tank is to move toward and then returns it. The Route class is very basic, and to save space here, I decided not to include it. Please refer to the full source code in the Source Code/Download area located on apress.com for more information on the Route class. (See Listing 10-45.)

Listing 10-45. Creating a Tank Route

```
Route CreateTankRoute1()
{
        // Around Pyramid
        Route TankRoute = null;
        int NumberWayPoints = 4;
        Vector3[] WayPoints = new Vector3[NumberWayPoints];
        WayPoints[0] = new Vector3(  7, 0, -10);
        WayPoints[1] = new Vector3( -7, 0, -10);
        WayPoints[2] = new Vector3( -7, 0, 5);
        WayPoints[3] = new Vector3(  7, 0, 5);
        TankRoute = new Route(NumberWayPoints, WayPoints);
        return TankRoute;
}
```

The CreateTankRoutes() function creates an array of routes and returns these routes in the TankRoutes array, along with the number of routes in the array. (See Listing 10-46.)

Listing 10-46. Creating a List of Tank Routes

```
int CreateTankRoutes(Route[] TankRoutes)
{
        int NumberRoutes = 6;
        TankRoutes[0] = CreateTankRoute1();
        TankRoutes[1] = CreateTankRoute2();
        TankRoutes[2] = CreateTankRoute3();
        TankRoutes[3] = CreateTankRoute4();
        TankRoutes[4] = CreateTankRoute5();
        TankRoutes[5] = CreateTankRoute6();
        return NumberRoutes;
}
```

The CreateGamePlayController() function creates an array of tank routes that the game controller uses in assigning paths to the enemy tanks and then creates the actual GamePlay Controller. (See Listing 10-47.)

Listing 10-47. Creating the GamePlay Controller

```
void CreateGamePlayController(Context iContext)
{
        int MaxNumberRoutes = 10;
        // Tanks
        int NumberTankRoutes = 0;
        Route[] TankRoutes = new Route[MaxNumberRoutes];
        NumberTankRoutes = CreateTankRoutes(TankRoutes);

        m_GamePlayController = new GamePlayController(iContext, m_ArenaObjectsSet, m_TankFleet,
m_Grid, NumberTankRoutes, TankRoutes);
}
```

Next, the code that updated and rendered the game elements that were in the onDrawFrame() function have been separated into code in UpdateScene() and RenderScene(). This is needed to implement the additional new code that helps run the game at a constant set frame rate and game speed that I will discuss later in this section.

The UpdateScene() function updates the elements of the game in terms of their position, orientation, status, etc. Some elements, such as the main menu, high score table, high score entry menu, and game over graphic, are only updated when the game is in a certain state and then after updating returns and does not update the rest of the elements. (See Listing 10-48.)

Listing 10-48. Updating the Game

```
void UpdateScene()
{
        m_Camera.UpdateCamera();

        // Main Menu
        if (m_GameState == GameState.MainMenu)
        {
                m_MainMenu.UpdateMenu(m_Camera);
                return;
        }

        // High Score Table
        if (m_GameState == GameState.HighScoreTable)
        {
                m_HighScoreTable.UpdateHighScoreTable(m_Camera);
                return;
        }

        // High Score Entry
        if (m_GameState == GameState.HighScoreEntry)
        {
                // Update HighScore Entry Table
                m_HighScoreEntryMenu.UpdateHighScoreEntryMenu(m_Camera);
                return;
        }

        // Game Over Screen
        if (m GameState == GameState.GameOverScreen)
        {
                // Update Game Over Screen Here
                UpdateGameOverBillBoard();
                m_GameOverBillBoard.UpdateObject3d(m_Camera);

                return;
        }

        // Check if Game has ended and go to
        if (m_Pyramid.GetObjectStats().GetHealth() <= 0)
        {
                m_GameState = GameState.GameOverScreen;
                m_GameOverStartTime = System.currentTimeMillis();

                // Game is over cannot continue current game.
                m_CanContinue = false;
        }

        // Process the Collisions in the Game
        ProcessCollisions();
```

```
///////////////////////// Update Objects
// Arena Objects
m_ArenaObjectsSet.UpdateArenaObjects();

// Tank Objects
m_TankFleet.UpdateTankFleet();

///////////////////////// Update and Draw Grid
UpdateGravityGrid();

// Player's Pyramid
m_Pyramid.UpdateObject3d();

// Player's Weapon
m_Weapon.UpdateWeapon();

///////////////////////// HUD
// Update HUD
UpdateHUD();
m_HUD.UpdateHUD(m_Camera);

// Update Game Play Controller
            m_GamePlayController.UpdateController(System.currentTimeMillis());
}
```

The RenderScene() function renders the game elements to the screen. No updating of the elements is done, only drawing the objects to the screen. (See Listing 10-49.)

Listing 10-49. Rendering the Game

```
void RenderScene()
{
        // Main Menu
        if (m_GameState == GameState.MainMenu)
        {
                m_MainMenu.RenderMenu(m_Camera, m_PointLight, false);
                return;
        }

        // High Score Table
        if (m_GameState == GameState.HighScoreTable)
        {
                m_HighScoreTable.RenderHighScoreTable(m_Camera, m_PointLight, false);
                return;
        }

        // High Score Entry
        if (m_GameState == GameState.HighScoreEntry)
        {
                m_HighScoreEntryMenu.RenderHighScoreEntryMenu(m_Camera, m_PointLight, false);
                return;
        }
```

```
        // Game Over Screen
        if (m_GameState == GameState.GameOverScreen)
        {
                // Update Game Over Screen Here
                m_GameOverBillBoard.DrawObject(m_Camera, m_PointLight);
        }
        /////////////////////////////// Draw Objects
        m_ArenaObjectsSet.RenderArenaObjects(m_Camera, m_PointLight,false);
        m_TankFleet.RenderTankFleet(m_Camera, m_PointLight,false);
        /////////////////////////////// Update and Draw Grid
        m_Grid.DrawGrid(m_Camera);

        // Player's Pyramid
        m_Pyramid.DrawObject(m_Camera, m_PointLight);

        // Player's Weapon
        m_Weapon.RenderWeapon(m_Camera, m_PointLight, false);
        ////////////////////////////// HUD
        // Render HUD
        m_HUD.RenderHUD(m_Camera, m_PointLight);
}
```

The CalculateFrameUpdateElapsedTime() function calculates the elapsed time since the last frame update and stores this value in m_ElapsedTime. (See Listing 10-50.)

Listing 10-50. Calculating the Elapsed Time for the Frame Update

```
void CalculateFrameUpdateElapsedTime()
{
        long Oldtime;

        // Elapsed Time Since Last in this function
        if (!m_TimeInit)
        {
        m_ElapsedTime = 0;
                m_CurrentTime = System.currentTimeMillis();
                m_TimeInit = true;
        }
        else
        {
                Oldtime = m_CurrentTime;
                m_CurrentTime = System.currentTimeMillis();
                m_ElapsedTime = m_CurrentTime - Oldtime;
        }
}
```

The FrameMove() function updates the game by calling UpdateScene() and, if needed, the ProcessCameraMove() function. The purpose of this function is to update the game at a constant and smooth frame rate as close to a rate of one game update every k_SecondsPerTick as possible. (See Listing 10-51.)

The FrameMove() function does the following:

1. The UpdateScene() and ProcessCameraMove() functions are called only if the elapsed time since the last update is greater than k_SecondsPerTick, which is the time for one update to occur.

2. After UpdateScene() is called, m_UpdateTimeCount is changed to reflect that an update has happened, by subtracting k_SecondsPerTick, which is the time for a single frame update or "tick."

3. If more frame updates have to occur in order to meet the goal of one game update every k_SecondsPerTick, which means that m_UpdateTimeCount ➤ k_SecondsPerTick, UpdateScene() is called repeatedly, until the elapsed time since the last game update is equal to or less than k_SecondsPerTick.

Listing 10-51. Updating the Game

```
void FrameMove()
{
    m_UpdateTimeCount += m_ElapsedTime;
    if (m_UpdateTimeCount > k_SecondsPerTick)
    {
    while(m_UpdateTimeCount > k_SecondsPerTick)
        {
                // Update Camera Position
            if (m_CameraMoved)
            {
                ProcessCameraMove();
            }
                // update the scene
                UpdateScene();

                m_UpdateTimeCount -= k_SecondsPerTick;
        }
      }
}
```

The onDrawFrame() function modifications (see Listing 10-52) involve the following:

1. New code for sound control is added. If m_SFXOn is true, the sound effects are turned on for the arena objects, tanks, and pyramid by calling TurnSFXOnOff(). Otherwise, the sound effects are turned off for these game elements.

2. The time that has elapsed since the last game update is calculated by calling the CalculateFrameUpdateElapsedTime() function.

3. The game is updated by calling FrameMove().

4. The game objects are rendered to the screen by calling RenderScene().

Listing 10-52. Modifying the onDrawFrame() Function

```
public void onDrawFrame(GL10 unused)
{
        GLES20.glClearColor(0.0f, 0.0f, 0.0f, 1.0f);
        GLES20.glClear( GLES20.GL_DEPTH_BUFFER_BIT | GLES20.GL_COLOR_BUFFER_BIT);

        // UPDATE SFX
        if (m_SFXOn)
        {
                TurnSFXOnOff(true);
        }
        else
        {
                TurnSFXOnOff(false);
        }
        // Did user touch screen
        if (m_ScreenTouched)
        {
                // Process Screen Touch
                CheckTouch();
                m_ScreenTouched = false;
        }
        CalculateFrameUpdateElapsedTime();
        FrameMove();
        RenderScene();
}
```

Now, run and play the game, and you should see something as in the following Figures 10-1 through 10-4.

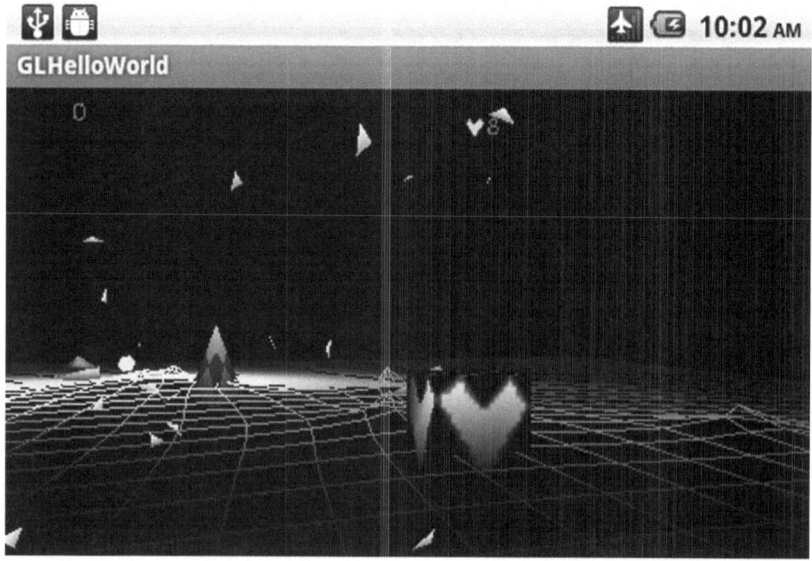

Figure 10-1. Heart arena object

Figure 10-2. *Two different tank types*

Figure 10-3. *Android arena object*

Figure 10-4. *Game over screen*

Summary

In this chapter, I have discussed the final Drone Grid game. I started with a discussion of the classes needed to manage our player's enemies. I then covered the GamePlayController class that was used to control important elements of how our enemies were presented. Next, I went over code and functions that were needed to save and load the state of our final game. The game over graphic and related game logic were then discussed. Finally, a hands-on example was presented that illustrated the classes, code, and concepts given in this chapter.

Chapter 11

The Android Native Development Kit (NDK)

This chapter covers the Android Native Development Kit, or NDK. I begin with an overview of the NDK, discussing what the NDK actually is, then examine the system and software requirements that must be met in order for you to use this kit in developing your Android programs. Then, the Java Native Interface, or JNI, is discussed, including how to use it to create functions that run in native machine code compiled from C/C++ code and can be called from Java functions from within the default Java framework we have used throughout this book. Next, a "Hello World" example using the JNI is presented that provides a step-by-step guide into creating a simple Android program that uses the JNI and native code written in the C language to output a string. Finally, another hands-on example is presented that takes pieces of existing code from our Drone Grid game and converts the existing Java code to native code.

NDK Overview

The NDK is a set of tools designed to work with existing Android development tools, such as Eclipse, to embed native machine code compiled from C and C++ code into Android programs. The NDK can be used to compile C/C++ code into a library that is then used by Eclipse to compile the final Android application. One important issue is that only Android operating system versions 1.5 (Cupcake) or higher can use the NDK to embed native code in Android applications.

NDK System Requirements

A complete Android SDK installation (including all dependencies) is required. Android 1.5 SDK or later version is required.

Supported operating systems include the following:

- Windows XP (32-bit) or Vista (32- or 64-bit)

- Mac OS X 10.4.8 or later (x86 only)

- Linux (32- or 64-bit; Ubuntu 8.04, or other Linux distributions using GLibc 2.7 or later)

Required development tools:

- For all development platforms, GNU Make 3.81 or later is required. Earlier versions of GNU Make might work but have not been tested.

- A recent version of awk (either GNU Awk or Nawk) is also required.

- For Windows, Cygwin 1.7 or higher is required. The NDK will not work with Cygwin 1.5 installations.

Note The Cygwin program can be downloaded from `www.cygwin.com`.

Android Platform Compatibility

Native code generated by the NDK targeting a specific CPU architecture requires a minimum Android operating system version depending on the CPU targeted.

- ARM, ARM-NEON targeted code requires Android 1.5 (API Level 3) and higher. Practically speaking, nearly 100% of current available Android devices are at least 1.5 or greater.

- x86 targeted code requires Android 2.3 (API Level 9) and higher.

- MIPS targeted code requires Android 2.3 (API Level 9) and higher.

Installing the Android NDK

The main Android NDK installation file is located on the official Android web site at `www.android.com`. The NDK is a zip file that you will have to download and uncompress onto your hard drive. (See Figure 11-1.)

Downloads

Platform	Package	Size (Bytes)	MD5 Checksum
Windows 32-bit	android-ndk-r9-windows-x86.zip	485200055	8895aec43f5141212c8dac6e9f07d5a8
	android-ndk-r9-windows-x86-legacy-toolchains.zip	292738221	ae3756d3773ec068fb653ff6fa411e35
Windows 64-bit	android-ndk-r9-windows-x86_64.zip	514321606	96c725d16ace7fd487bf1bc1427af3a0
	android-ndk-r9-windows-x86_64-legacy-toolchains.zip	312340413	707d1eaa6f5d427ad439c764c8bd68d2
Mac OS X 32-bit	android-ndk-r9-darwin-x86.tar.bz2	446858202	781da0e6bb5b072512e67b879b56a74c
	android-ndk-r9-darwin-x86-legacy-toolchains.tar.bz2	264053696	9fd7f76a1f1f59386a34b019dcd20976
Mac OS X 64-bit	android-ndk-r9-darwin-x86_64.tar.bz2	454408117	ff27c8b9efc8260d9f883dc42d08f651
	android-ndk-r9-darwin-x86_64-legacy-toolchains.tar.bz2	271922968	251c21defcf90a2f0e8283bab90ed861
Linux 32-bit (x86)	android-ndk-r9-linux-x86.tar.bz2	419862465	beadafdc187461c057d513c40f0ac33b
	android-ndk-r9-linux-x86-legacy-toolchains.tar.bz2	241172797	957c415de9d7c7ce1c2377ec4d3d60f1
Linux 64-bit (x86)	android-ndk-r9-linux-x86_64.tar.bz2	425113267	0ccfd9960526e61d1527155fa6f84ac0
	android-ndk-r9-linux-x86_64-legacy-toolchains.tar.bz2	244427866	3976a8237d75526b8a0f275375dd68b5

Figure 11-1. Android NDK downloads

If you are developing on a Windows platform, you will have to download and install Cygwin, which is a Unix-style command-line interface that allows you to execute Unix commands on your PC. (See Figure 11-2.)

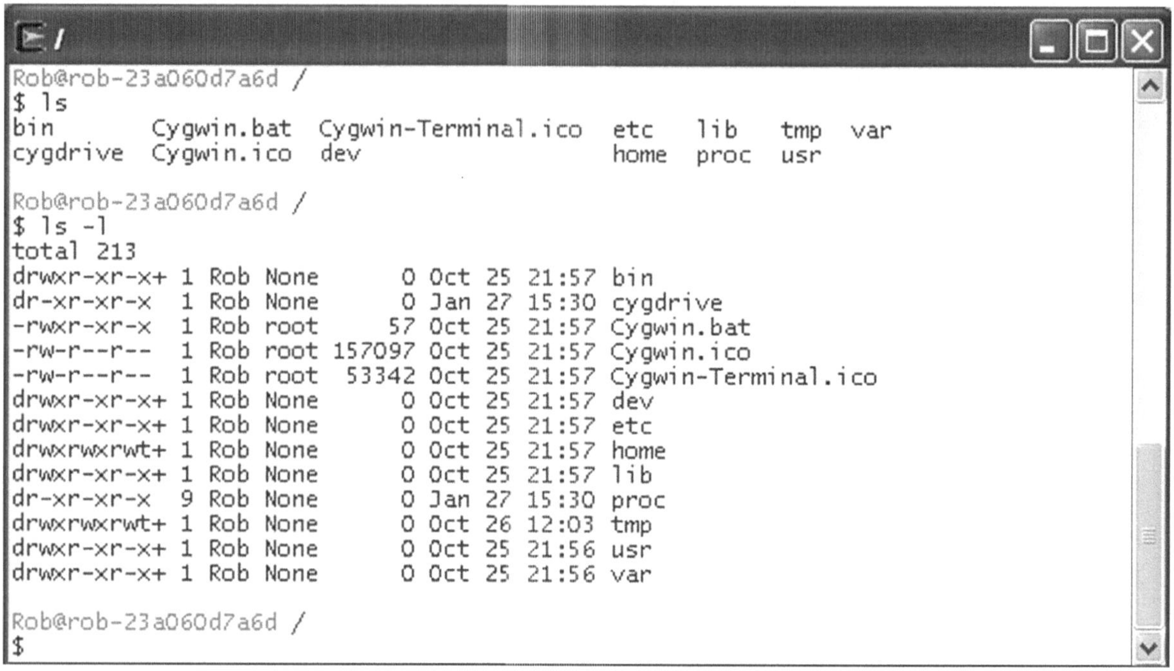

```
Rob@rob-23a060d7a6d /
$ ls
bin          Cygwin.bat  Cygwin-Terminal.ico   etc    lib    tmp   var
cygdrive  Cygwin.ico   dev                           home  proc   usr

Rob@rob-23a060d7a6d /
$ ls -l
total 213
drwxr-xr-x+ 1 Rob None         0 Oct 25 21:57 bin
dr-xr-xr-x  1 Rob None         0 Jan 27 15:30 cygdrive
-rwxr-xr-x  1 Rob root        57 Oct 25 21:57 Cygwin.bat
-rw-r--r--  1 Rob root 157097 Oct 25 21:57 Cygwin.ico
-rw-r--r--  1 Rob root  53342 Oct 25 21:57 Cygwin-Terminal.ico
drwxr-xr-x+ 1 Rob None         0 Oct 25 21:57 dev
drwxr-xr-x+ 1 Rob None         0 Oct 25 21:57 etc
drwxrwxrwt+ 1 Rob None         0 Oct 25 21:57 home
drwxr-xr-x+ 1 Rob None         0 Oct 25 21:57 lib
dr-xr-xr-x  9 Rob None         0 Jan 27 15:30 proc
drwxrwxrwt+ 1 Rob None         0 Oct 26 12:03 tmp
drwxr-xr-x+ 1 Rob None         0 Oct 25 21:56 usr
drwxr-xr-x+ 1 Rob None         0 Oct 25 21:56 var

Rob@rob-23a060d7a6d /
$
```

Figure 11-2. Cygwin Unix command shell for Windows

Ways to Use the Android NDK

There are two ways to use the Android NDK.

- Use the Android Java framework and use the Java Native Interface or JNI to call native code from your Java-based Android program.

- Use the NativeActivity class provided by the Android SDK to replace normally Java language life cycle callbacks such as onCreate(), onPause(), onResume(), etc., with native code written in C/C++. However, native activities must be run on Android operating system versions 2.3 (API Level 9) and higher. Also, some Android framework services cannot be accessed natively.

In this chapter, I will cover using the JNI to access native code from the Java framework.

Java Native Interface Overview

The JNI allows Java code that runs within the Android virtual machine to operate with applications and libraries written in other programming languages, such as C, C++, and assembly. The Java Interface Pointer is discussed in this section. Native C/C++ coding methods are treated, including the variable types used with the JNI, the required naming conventions for native C/C++ functions, and the required input parameters for these functions. The procedure to integrate native functions into Java code is given. Examples of how to use native code from Java and how to use Java functions from native code are also presented in this section.

The Java Interface Pointer

Native code in C/C++ accesses the Java Virtual Machine through JNI functions that are accessed through the Java Interface Pointer. The JNI Interface Pointer is a pointer to an array of pointers that point to JNI functions. (See Figure 11-3.)

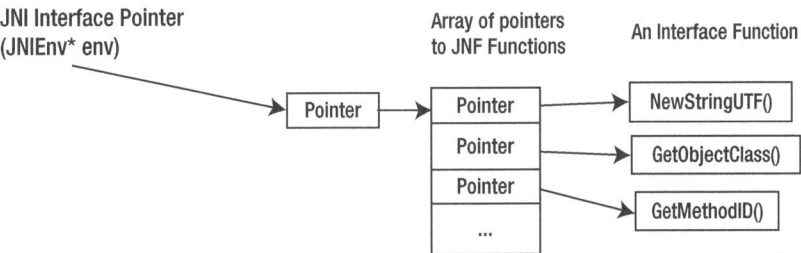

Figure 11-3. *The Java Interface Pointer*

Loading and Linking Native C/C++ Methods

In order to use native classes in your Android Java code (see Listing 11-1), you will have to

1. Load the compiled library by using the `System.loadLibrary()` function.

2. Declare the native class that is defined in the C/C++ source code as native in the Android Java code by using the native keyword in a function declaration.

Listing 11-1. *Loading and Linking Native C/C++ Methods from Android Java Code*

```
package robs.gldemo.robsgl20tutorial;
class GLES20TriangleRenderer implements GLSurfaceView.Renderer
{
        /* this is used to load the 'hello-jni' library on application
        * startup. The library has already been unpacked into
        * /data/data/com.example.hellojni/lib/libhello-jni.so at
        * installation time by the package manager.
        */
        static {
                System.loadLibrary("hello-jni");
        }
        public native String  RobsstringFromJNI();
}
```

Naming Native Functions

The native function declared in the Java code must match the function name declared in the native C/C++ code based on certain formatting, as follows:

1. The function starts with "Java."

2. It is followed by the package name "`robs_gldemo_robsgl20tutorial`" from the example in Listing 11-1.

3. This is followed by the class name "`GLES20TriangleRenderer`" from the example in Listing 11-1.

4. Next comes the function name "`RobsstringFromJNI`" from the example in Listing 11-1.

See the full function name in Listing 11-2.

Listing 11-2. The Native `RobsstringFromJNI()` Function

```
jstring
Java_robs_gldemo_robsgl20tutorial_GLES20TriangleRenderer_RobsstringFromJNI(JNIEnv* env, jobject thiz
)
{
    return (*env)->NewStringUTF(env, "Rob's String Text Message!");
}
```

Native Function Parameters

The parameter list for a native function always starts with a pointer to JNIEnv, which is the Java Interface Pointer, for example, env in our native function from Listing 11-2.

```
JNIEnv* env
```

The second parameter is a reference to the object, if the native function is nonstatic, for example, thiz in our native function example from Listing 11-2.

```
 jobject thiz
```

However, if the native function is static, the second parameter is a reference to its Java class.

C vs. C++ Native Function Formats

The function in Listing 11-2 is the C native function that utilizes the Java Native Interface. The C++ version is slightly different, but the underlying mechanisms are the same (see Listing 11-3). The main differences are

1. The extern "C" specification

2. The change from (*env)-> to env-> for accessing the JNI functions

3. The removal of env as the first parameter of the JNI function call

Listing 11-3. The C++ Equivalent Native Function

```
extern "C" /* specify the C calling convention */
jstring
Java_robs_gldemo_robsgl20tutorial_GLES20TriangleRenderer_RobsstringFromJNI(JNIEnv* env, jobject thiz
)
{
    return env->NewStringUTF("Rob's String Text Message!");
}
```

Native Types

The JNI native data types and their Java equivalents include

jboolean: This native type is equivalent to the boolean Java type and is unsigned 8 bits in size.

jbyte: This native type is equivalent to the byte Java type and is signed 8 bits in size.

jchar: This native type is equivalent to the char Java type and is unsigned 16 bits in size.

jshort: This native type is equivalent to the short Java type and is signed 16 bits in size.

jint: This native type is equivalent to the int Java type and is signed 32 bits in size.

jlong: This native type is equivalent to the long Java type and is signed 64 bits in size.

jfloat: This native type is equivalent to the float Java type and is 32 bits in size.

jdouble: This native type is equivalent to the double Java type and is 64 bits in size.

Reference Types

The JNI includes some reference types that correspond to various Java objects. See Figure 11-4 for a list of these reference types in a hierarchical view.

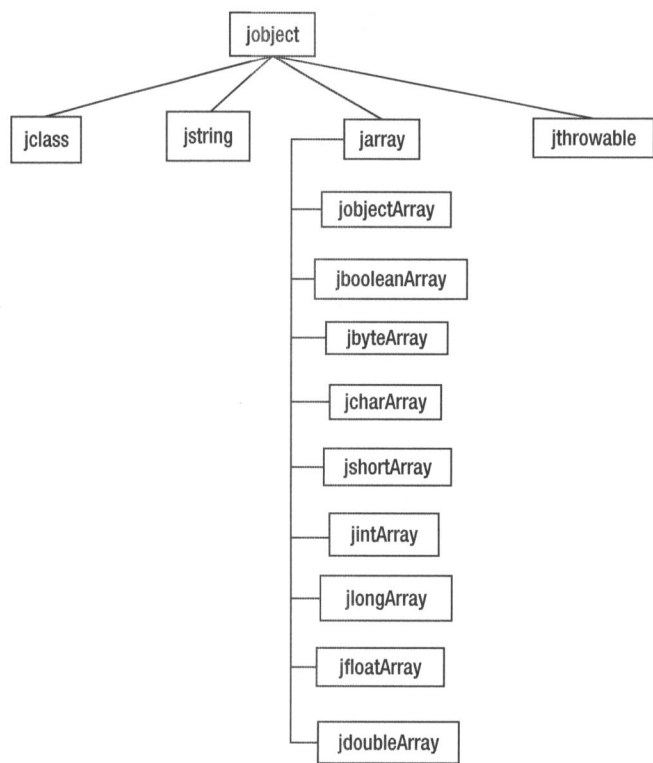

Figure 11-4. Reference type hierarchy

JNI Signature Types

The JNI uses the Java Virtual Machine's representation of signature types that are used to define a specific function, including its return value type and the types of its input parameters. The signature types are as follows:

Z boolean type

B byte type

C char type

S short type

I int type

J long type

F float type

D double type

L fully-qualified-class; fully qualified class

[**type** type[] array

(arg-types) ret-type (method) return type

V void

For example, the Java method

```
long JavaMethod1(int number, String str, int[] intarray1);
```

has the following type signature:

```
(ILjava/lang/String;[I)J
```

Another example is the AddRotation() function from the Orientation class.

```
void AddRotation(float AngleIncrementDegrees)
```

The signature type for this function would be

```
(F)V
```

The F would represent the float input parameter, and the V would represent the void return type.

Calling Native Code from Java and Accessing Java Methods from Native Code

To call a native function from Java code, you would use the native function name without the mangled prefix with the extra identifying information. For example, to call the native function AddRotationNative() shown in Listing 11-5, you would use the code shown in Listing 11-4.

Listing 11-4. Calling Native Code from Java

```
void AddRotationToObject(Orientation O, float AngleAmount)
{
        AddRotationNative(O, AngleAmount);
}
```

The native function AddRotationNative()shown in Listing 11-5 calls the AddRotation() function for the Orientation object that is passed into the function and held in the Orient variable.

The AddRotationNative() function does the following:

1. Gets the class of the Java object Orient by calling the GetObjectClass() JNI function and assigns it to the OrientationClass variable

2. Gets the method id of a specific function by calling the GetMethodID() JNI function with parameters including the function name, which is "AddRotation"; the function signature type, which is "(F)V"; and the Java class object that contains the function, which is OrientationClass. This method id is assigned to the MethodID variable.

3.   Calls the Orient Java object's AddRotation() function with parameter
       RotationAngle by calling the CallVoidMethod() JNI function

Listing 11-5. Accessing a Java Method from a Native Code

```
Java_robs_gldemo_robsgl2Otutorial_Physics_AddRotationNative(JNIEnv* env,
                                                            jobject thiz,
                                                            jobject Orient,
                                                            jfloat RotationAngle)
{
       /*
       GetObjectClass
       jclass GetObjectClass(JNIEnv *env, jobject obj);
       */
       jclass OrientationClass = (*env)->GetObjectClass(env, Orient);

       /*
       GetMethodID
       jmethodID GetMethodID(JNIEnv *env, jclass clazz, const char *name, const char *sig);
       */
       jmethodID  MethodID = (*env)->GetMethodID(env,
                                          OrientationClass,
                                          "AddRotation",
                                                  "(F)V");

       /*
       NativeType Call<type>Method(JNIEnv *env, jobject obj, jmethodID methodID, ...);
       */
       (*env)->CallVoidMethod(env, Orient, MethodID, RotationAngle);
}
```

JNI Functions

There are many more JNI functions besides those discussed in Listing 11-5. For example, if
the Java function we want to call returns a double numeric value, we would have to call the
CallDoubleMethod() function instead of the CallVoidMethod for a function that returns void.
We won't try and discuss every JNI function here, because this is not intended to be a JNI reference
manual. If you want more information on the complete list of JNI functions supported, please go to
http://docs.oracle.com/javase/6/docs/technotes/guides/jni/spec/functions.html#wp9502.

> **Note**   The main web site for JNI specifications is http://docs.oracle.com/javase/6/docs/
> technotes/guides/jni/spec/jniTOC.html.

Android JNI Makefile

The Android makefile (`Android.mk`) is a file that describes your native code that you want to compile to the NDK build system.

The `LOCAL_PATH` variable holds the location of the source files. The value of `my-dir` is already defined by the NDK build system to point to the current directory that contains the `Android.mk` makefile. What you will do is put this makefile in the JNI directory, along with all the C source code files that you want to compile.

```
LOCAL_PATH := $(call my-dir)
```

The `CLEAR_VARS` variable is already defined by the NDK build system and points to a makefile that will clear many of the local variables that are used in the build system.

```
include $(CLEAR_VARS)
```

The `LOCAL_MODULE` variable sets the library name that will be generated from the native source code files. The library name format will be the prefix "`lib`" + "`hello-jni`" + the "`.so`" suffix. However, if the library name already begins with "`lib`", the prefix "`lib`" is not added to the final file name.

```
LOCAL_MODULE    := hello-jni
```

The `LOCAL_SRC_FILES` variable holds the names of the C/C++ source files that the NDK build system will compile and create a final library from.

```
LOCAL_SRC_FILES := hello-jni.c
```

The `BUILD_SHARED_LIBRARY` variable is defined by the NDK build system and points to a makefile that gathers and processes all the information needed for building the final library.

```
include $(BUILD_SHARED_LIBRARY)
```

The complete makefile is shown in Listing 11-6.

Listing 11-6. The Android JNI Makefile

```
LOCAL_PATH := $(call my-dir)

include $(CLEAR_VARS)

LOCAL_MODULE    := hello-jni
LOCAL_SRC_FILES := hello-jni.c

include $(BUILD_SHARED_LIBRARY)
```

Hands-on Example: "Hello World from JNI and Native Code"

In this hands-on example, a simple "Hello World" example is discussed, in which the actual string "Hello World from JNI and Native Code" is generated from native C code and returned to the Java caller where it is then printed out to the log window.

First, we have to create the `jni` directory for the Android project. Select the main project directory that you want to create the `jni` directory in. Go to File ➤ New ➤ Folder to bring up the New Folder dialog window. Enter the file name "jni" in the Folder name edit box and click the Finish button to create a new directory named `jni`. (See Figure 11-5.)

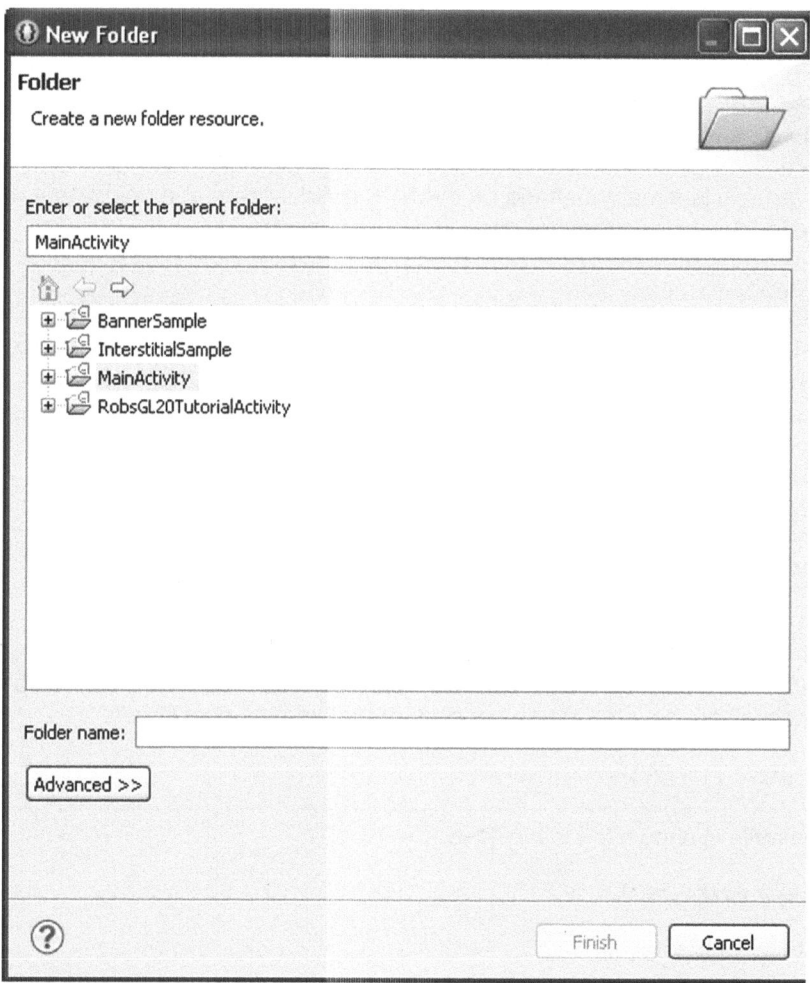

Figure 11-5. *New Folder window dialog*

The `jni` directory should appear under the main project directory. (See Figure 11-6.)

Figure 11-6. *Creating the* jni *directory*

Next, create a new file under the jni directory by selecting File ➤ New ➤ File to bring up a New File window dialog. (See Figure 11-7.)

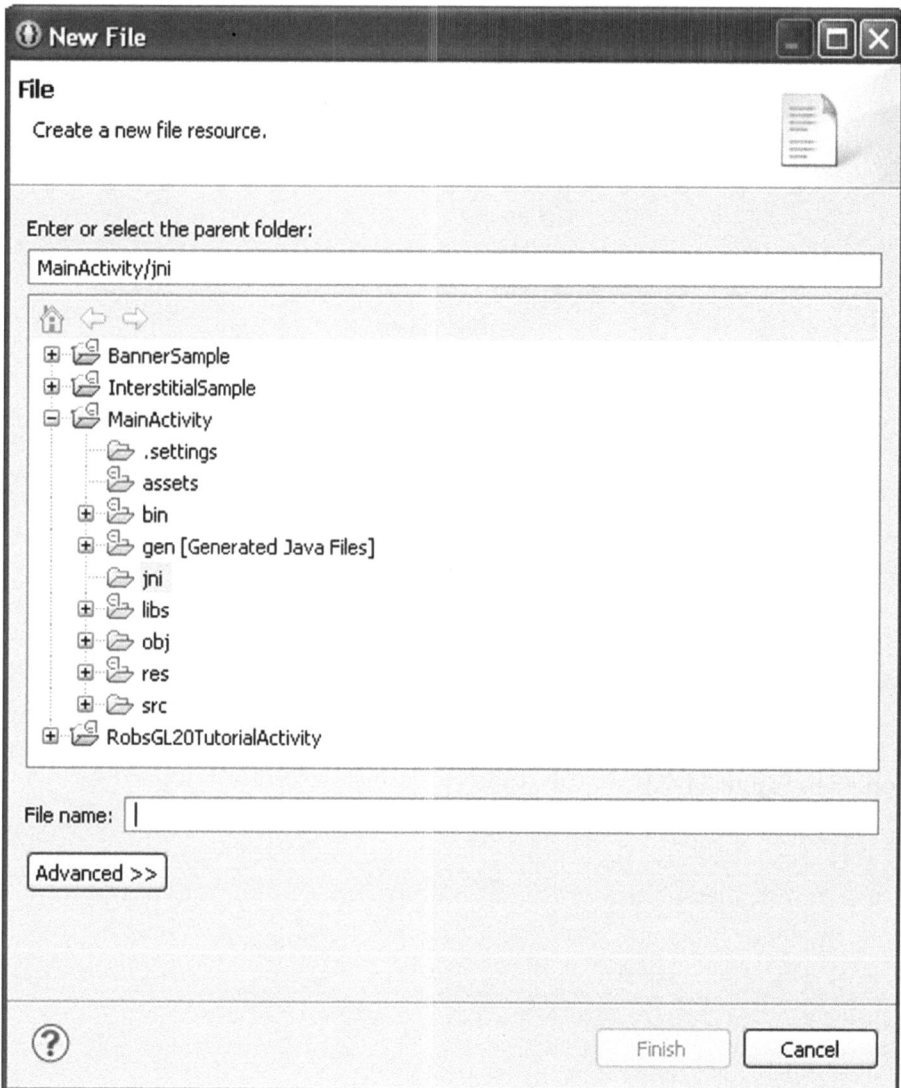

Figure 11-7. New File window dialog

Type in "Android.mk" for the file name, and click the Finish button to create the new file. Double-click the file in the Package Explorer window to bring up the text in the Eclipse source code area. Copy the makefile code from Listing 11-6 to the new file and save it by selecting File ➤ Save All.

Next, repeat the previous steps to create a new file for the C source code file, which will be named "hello-jni.c." Copy the source code shown in Listing 11-7 to `hello-jni.c`.

Listing 11-7. Hello-Jni.c Source Code

```
#include <string.h>
#include <jni.h>

// package = com.robsexample.glhelloworld;
// class = MyGLRenderer
jstring
Java_com_robsexample_glhelloworld_MyGLRenderer_RobsstringFromJNI(JNIEnv* env,
                                                                  jobject thiz )
{
    return (*env)->NewStringUTF(env, "Hello World from JNI and Native Code.");
}
```

The native C function `RobsstringFromJNI()` is shown in Listing 11-7. The function creates a new Java string by calling the `NewStringUTF()` function and returns the string to the Java caller.

Next, the native code must be compiled using the NDK build system. In order to do this, we have to start up the Cygwin Unix emulator, which allows you to run Unix commands on your Windows PC, if you are using a PC for your Android development.

You must navigate using Unix commands to the `jni` directory that you created previously. Use the "cd .." command to change the directory to one directory level up and "cd foldername" to change the current directory to the folder name. Use the "ls" command to list the files and folders in the current directory. Use the "pwd" command to get the current directory path you are in.

You will have to go to the root directory and change the directory to the `cygdrive/` folder. Then change the directory to the drive your Android project is stored on, and go to the specific folder your native source files are in. Once you are in the `jni` directory that contains your makefile and the source code, you have to run the `ndk-build` script from the Android NDK you downloaded and unzipped to your hard drive. For example, let's say your files are in the /cygdrive/c/ AndroidWorkSpaces/WorkSpace1/MainActivity/jni directory. You can execute the `ndk-build` script while in the current directory by typing in the full path to the script, such as /cygdrive/c/ AndroidNDK/andoird-ndk-r9/ndk-build. The build script will then execute and produce the output shown in Figure 11-8.

```
Rob@rob-23a060d7a6d /cygdrive/c/AndroidWorkSpaces/WorkSpace1/MainActivity/jni
$ /cygdrive/c/AndroidNDK/android-ndk-r9/ndk-build
Android NDK: WARNING: APP_PLATFORM android-14 is larger than android:minSdkVersi
on 8 in /cygdrive/c/AndroidWorkSpaces/WorkSpace1/MainActivity/AndroidManifest.xm
l
Compile thumb  : hello-jni <= hello-jni.c
SharedLibrary  : libhello-jni.so
Install        : libhello-jni.so => libs/armeabi/libhello-jni.so
```

Figure 11-8. *Running the ndk-build script in the jni directory*

The ndk-build script will process your native code files and package them into a shared library called libhello-jni.so, which is placed in the libs/armeabi directory. (See Figure 11-9.)

Figure 11-9. *The shared library generated by the* ndk-build *script*

Modifying the MyGLRenderer Class

In order to use the compiled native code, we have to make some modifications to the MyGLRenderer class from our hands-on example from the previous chapter.

The shared library with the native code has to be loaded using the loadLibrary() function.

```
static {
        System.loadLibrary("hello-jni");
}
```

The native C function in the library has to be declared with the native keyword, in order to be recognized and used.

```
public native String  RobsstringFromJNI();
```

In the onDrawFrame() function, the String variable TestJNIString is assigned the return value from calling the RobsstringFromJNI() function. This return value is incorporated in a debug log statement. (See Listing 11-8.)

Listing 11-8. Modifying the onDrawFrame() Function

```
String  TestJNIString = RobsstringFromJNI();
Log.e("RENDERER" , "JNI STRING = " + TestJNIString);
```

Run the program. You should see debug statements in your Log window. The final output of the log statement is shown in Figure 11-10.

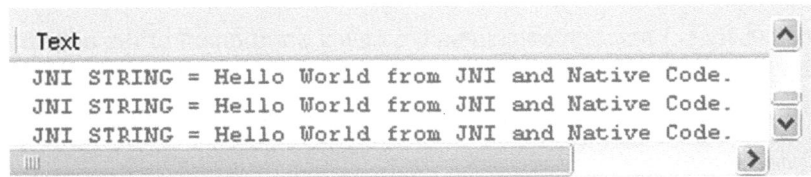

Figure 11-10. JNI test results in Log window

Hands-on Example: Adding Native Functions to the Drone Grid Game Case Study

This hands-on example will demonstrate how to convert parts of our Drone Grid game from Java code to native C code and how to call Java functions from native C code.

Calculating Gravity in Native Code

The gravity calculations can be modified in our Drone Grid game so that the actual gravity calculation is compiled in C and executed in native code on the Android. To do this, we have to modify the `hello-jni.c` file and the Physics class.

Modifying the `hello-jni.c` File

The `hello-jni.c` file must be modified to include C source code.

The Gravity variable holds the value of the acceleration of gravity for our 3D game world.

```
float Gravity = 0.010f;
```

The ApplyGravityToObjectNative() function calculates the new acceleration of an object after the acceleration from gravity in our game is applied. This new acceleration is then returned. (See Listing 11-9.)

Listing 11-9. Calculating the New Y Acceleration of an Object

```
jfloat
Java_com_robsexample_glhelloworld_Physics_ApplyGravityToObjectNative(JNIEnv* env,
                                                          jobject thiz,
                                                          jfloat YAccel)
{
      YAccel = YAccel - Gravity;
      return YAccel;
}
```

Modifying the Physics Class

Next, the Physics class has to be modified to call the native class.

The `ApplyGravityToObjectNative()` function has to be declared as a native function in the Physics class.

```
native float ApplyGravityToObjectNative(float YAccel);
```

The `ApplyGravityToObject()` function calculates the new y component of the acceleration by calling the native function `ApplyGravityToObjectNative()` with the current y acceleration of the object. (See Listing 11-10.)

Listing 11-10. Calling the Native Gravity Calculation Function

```
void ApplyGravityToObject()
{
        // Do Native Apply Gravity
        float YAccel = m_Acceleration.y;

        m_Acceleration.y = ApplyGravityToObjectNative(YAccel);
}
```

Rotating Objects from Native Code

Demonstrating the rotation of objects from native C code using Java functions requires modifications to the `hello-jni.c` file, the Physics class, and the MyGLRenderer class.

Modifying the `hello-jni.c` File

The `hello-jni.c` file has to be modified to add the native class `AddRotationNative()`.

The `AddRotationNative()` function is almost identical to the function we discussed in Listing 11-5. The difference is that the full function name involves a different package and class. What `AddRotationNative()` does is add `RotationAngle` degrees to the rotation of the object whose orientation is represented by the input `Orient` parameter. This is done by calling the actual Java language method "AddRotation()". (See Listing 11-11.)

Listing 11-11. Adding a Rotation

```
Java_com_robsexample_glhelloworld_Physics_AddRotationNative(JNIEnv* env,
                                                            jobject thiz,
                                                            jobject Orient,
                                                            jfloat RotationAngle)
{
        /*
        GetObjectClass
        jclass GetObjectClass(JNIEnv *env, jobject obj);
        */
        jclass OrientationClass = (*env)->GetObjectClass(env, Orient);
```

```
/*
GetMethodID
jmethodID GetMethodID(JNIEnv *env, jclass clazz, const char *name, const char *sig);
*/
jmethodID  MethodID = (*env)->GetMethodID(env, OrientationClass,"AddRotation", "(F)V");

/*
NativeType Call<type>Method(JNIEnv *env, jobject obj, jmethodID methodID, ...);
*/
(*env)->CallVoidMethod(env, Orient, MethodID, RotationAngle);
}
```

Modifying the Physics Class

The Physics class must also be modified to use the native rotation function.

The AddRotationNative() function has to be declared with the native keyword in the Physics class.

```
native void  AddRotationNative(Orientation O, float RotationAngle);
```

The AddRotationToObject() function calls the AddRotationNative() function to perform the rotation on the object and serves as a Java wrapper interface for the native function. (See Listing 11-12.)

Listing 11-12. Wrapper Function for AddRotationNative()

```
void AddRotationToObject(Orientation O, float AngleAmount)
{
        AddRotationNative(O, AngleAmount);
}
```

The UpdatePhysicsObject() function is modified by adding the Java function AddRotationToObject() that calls the native C function that rotates the object as part of the physics update. The old AddRotation() function is also commented out. (See Listing 11-13.)

Listing 11-13. Modifying the UpdatePhysicsObject() Function

```
void UpdatePhysicsObject(Orientation orientation)
{
        // 0. Apply Gravity if needed
        if (m_ApplyGravity)
        {
                ApplyGravityToObject();
        }

        // 1. Update Linear Velocity
        //////////////////////////////////////////////////////////////////////
        m_Acceleration.x = TestSetLimitValue(m_Acceleration.x, m_MaxAcceleration.x);
        m_Acceleration.y = TestSetLimitValue(m_Acceleration.y, m_MaxAcceleration.y);
        m_Acceleration.z = TestSetLimitValue(m_Acceleration.z, m_MaxAcceleration.z);
```

```
        m_Velocity.Add(m_Acceleration);
        m_Velocity.x = TestSetLimitValue(m_Velocity.x, m_MaxVelocity.x);
        m_Velocity.y = TestSetLimitValue(m_Velocity.y, m_MaxVelocity.y);
        m_Velocity.z = TestSetLimitValue(m_Velocity.z, m_MaxVelocity.z);

        // 2. Update Angular Velocity
        ////////////////////////////////////////////////////////////////////////
        m_AngularAcceleration = TestSetLimitValue(m_AngularAcceleration, m_MaxAngularAcceleration);

        m_AngularVelocity += m_AngularAcceleration;
        m_AngularVelocity = TestSetLimitValue(m_AngularVelocity,m_MaxAngularVelocity);

        // 3. Reset Forces acting on Object
        //     Rebuild forces acting on object for each update
        ////////////////////////////////////////////////////////////////////////
        m_Acceleration.Clear();
        m_AngularAcceleration = 0;

        //4. Update Object Linear Position
        ////////////////////////////////////////////////////////////////////////
        Vector3 pos = orientation.GetPosition();
        pos.Add(m_Velocity);

        // Check for object hitting ground if gravity is on.
        if (m_ApplyGravity)
        {
                if ((pos.y < m_GroundLevel)&& (m_Velocity.y < 0))
                {
                        if (Math.abs(m_Velocity.y) > Math.abs(m_Gravity))
                        {
                                m_JustHitGround = true;
                        }
                        pos.y = m_GroundLevel;
                        m_Velocity.y = 0;
                }
        }

        //5. Update Object Angular Position
        ////////////////////////////////////////////////////////////////////////
        // Add Rotation to Rotation Matrix
        //orientation.AddRotation(m_AngularVelocity);

        // Call Native Method
        AddRotationToObject(orientation, m_AngularVelocity);
}
```

Modifying the MyGLRenderer Class

Finally, the MyGLRenderer class has to be modified.

The CreateArenaObjectsSet() function must be modified to apply rotational forces to the arena objects, in order to demonstrate the use of native functions in rotating objects. The value of the rotational force to apply to the arena objects is held in the RotationalForce variable and is set to 5000.

The `ApplyRotationalForce()` function is used to apply the actual force to the arena objects. (See Listing 11-14.)

Listing 11-14. Modifying the Arena Objects Set Creation Function

```
void CreateArenaObjectsSet(Context iContext)
{
        m_ArenaObjectsSet = new ArenaObjectSet(iContext);

        // Cube
        float RotationalForce = 5000;
        float MaxVelocity = 0.1f;

        ArenaObject3d Obj = CreateArenaObjectCube1(iContext);
        Obj.SetArenaObjectID("cube1");
        Obj.GetObjectStats().SetDamageValue(10);
        Obj.GetObjectPhysics().GetMaxVelocity().Set(MaxVelocity, 1, MaxVelocity);
        Obj.GetObjectPhysics().ApplyRotationalForce(RotationalForce, 1);

        // Add new Object
        boolean result = m_ArenaObjectsSet.AddNewArenaObject(Obj);

        /////////////////////////////////////////////////
        Obj = CreateArenaObjectCube2(iContext);
        Obj.SetArenaObjectID("cube2");
        Obj.GetObjectStats().SetDamageValue(10);
        Obj.GetObjectPhysics().GetMaxVelocity().Set(MaxVelocity, 1, MaxVelocity);
        Obj.GetObjectPhysics().ApplyRotationalForce(RotationalForce, 1);

        // Add new Object
        result = m_ArenaObjectsSet.AddNewArenaObject(Obj);
}
```

Calculating the Reaction Force for a Collision from Native Code

In order to calculate the reaction force for a collision, modifications must be made to the `hello-jni.c` file and the Physics class.

Modifying the `hello-jni.c` File

The `hello-jni.c` file has to be modified by adding two functions.

The `DotProduct()` function calculates the dot product between two vectors (x1,y1,z1) and (x2,y2,z2) and returns it. (See Listing 11-15.)

Listing 11-15. Calculating the Dot Product of Two Vectors

```
float DotProduct(float x1, float y1, float z1,
                 float x2, float y2, float z2)
{
        return ((x1 * x2) + (y1 * y2) + (z1 * z2));
}
```

The CalculateCollisionImpulseNative() function calculates the collision reaction force from two objects colliding with each other and returns the value. (See Listing 11-16.)

Listing 11-16. Calculating the Reaction Force for a Collision

```
jfloat
Java_com_robsexample_glhelloworld_Physics_CalculateCollisionImpulseNative(JNIEnv* env,
                                                      jobject thiz,
                                                      jfloat CoefficientOfRestitution,
                                                      jfloat Mass1,
                                                      jfloat Mass2,
                                                      jfloat RelativeVelocityX,
                                                      jfloat RelativeVelocityY,
                                                      jfloat RelativeVelocityZ,
                                                      jfloat CollisionNormalX,
                                                      jfloat CollisionNormalY,
                                                      jfloat CollisionNormalZ)
{
        // 1. Calculate the impulse along the line of action of the Collision Normal
        //float Impulse = (-(1+CoefficientOfRestitution) * (RelativeVelocity.
DotProduct(CollisionNormal))) /
        //                              (1/Mass1 + 1/Mass2);

        float RelativeVelocityDotCollisionNormal = DotProduct(RelativeVelocityX, RelativeVelocityY,
RelativeVelocityZ, CollisionNormalX, CollisionNormalY, CollisionNormalZ);
        float Impulse = (-(1+CoefficientOfRestitution) *
RelativeVelocityDotCollisionNormal)/(1/Mass1 + 1/Mass2);
        return Impulse;
}
```

Modifying the Physics Class

The Physics class must also be modified to implement the reaction force calculation.

The CalculateCollisionImpulseNative() function has to be declared as native in order to be used.

```
native float CalculateCollisionImpulseNative(float CoefficientOfRestitution,
                              float Mass1,float Mass2,
                              float RelativeVelocityX,  float RelativeVelocityY, float
RelativeVelocityZ,
                              float CollisionNormalX, float CollisionNormalY, float
CollisionNormalZ);
```

The ApplyLinearImpulse() function has to be modified so that it calls the CalculateCollisionImpulseNative() function to calculate the reaction force of the collision. The existing calculation for the reaction force is commented out. (See Listing 11-17.)

Listing 11-17. Modifying the ApplyLinearImpulse() Function

```
void ApplyLinearImpulse(Object3d body1, Object3d body2)
{
        float m_Impulse = 0;

        /*
        // 1. Calculate the impulse along the line of action of the Collision Normal
        m_Impulse = (-(1+m_CoefficientOfRestitution) *
(m_RelativeVelocity.DotProduct(m_CollisionNormal))) / ((1/body1.GetObjectPhysics().GetMass() +
1/body2.GetObjectPhysics().GetMass()));
        */
        m_Impulse = CalculateCollisionImpulseNative(m_CoefficientOfRestitution,
                                        body1.GetObjectPhysics().GetMass(),
                                        body2.GetObjectPhysics().GetMass(),
                                                m_RelativeVelocity.x,
                                        m_RelativeVelocity.y,
                                        m_RelativeVelocity.z,
                                                m_CollisionNormal.x,
                                                m_CollisionNormal.y,
                                                m_CollisionNormal.z);

        // 2. Apply Translational Force to bodies
        // f = ma;
        // f/m = a;
        Vector3 Force1 = Vector3.Multiply( m_Impulse, m_CollisionNormal);
        Vector3 Force2 = Vector3.Multiply(-m_Impulse, m_CollisionNormal);

        body1.GetObjectPhysics().ApplyTranslationalForce(Force1);
        body2.GetObjectPhysics().ApplyTranslationalForce(Force2);
}
```

Run the program. The natively calculated gravity should pull toward the ground objects such as the tank in Figure 11-11. The arena objects should rotate as in Figure 11-12. Collision forces acting on objects after a collision should work to deflect objects away from one another, as shown in Figure 11-13.

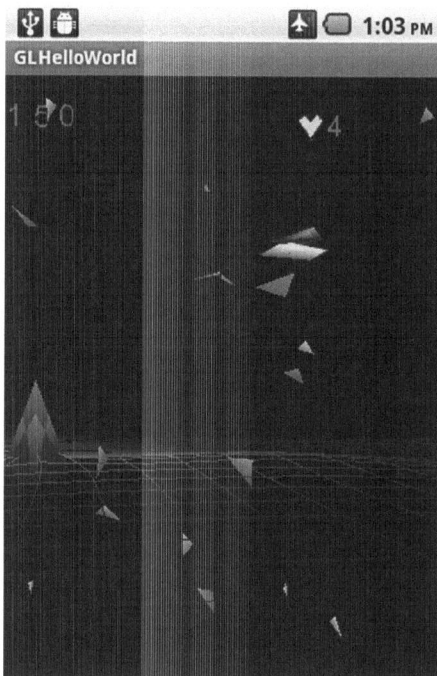

Figure 11-11. *Tank falling from natively calculated gravity*

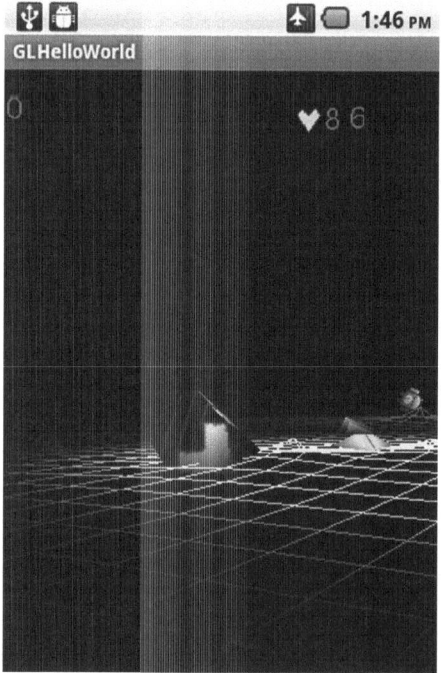

Figure 11-12. *Arena object turning from natively called rotation*

Figure 11-13. *Collision between player's ammunition and an arena object with reaction force calculated natively*

Summary

In this chapter, I covered the Android Native Development Kit (NDK). I started with an overview of what exactly the NDK is and what system requirements, software requirements, and actual Android hardware requirements are needed to use the NDK. Then, the Java Native Interface, or JNI, was covered and used to demonstrate how Java functions can call native functions written in C and how native functions written in C can be used to call Java functions. A simple "Hello World from JNI and Native Code" hands-on example was then introduced, and you were taken through a step-by-step implementation of native code into an existing Java program. Finally, another hands-on example was presented that demonstrated how to integrate native code into our existing Drone Grid game case study.

Chapter 12

Publishing and Marketing Your Final Game

In this chapter, I cover the publishing and marketing of your final game. I start with a discussion of how to create the final game distribution file that users will install. I then cover how to test the distribution file by copying and then installing it on an actual Android device. Next, I cover a list of Android marketplaces in which you can upload your game distribution file for sale and/or download. Then, numerous ad networks that support Android are presented for those who want to make money from their game through advertisements. A list of game sites that review Android games is provided. Finally, other helpful web sites for the Android game developer are listed.

Creating the Final Distribution File

The final distribution file that you will submit for your users to download and install is an .apk file generated from the Eclipse program with Android Development Tools installed.

To begin creating an .apk distribution file, select File ➤ Export from Eclipse, as shown in Figure 12-1.

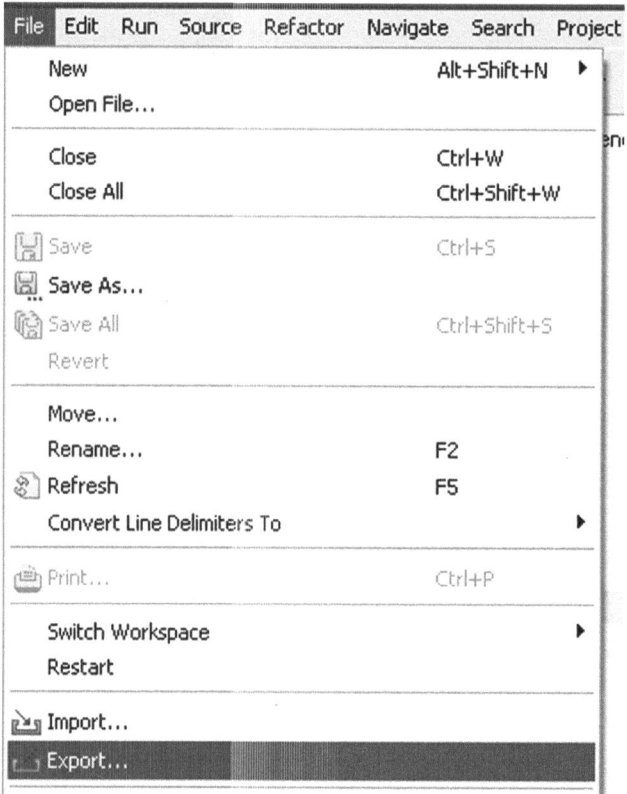

Figure 12-1. Selecting Export from the File menu

The Export window dialog should pop up. Under the Android folder, select Export Android Application and click the Next button. (See Figure 12-2.)

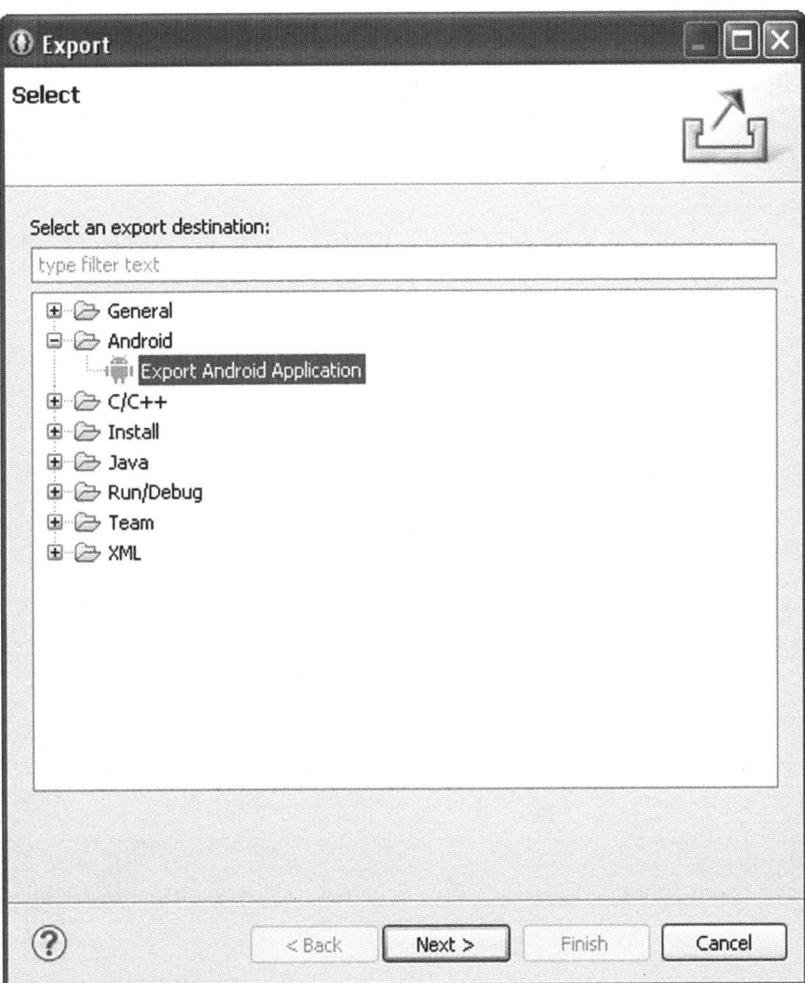

Figure 12-2. *Exporting an Android application*

Next, the Export Android Application window should pop up. Click the Browse button to select an Android project to export and turn into an .apk distribution file. Once you select a project, it is checked for any errors that might hinder the packaging of the project. Then, click the Next button to move to the next screen. (See Figure 12-3.)

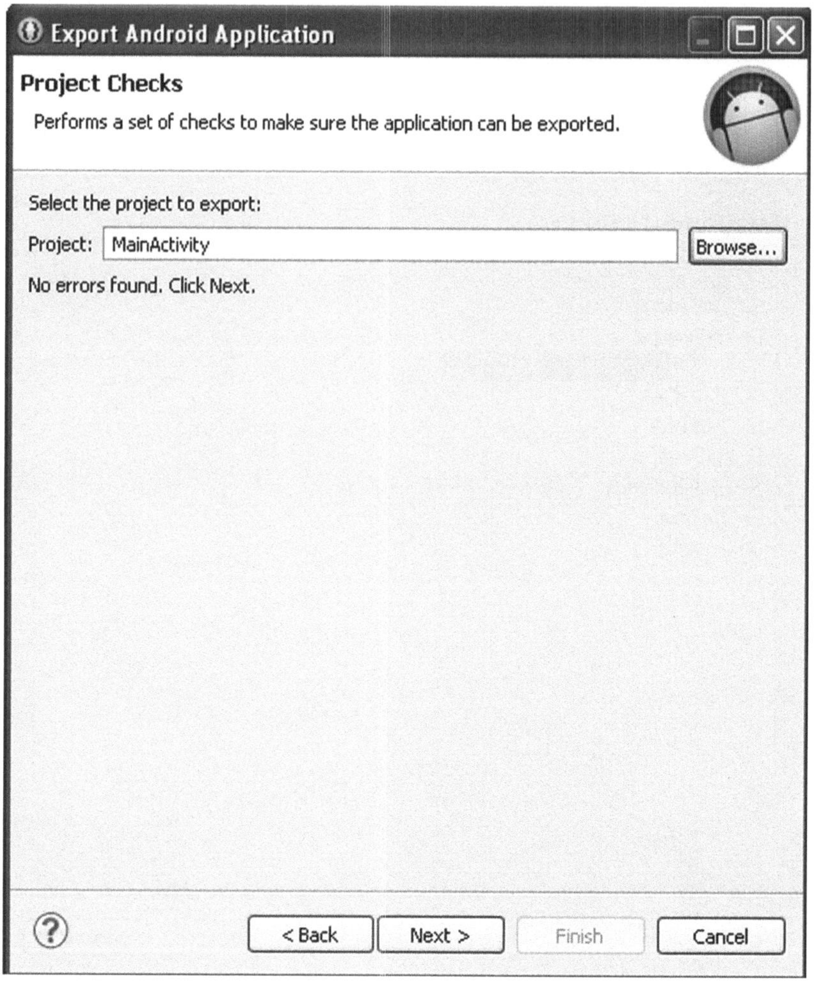

Figure 12-3. Select application to export

The Keystore selection window should come up. Select the Create new keystore radio button. Click the Browse button and choose the directory you want to store the new keystore file in. Type in a password in the Password box and confirm the password in the Confirm box. Click the Next button to continue. (See Figure 12-4.)

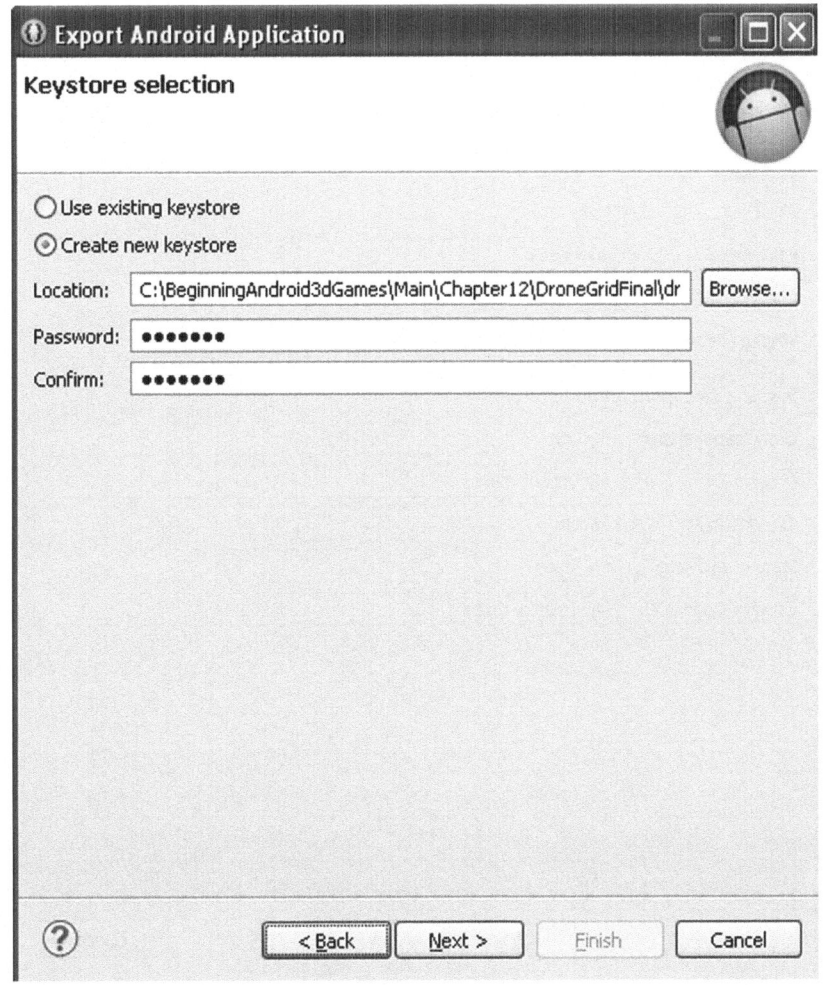

Figure 12-4. Create a new keystore selection

The Key Creation window should now be displayed. Filling out this form will create a key that will be used to sign your application. Fill out the form and click the Next button. (See Figure 12-5.)

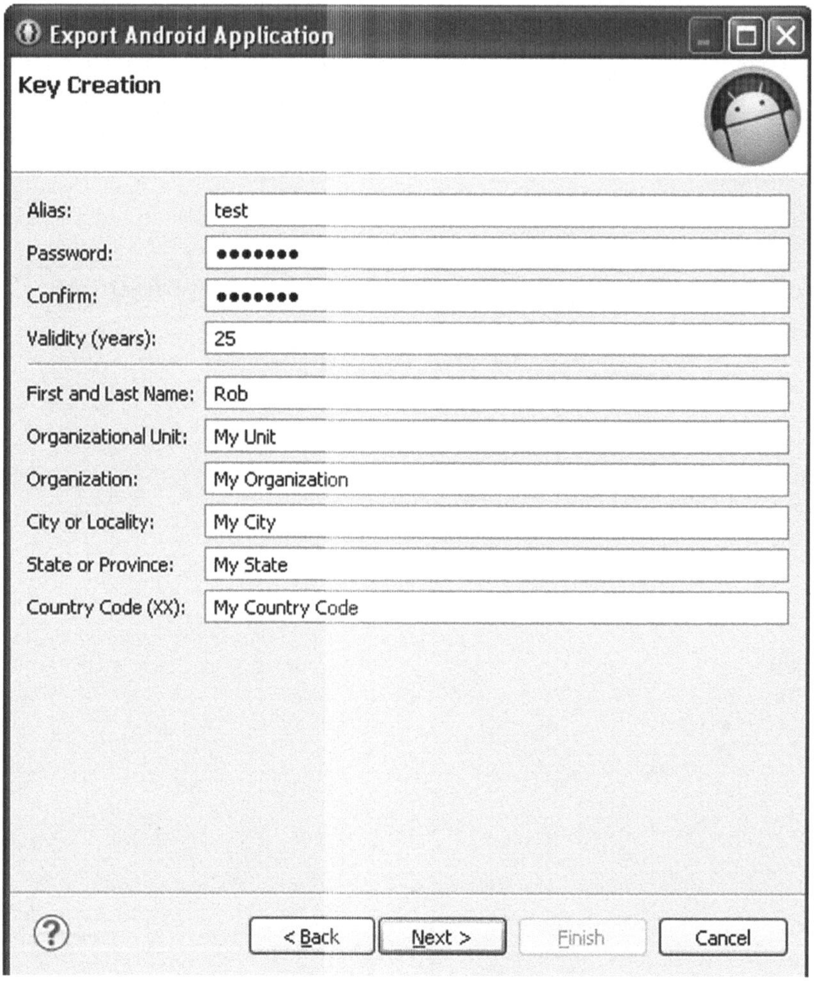

Figure 12-5. Key Creation window

> **Note**   It's recommended that you back up the keystore file to a safe location. You will have to use this
> keystore file if you want to update the games that are currently published using this keystore file.

The Destination and key/certificate checks window should appear. Click the Browse button to enter
a directory and file name for your distribution .apk file. Click the Finish button to start creating your
final distribution .apk file. (See Figure 12-6.)

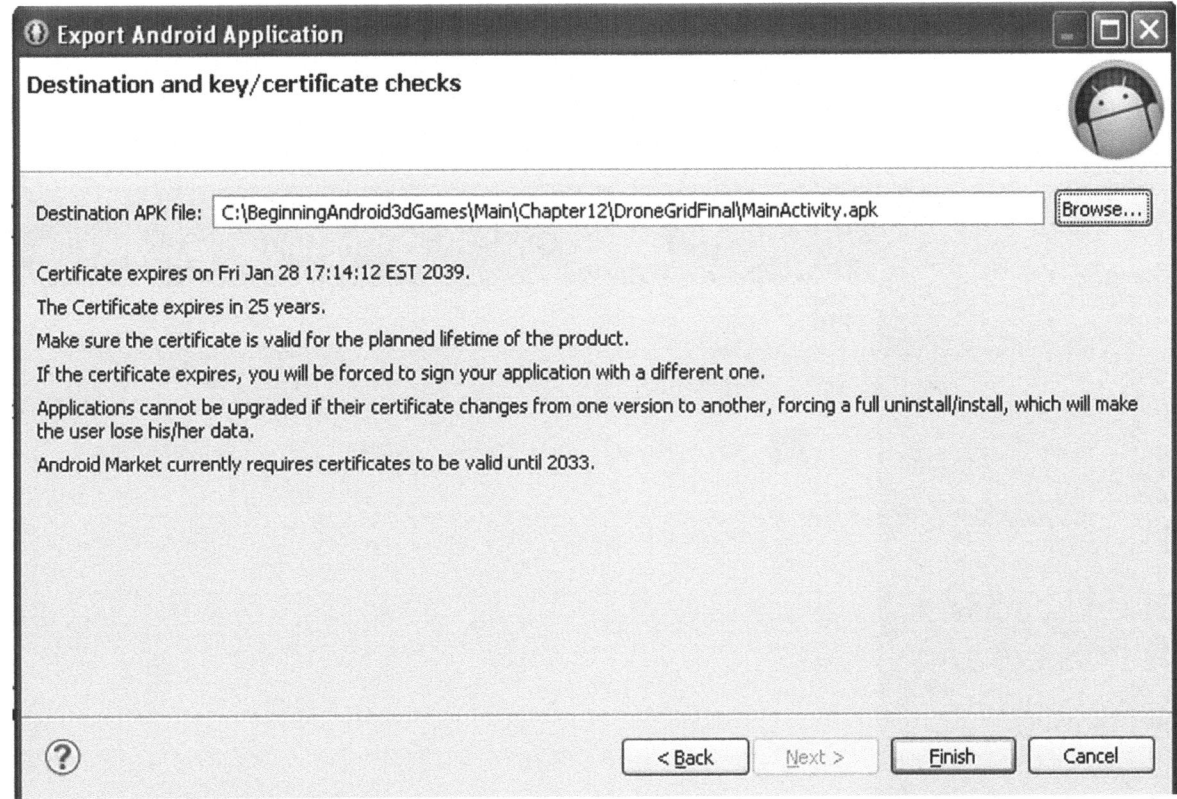

Figure 12-6. Creating the final .apk file

Testing the Distribution .apk File

Now let's test the distribution .apk file you just created, by installing it on an Android device. First, the .apk file has to be copied to the actual Android device. There are many ways to do this, depending on what version of the Android operating system you are using, what software (such as FTP) you have installed on your Android, and what network connections you have set up. I will demonstrate the copying method that will work on all the Android operating systems, regardless of what file transfer software you have installed or what network connections you have set up. To do this, we can use the Android Debug Bridge (adb) push command to put the file on the device that is connected to your computer via USB cable. The general form of the command is as follows:

```
adb push Filename LocationOnAndroidDevice
```

The specific command to put the MainActivity.apk file on the Android device at the location/sdcard/Download using the adb located at C:\Android\adt-bundle-windows-x86\sdk\ platform-tools and assuming we are in the same directory that the .apk is in is

```
C:\Android\adt-bundle-windows-x86\sdk\platform-tools\adb push MainActivity.apk /sdcard/Download
```

The C: refers to the drive letter that your Android SDK is installed on and may differ according to where you have stored your SDK and the specific operating system you are using. After executing the command, the MainActivity.apk file should now be on your Android device at /sdcard/Download, assuming this directory already exists. (See Figure 12-7.)

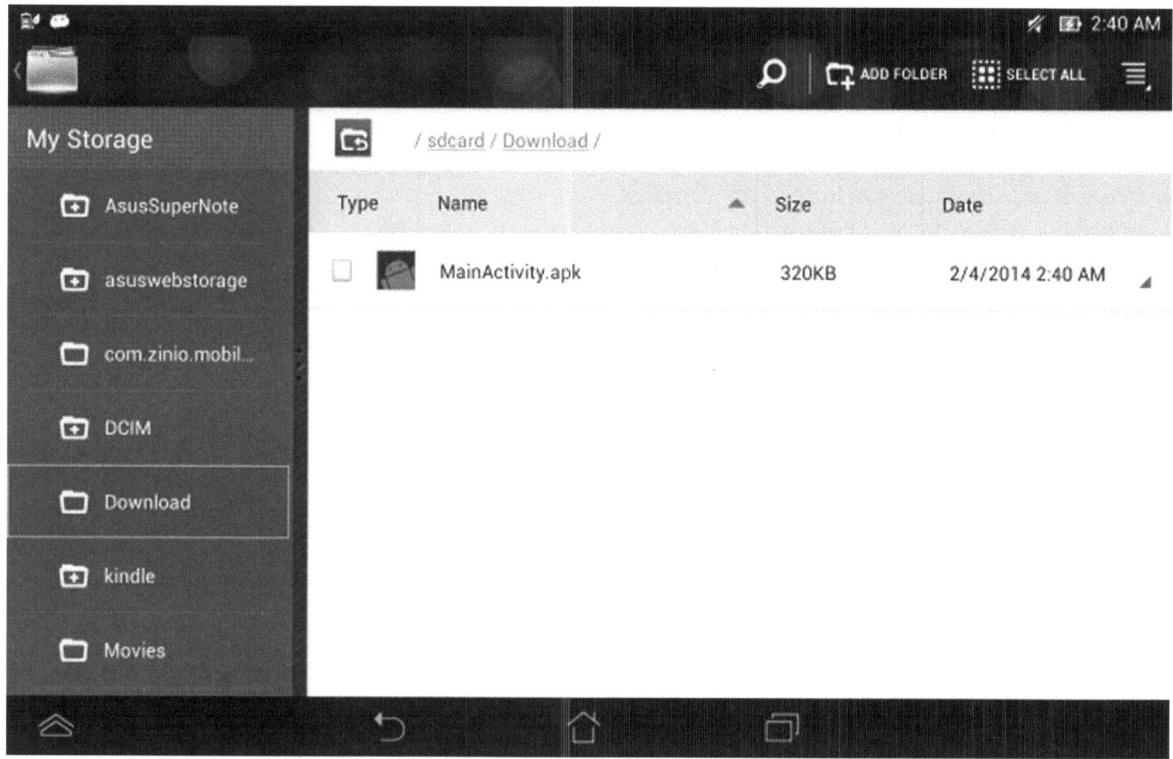

Figure 12-7. Copying MainActivity.apk on an Android device

Before you try to install the .apk file, you will have to go to the Settings ➤ Applications section and check the box under Unknown sources that allows the installation of apps from unknown sources, if you are using older Android operating system versions such as 2.2. On later versions of the Android operating system, you will have to look under Settings ➤ Security. (See Figure 12-8.)

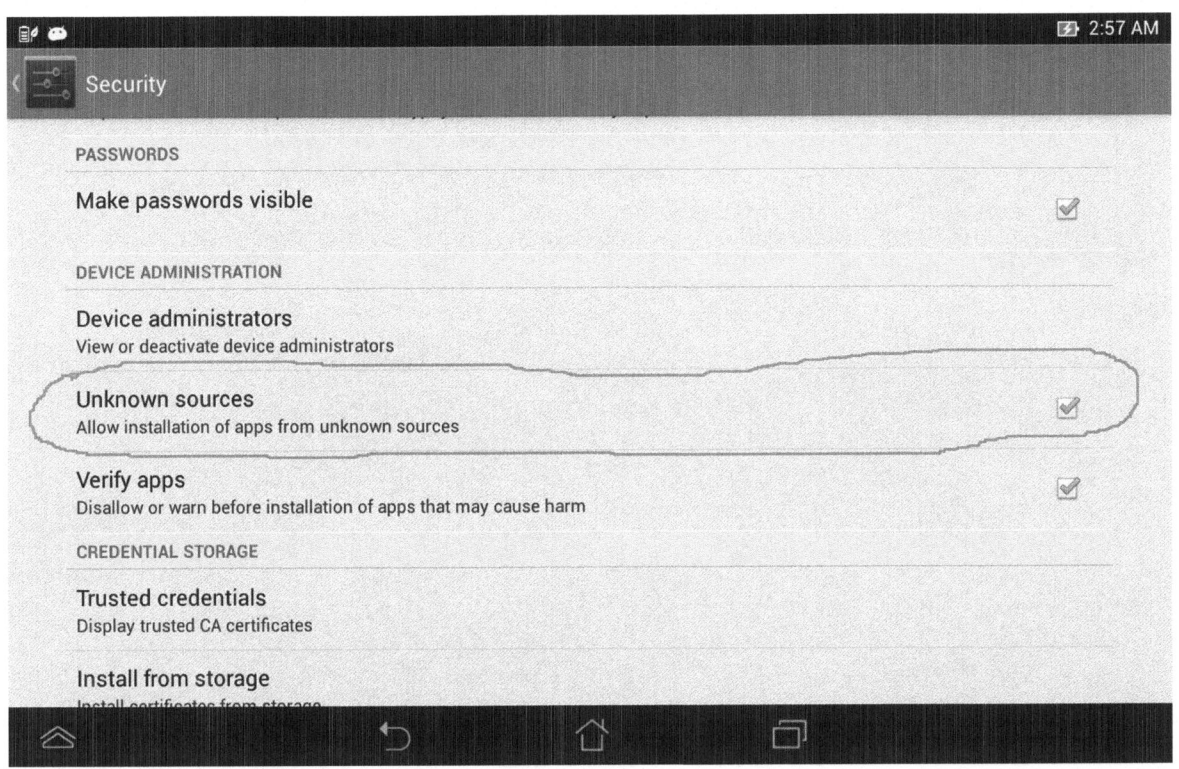

Figure 12-8. Allowing installation of .apk files from unknown sources

Go back to the Android's file manager program and navigate back to the directory where you copied the .apk file. Click the .apk file to start the installation process. A screen should come up asking if you want to install this application. Click the Install button. (See Figure 12-9.)

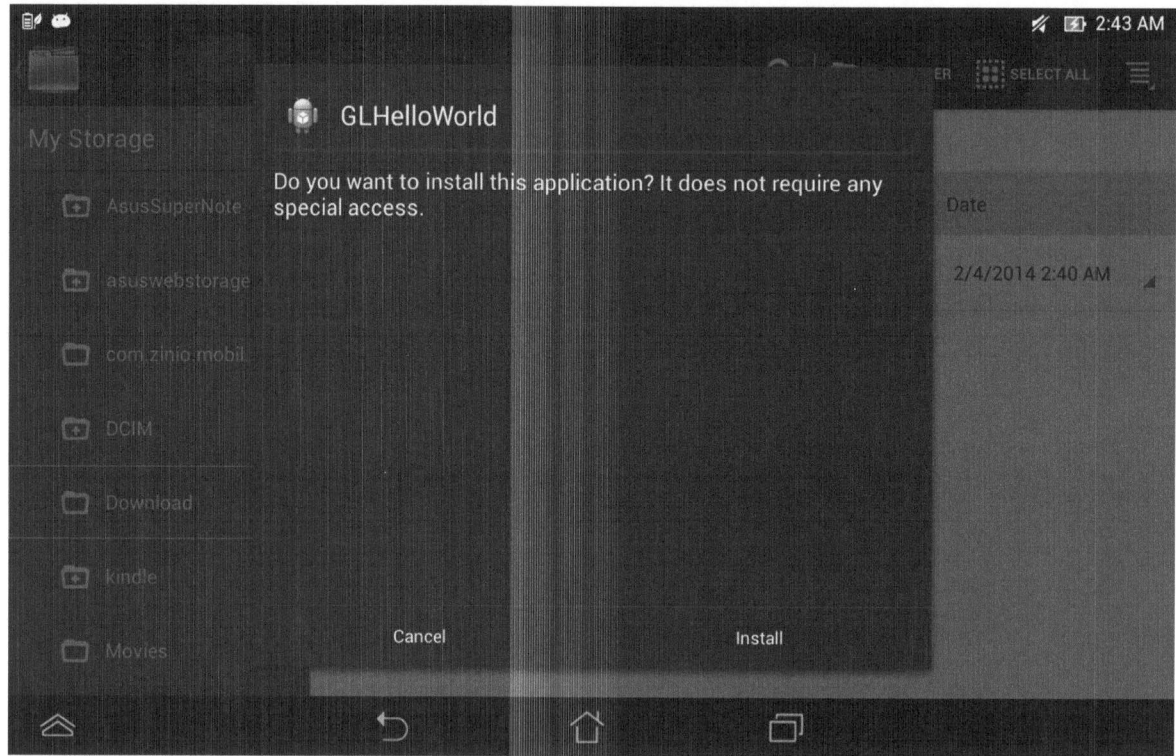

Figure 12-9. Installing the .apk

After the installation is complete, another screen should appear, verifying that the .apk has been successfully installed. (See Figure 12-10.)

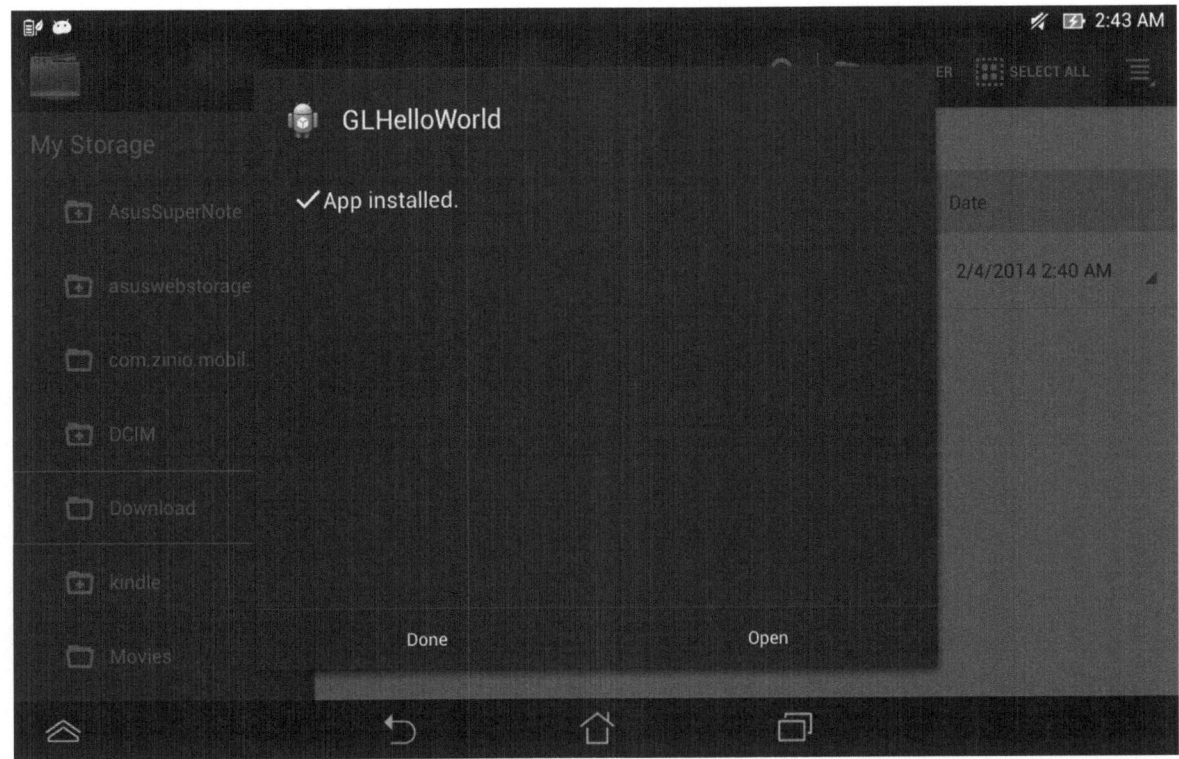

Figure 12-10. App installed verification screen

Click the Open button to start the Drone Grid game. (See Figure 12-11.)

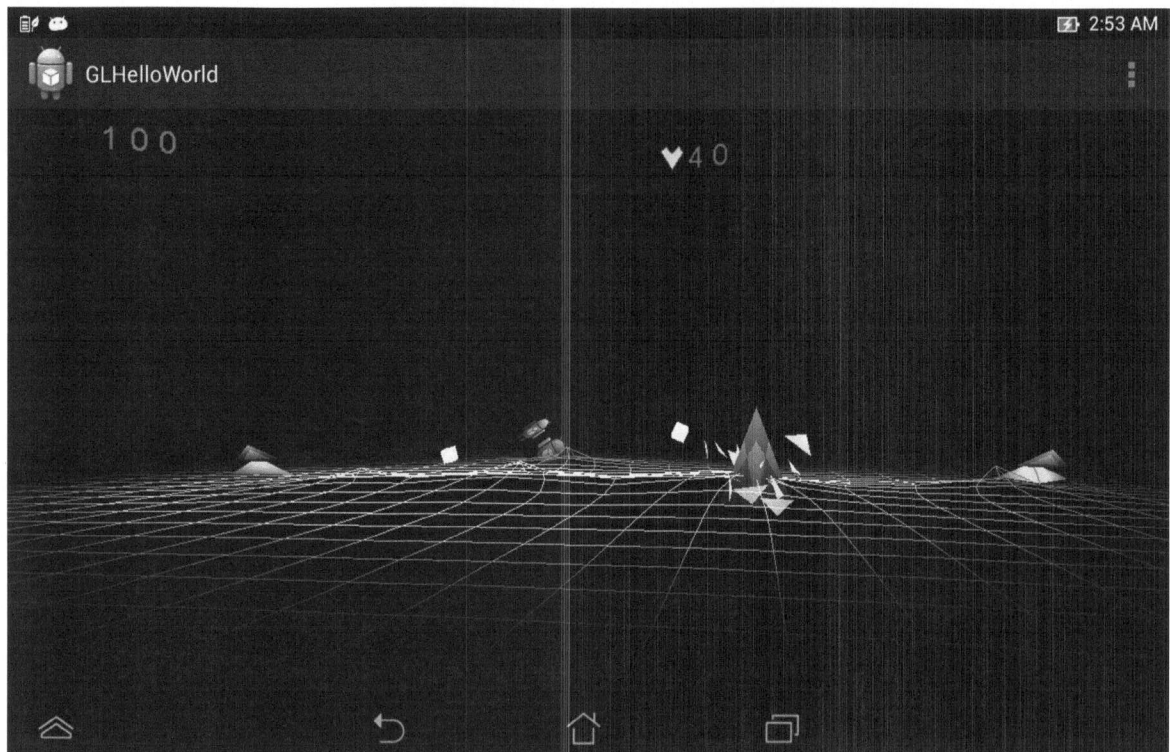

Figure 12-11. The game running from the newly installed .apk file

Now that your game is installed and running on an Android device, it is time to start marketing your game.

List of Android Marketplaces and Policies

This section lists some of the Android marketplaces in which you can upload your application for download by other users. Each marketplace has its own policies, which can change frequently, depending on market conditions. For example, Google recently tightened its restrictions on what types of ads can be used in applications sold in its Google Play market. Amazon has recently eliminated its yearly fee for selling games and apps in its store on Amazon.com.

Google Play

Google Play is the main Android market. The link to sign up for a Google publisher account is

https://play.google.com/apps/publish

There is a $25 one-time registration fee that you will have to pay using Google Wallet. If you want to sell items, you will also need a Google Wallet Merchant account. You can apply for a Google Wallet Merchant account from within the Google Developers Console by navigating to the Financial Reports ➤ Set Up a Merchant Account Now tab. This should take you to the Google Wallet site to sign up as a merchant.

You can publish and unpublish your application or game quickly through the Developers Console. Google does not screen your Android program before it makes it publicly available.

A full description of Google's policies is located at

http://play.google.com/about/developer-content-policy.html

> **Note**   If you decide to include ads in your game, make sure that they comply with Google Play's ad policy, or you risk having your games banned or your account made inactive permanently.

Amazon Appstore for Android

Amazon runs an Android application and game store in which you can sell your game or provide it for a free download. The web link to sign up for a developer's account is

https://developer.amazon.com/welcome.html

There is no fee for registration. If you sell programs through Amazon's Appstore, you will receive 70% of the list price of the item.

You will be required to submit your game to Amazon for testing and verification before it is available for download or purchase. Review usually takes a few days.

Samsung Apps Store

Samsung operates its own Android store in which you can upload your apps and games for sale or for a free download. The web link to log in as a developer is

http://seller.samsungapps.com/login/signIn.as

There is no fee for registration. If you sell games through the store, you will receive 70% of the list price of the item.

You will have to submit your game for review by Samsung before it is available for sale or download. Review usually takes a week but depends on the number of devices you choose to test your game against. Samsung tests your game specifically with different models of Samsung phones and tablets, depending on which models you select.

Aptoide

Aptoide is different from previously discussed stores in that each developer or publisher manages his/her own store and the user must download the Aptoide client and install it in order to download and install Android software from these stores. The official web site is

```
www.aptoide.com
```

Here is the official web site description: "Aptoide is a website where you can download free apps to mobile Android devices through a software client, Aptoide. In Aptoide you can also upload Android apps to share with others."

Appitalism

Appitalism is an app store similar to Google Play in which the developer can sell or upload free apps for distribution. The official site is

```
www.appitalism.com
```

There is no registration fee.

In terms of profit, 70% of the price of an item is returned to the developer.

GetJar

GetJar allows you to publish your game or app on its site for free. The main web site is

```
www.getjar.mobi
```

The developer login link is

```
http://developer.getjar.mobi
```

GetJar claims more than 3 million downloads per day from its web site. However, GetJar does not accept paid apps.

SlideMe

SlideMe is an Android application and game store in which you can upload your free and paid Android games for distribution and/or sale. The official web site is

```
http://slideme.org/
```

There are no developer fees.

Soc.Io Mall

Soc.Io Mall is an Android application and game store that accepts free and paid apps. The official developer web site is

```
https://developer.soc.io/home
```

There is no fee for submitting an application or game.

Your Own WebSite.Com

Remember that with Android, you can just publish your final distribution file on your own web site. However, if you wish to receive money for it, you will probably have to rely on other entities such as payment processors to process credit and debit card transactions or ad networks that pay for clicks on their ads generated by the users of your programs.

List of Android Ad Networks

One way of making money from your game or app is to use an Android ad network that will pay you based on the amount of clicks that your users make on the ads the network places in your game. Each ad network usually has its own Software Development Kit (SDK) that you will have to integrate into your game. The SDK usually consists of an Android library in the form of a `.jar` file and code that uses the functions in this library to display ads. Different ad networks have different styles of ads to choose from. This section first lists a few of the more prominent ones in the Android developer community and then lists other ad networks and marketing companies that might be useful to Android developers for monetizing and promoting their games.

AppFlood

AppFlood is an ad system from PapayaMobile that is based in Beijing, China, with offices in San Francisco, in the United States, and London, England. It has a web site at

```
http://appflood.com/
```

It has the following types of ads available:

- Interstitials: These are full screen ads that are generally displayed at a natural break point in the game, such as the end of a level or the end of the game. (See Figure 12-12.)

Figure 12-12. AppFlood interstitial ad

■ App lists: These ads mimic the look and feel of a typical Android app/game store. (See Figure 12-13.)

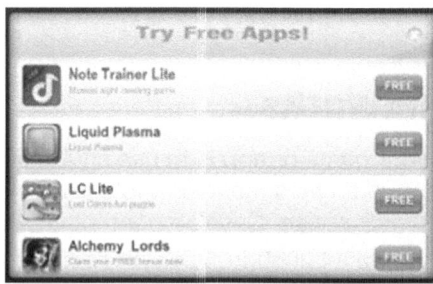

Figure 12-13. AppFlood app list

■ More games ad: These ads display one big game ad, along with four smaller ones. (See Figure 12-14.)

Figure 12-14. AppFlood more games ad

■ Notification ads: These ad types are notifications that are pushed to the user's Android phone.

■ Icon ads: These ad types put an icon on the screen of the user's phone. Be aware that this type of ad is annoying to many users and may not be compliant with Google Play's latest ad policies.

Appwiz

Appwiz is an ad network that was founded in 2012 and has a web site at

`www.appwiz.com`

The types of ads it offers developers are

- Search icon: A search icon is placed on the user's home screen. Note, however, that this type of ad is very annoying to many users.

- Bookmark: A bookmark is placed in the user's web browser.

- Offer wall: A full-screen ad that dynamically optimizes between other subformats that Appwiz offers, such as AppWall, SmartWall, dialog ads, video ads, and Rich Media.

- Premium ad: A shortcut placed on the home screen that links to free apps and hot deals.

LeadBolt

LeadBolt is an ad network founded in 2010 and located in Sydney, Australia. Its web site is

`www.leadbolt.com`

Ad types available are

- Banners
- Push notifications
- Home screen icons
- Browser bookmarks
- Interstitials

AppBucks

AppBucks is an ad network located in Fort Myers, Florida, in the United States. The company's web site is

`www.app-bucks.com`

The types of ads available from AppBucks are

- Interstitial: This type of ad fills up the entire screen and is generally used at key points in the game, to get the user's attention, such as the end of a level. See Figure 12-15 for an example of an interstitial ad from AppBucks.

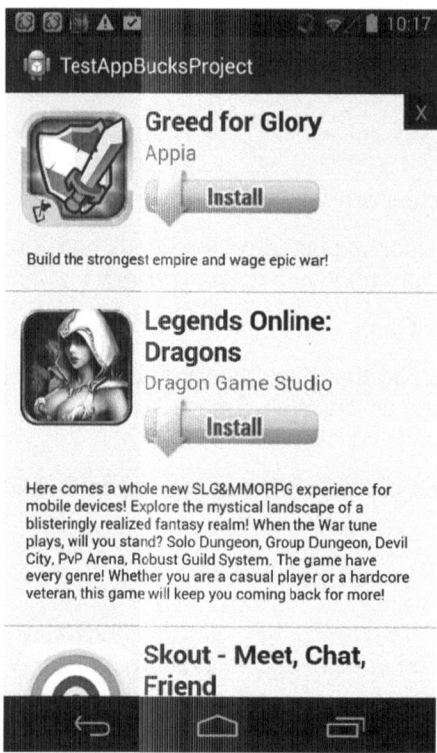

Figure 12-15. AppBucks interstitial ad

- Slider ad: This type of ad, which works well with wallpaper and service-orientated apps, slides out from the side of the screen. (See Figure 12-16.)

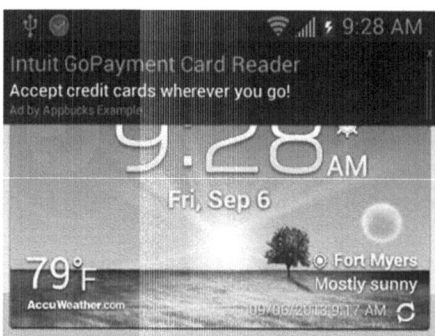

Figure 12-16. AppBucks slider ad

- Banner ad: The banner ad displays a banner, usually across the top or bottom of the screen.

MobileCore

MobileCore is an ad network based in Tel Aviv, Israel, that was founded in 2009. The company's web site is

www.mobilecore.com

The types of ads available from MobileCore are

- AppWall ad: Half-screen or full-screen ads that offer other apps or deals. The developer will get paid for each click or install generated from the AppWall.

- Slider ad: Ads that slide out from the side of the screen.

AdMob

AdMob is run by Google and is probably the safest to use if you want your apps to be compliant with Google's marketplace policies. Violations of those policies can get your game or app banned and/or your account frozen. The web site for AdMob is

www.google.com/ads/admob/

AdMob has the following types of ads:

- Banner ads: These ads take up a small portion of the screen and allow the user to click through to a more detailed information page or web site. (See Figure 12-17.)

Figure 12-17. AdMob banner ad

■ Interstitial ads: Interstitial ads are large full-screen ads that are designed to grab attention. (See Figure 12-18.)

Figure 12-18. AdMob interstitial ad

StartApp

StartApp is a mobile advertising platform started in 2010. Headquartered in the United States in New York, the company's web site is

www.startapp.com

Types of ads offered are

- Interstitial ads: Full-screen ads appear at any point the developer chooses. (See Figure 12-19.)

Figure 12-19. StartApp interstitial ad

■ Banner ads: 3D banner ads. (See Figure 12-20.)

Figure 12-20. StartApp 3D banner ad

■ Exit ads: Show an ad that appears when the user exits the application by clicking the back button or home button. (See Figure 12-21.)

Figure 12-21. StartApp exit ad

■ Search box: Shows a useful sliding search box from within the app.
   (See Figure 12-22.)

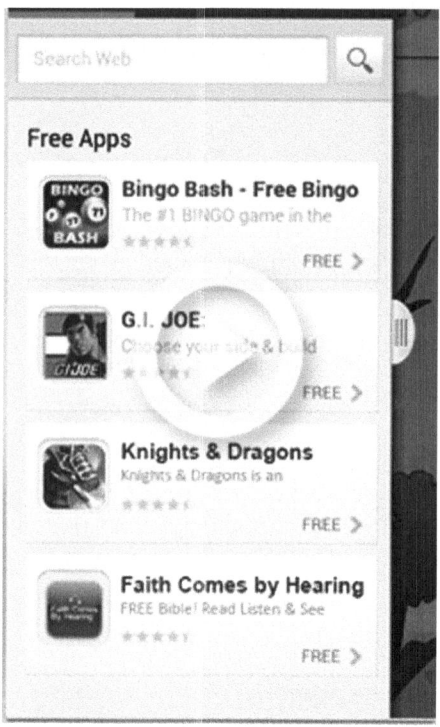

Figure 12-22. StartApp search box

■   Splash screen: Shows an ad while your program is loading. (See Figure 12-23.)

Figure 12-23. StartApp splash screen

Other Ad Network and Marketing-Related Companies

The following list covers ad networks and marketing-related companies that may be helpful both in marketing your game and earning you money from ads placed in the game.

Aarki (http://aarki.com): Aarki is a mobile advertising provider based in Silicon Valley, California.

AdColony (http://adcolony.com): Launched in 2011, AdColony is a leading mobile video advertising and monetization platform that plays crystal-clear HD video at lightning speed and drives deep engagement with content.

Adfonic (http://adfonic.com): Adfonic is a mobile advertising buying platform that is headquartered in London.

AdIQuity (http://adiquity.com): AdIQuity is a global mobile ad platform that helps mobile publishers and app developers to earn high revenue from their mobile inventory. It also helps ad agencies, ad networks, and other media buyers to acquire quality global mobile traffic.

AdMarvel (www.admarvel.com): AdMarvel is a mobile ad optimization used by the world's largest publishers, agencies, and carriers.

Admoda (www.admoda.com): Admoda is a mobile ad network. Its primary focus is providing traffic to the performance-based and affiliate marketing sector.

Applifier (www.applifier.com): Applifier helps game and app publishers of all sizes grow their applications through cross-promotion.

Apprupt (www.apprupt.com): Apprupt consists of mobile marketing specialists.

Avocarrot (www.avocarrot.com): Avocarrot is a unique mobile ad network specializing in high-engagement native advertising. Choose from a range of customizable ad-units to create a seamless user experience that results in higher revenues.

BuzzCity (www.buzzcity.com): BuzzCity is a global advertising network.

ChartBoost (www.chartboost.com): ChartBoost is a mobile game service specializing in games.

Epom (http://epom.com): Founded in 2010, Epom specializes in ad serving and ad management.

4th Screen Advertising (www.4th-screen.com): Established in 2006 and now part of the Opera Software group, 4th Screen Advertising is Europe's leading premium mobile advertising sales agency.

Hunt Mobile Ads (www.huntmads.com): Hunt Mobile is the leading independent mobile advertising company targeted to Spanish-speaking market, including all Latin American and the US Hispanic market, and offers solutions to discover, build brands, and capitalize on the mobile Internet sector.

InMobi (www.inmobi.com): InMobi is a performance-based mobile ad network backed by Soft Bank and Kleiner Perkins Caufield & Byers. The company was founded in 2007 in India and has offices in several countries.

Inneractive (http://inner-active.com): Inneractive is a global programmatic ad stack for mobile publishers, focusing on video, hyper-local, and in-app search advertising.

Jampp (www.jampp.com): Jampp is a leading data-driven mobile app marketing platform that connects to a large number of mobile ad networks and real-time bidding exchanges.

Kiip (www.kiip.com): Kiip provides real rewards for virtual achievements.

Komli Mobile (www.komlimobile.com): Komli Mobile is a leading global mobile advertising and publishing network.

Leanmarket (www.lean.com): Leanmarket specializes in marketing efficiency issues.

LoopMe Media (http://loopme.biz): LoopMe is the leading global pioneer in social ad discovery on smartphones and tablets. LoopMe enables consumers to give feedback on ads ("like," "stop," and "share"), which increases click interactions, branded engagement, and value through social endorsement.

MdotM (www.mdotm.com): MdotM is a mobile marketing services company.

Medialets (http://medialets.com): Medialets is a mobile advertising company. Medialets's mobile and tablet ad-serving platform, Servo, provides advanced measurement technology and analytics and simplified campaign management.

Millennial Media (`www.millennialmedia.com`): Millennial Media is a mobile marketing and advertising company.

MKmob (`www.mkmob.com/`): MKmob is a global mobile ad network.

MMedia (`http://mmedia.com`): MMedia is a mobile advertising and monetization network.

Mobbnet (`www.mobbnet.com`): Mobbnet is a global advertising network.

Mobfox (`www.mobfox.com`): MobFox is a mobile advertising network that operates across iPhone, Android, Blackberry, Windows Mobile, and mobile web sites.

Mobgold (`www.mobgold.com`): MobGold helps advertisers to reach targeted users on various mobile devices and publishers to monetize their mobile traffic.

MobileFuse (`www.mobilefuse.com`): MobileFuse is a mobile ad network composed of strategically selected premium sites and applications, with a reach of 85 million uniques.

Mobile Theory (`http://mobiletheory.com`): Mobile Theory offers mobile advertising and services.

Mocean Mobile (`www.moceanmobile.com`): Mocean Mobile Marketplace (MMM) is the world's largest mobile ad marketplace.

Mojiva (`www.mojiva.com`): Mojiva is a mobile ad network focused on smartphones and tablets. It is most known for Mojiva tab, which is an ad network specifically designed for tablets.

MoPub (`www.mopub.com`): MoPub is a hosted ad-serving solution built specifically for mobile publishers.

Nexage (`www.nexage.com`): Nexage strengthens publishers' and developers' mobile ad businesses with solutions that increase mobile ad revenue and decrease operational costs.

OnMOBi (`http://on-mobi.com`): OnMOBi is an advertising network focusing exclusively on games and finance.

Placeplay (`www.placeplay.com`): Placeplay is a mobile advertising network targeting iOS and Android.

Playhaven (`www.playhaven.com`): Playhaven is a mobile advertising company focusing on games.

Pontiflex (`www.pontiflex.com`): Pontiflex is a mobile advertising company specializing in sign up–style ads.

Revmob (`www.revmobmobileadnetwork.com`): Revmob provides mobile advertising for Android and iOS.

SellAring (`www.sellaring.com`): SellAring provides mobile advertising that specializes in audio ads that replace existing ringtones.

SendDroid (`http://senddroid.com`): SendDroid is a mobile advertising company specializing in Android push notification ads.

SessionM (www.sessionm.com): SessionM is a mobile ad company focusing on games.

Smaato (www.smaato.com): Smaato is the leading global mobile advertising exchange. Smaato's SMX platform is the leading global mobile real-time bidding ad exchange, helping mobile app developers and publishers increase ad revenues worldwide.

Sofialys (www.sofialys.com): Sofialys delivers mobile advertising and marketing solutions, including an ad server and mobile ad network.

SponsorPay (www.sponsorpay.com): SponsorPay is an ad monetization company.

StrikeAd (www.strikead.com): StrikeAd is a US- and UK-based mobile advertising company.

Tapgage (www.tapgage.com): Tapgage is a mobile interstitial ad network that helps app developers and publishers monetize their apps and web sites.

TapIt! (www.tapit.com): TapIt! provides mobile advertising.

Tapjoy (www.tapjoy.com): Tapjoy is a mobile advertising company that allows the user to install an application in place of an in-game payment.

ThinkNear (www.thinknear.com): ThinkNear is a mobile advertising company that specializes in location-based ads.

Todacell (www.todacell.com): Todacell is a premium mobile advertising company.

Trademob (www.trademob.com): Based in Europe, Trademob provides mobile app marketing.

Vserv (www.vserv.mobi): Vserv is a mobile advertising exchange focusing on emerging markets.

Wapstart (wapstart.ru/en): Wapstart is a Russian mobile advertising company.

Webmoblink (www.webmoblink.com): Webmoblink is a leading mobile advertising network that targets Latin America (Spanish and Portuguese) and US Hispanic markets.

Widespace (www.widespace.com): Widespace is a premium mobile ad network based in Europe.

XAd (www.xad.com): XAd provides location-based mobile advertising.

Ybrant Mobile (www.ybrantmobile.com): Ybrant Mobile provides mobile advertising with targeted ad campaigns.

YOC Mobile Advertising (http://group.yoc.com): YOC Mobile Advertising is Europe's largest premium mobile ad network, with a strong presence in five key markets: UK, Germany, France, Spain, and Austria.

YOOSE (www.yoose.com): YOOSE is a mobile ad network concentrating on location-specific ads.

Zumobi (www.zumobi.com): Zumobi is a mobile media and advertising company.

List of Android Game Review Web Sites

This section lists web sites that review Android games. Android game-review sites are excellent places to get free publicity for your game. Some of these sites are exclusively dedicated to Android, and others are multi-platform with an Android section.

AndDev: www.anddev.org

Android and Me: http://androidandme.com

Android App Log: www.androidapplog.com

Android Appdictions: www.androidappdictions.com

Android Apps: http://android-apps.com

Android Apps: www.androidapps.com

Android Apps: www.androidapps.org

Android Apps Gallery: www.androidappsgallery.com

Android Apps Reviews: www.androidapps-reviews.com

Android Authority: www.androidauthority.com

Android Bloke: www.androidbloke.co.uk

Android Central: www.androidcentral.com

Android Community: http://androidcommunity.com

Android Encyclopedia: www.androidencyclopedia.com

Android Etvous: www.androidetvous.com

Android Forums: http://androidforums.com

Android France: http://forum.android-france.fr

Android Games: www.android-games.com

Android Games: www.android-games.fr

Android Games Review: www.androidgamesreview.com

Androidgen: www.androidgen.fr

Android Guys: www.androidguys.com

Android Headlines: www.androidheadlines.com

Androidki: http://androidki.com

Android Lab: www.androidlab.it

Android Market Apps: www.androidmarketapps.com

Android MT: www.android-mt.com

AndroidNG: www.androidng.com

Android Phone Themes: www.androidphonethemes.com

Android Pimps: http://androidpimps.com

Android Pit: www.androidpit.com

Android Pit (France): www.androidpit.fr

Android Police: www.androidpolice.com

Android Preview Source: www.androidappreviewsource.com

Android RunDown: www.androidrundown.com

Android Shock: www.androidshock.com

Android Social Media: www.androidsocialmedia.com

Android Spin: http://androidspin.com

Android Tablets: www.androidtablets.net

Android Tapp: www.androidtapp.com

Android Techie: www.androidtechie.com

Android Video Reviews: www.androidvideoreview.net

Android Viral: www.androidviral.com

Android World: www.androidworld.it

Android Zoom: www.androidzoom.com

Andro Lib: www.androlib.com

Andronica: www.andronica.com

Apkfile: http://androidgamesapps.apkfile.us

App Brain: www.appbrain.com

App Eggs: www.appeggs.com

Appgefahren: www.appgefahren.de

Application Android: www.applicationandroid.com

Applorer: www.applorer.com

App Modo: www.appmodo.com

App Review Central: www.appreviewcentral.net

Apps 400: www.apps400.com

Appsplit: http://appsplit.com

App Storm: http://android.appstorm.net

Apps to Use: www.appstouse.com

Apps Zoom: www.appszoom.com

Ask Your Android: www.askyourandroid.com

Attdroids: www.attdroids.com

Best Android Apps Review: www.bestandroidappsreview.com

Best Android Game Award: www.bestandroidgameaward.com

Best Apps: http://best-apps.t3.com

Best Droid Games: www.bestandroidgames.net

Cnet: http://reviews.cnet.com

Crazy Mikes Apps: www.crazymikesapps.com

Daily App Show: www.dailyappshow.com

Droid Android Games: www.droidandroidgames.com

Droid App of the Day: http://droidappoftheday.com

DroidForums: www.droidforums.net

DroidGamers: www.droidgamers.com

Droid Gaming: www.droidgaming.net

Droid Idol: www.droididol.com

Droid Life: www.droid-life.com

Droidologist: www.droidologist.com

Droid Review Central: www.droidreviewcentral.com

Droid Soft: www.droidsoft.fr

El Android Elibre: www.elandroidelibre.com

Euro Droid: www.eurodroid.com

Euro Gamer: www.eurogamer.net

Everything Android: www.everythingandroid.org

Frandroid: www.frandroid.com

Game Loft: www.gameloft.com/android-games

Game Play Today: www.gameplaytoday.com

GamePro: www.gamepro.de

Gamerpond: www.gamerpond.com

Game Spot: www.gamespot.com

GameZebo: www.gamezebo.com

Get Android Stuff: http://getandroidstuff.com

GiggleApps: www.giggleapps.com

Hardcore Droid: www.hardcoredroid.com

Hooked On Android: www.hookedondroid.com

HTC Desire Games: www.htcdesireforum.com/htc-desire-games

IGN: www.ign.com/games/reviews/android

IosRPG: www.iosrpg.com

Jeuxandroid: www.jeuxandroid.org

Know Your Mobile: www.knowyourmobile.com

Kotaku: http://kotaku.com

Latest Android Apps: www.latestandroidapps.net

Life of Android: www.lifeofandroid.com

MobiFlip: www.mobiFlip.de

Mobile Apps Gallery: www.mobileappsgallery.com

Mobiles 24: http://forum.mobiles24.com

Mobilism: www.mobilism.org

N-Droid: www.n-droid.de

New Apps Review: www.newappsreview.com

OmgDroid: www.omgdroid.com

100 Best Android Apps: www.100bestandroidapps.com

101 Best Android Apps: www.101bestandroidapps.com

148 Apps: www.148apps.com

PhanDroid: www.phandroid.com

PhoneDog: www.phonedog.com

Play Android: www.playandroid.com

Play Droid: http://playdroid.blogspot.com

Pocket Gamer: www.pocketgamer.co.uk

Pocket Lint: www.pocket-lint.com

Pocket Tactics: www.pockettactics.com

Rpg Watch: www.rpgwatch.com

Samsung Galaxy S Forum: www.samsunggalaxysforum.com

Screw Attack: www.screwattack.com

Slide To Play: www.slidetoplay.com

SmartKeitai: www.smartkeitai.com

Smart Phone Daily: www.smartphonedaily.co.uk

Tablette: http://tablette.com

Talk Android: www.talkandroid.com

Tapscape: www.tapscape.com

Tap Zone: www.tapzone.info

Tech Hive: www.techhive.com

The Android Galaxy: www.theandroidgalaxy.com

The Android Site: www.theandroidsite.com

Tips 4 Tech: www.tips4tech.net

Top Best Free Apps: http://topbestfreeapps.com

Touch Arcade: www.toucharcade.com

24 Android: www.24android.com

List of Other Helpful Sites for Android Developers

The following list contains other helpful sites for Android developers. Among these sites are ones that provide free graphics and graphics-related tools.

- **Open Clip Art** (www.openclipart.org): Contains public domain and royalty-free graphics.
- **Vector Open Stock** (www.vectoropenstock.com): Contains free vector clip art.
- **Blender 3D Renderer** (www.blender.org): Free 3D model builder and renderer available for Mac OS X, Linux, and Windows.
- **Making Money with Android** (www.makingmoneywithandroid.com): Site concentrating on making money with Android. Has a forum with lots of discussion on the best ad networks available for Android.

Summary

In this chapter, I discussed the publishing and marketing of your game. I started out by covering how to create the final distribution file for your game and how to test this final distribution file on an actual Android device. Next, I covered some of the available Android marketplaces in which you can sell your game or provide it for a free download. Then, I presented a list of ad networks with which you can make money by letting these ad networks place ads inside your game for users to view and click. Next, I provided a list of game review web sites from which you might be able to receive free publicity for your game. Finally, a list of other helpful web sites was given.

Index

B

E

F

W, X, Y, Z

Get the eBook for only $10!

> Now you can take the weightless companion with you anywhere, anytime. Your purchase of this book entitles you to 3 electronic versions for only $10.

This Apress title will prove so indispensible that you'll want to carry it with you everywhere, which is why we are offering the eBook in 3 formats for only $10 if you have already purchased the print book.

Convenient and fully searchable, the PDF version enables you to easily find and copy code—or perform examples by quickly toggling between instructions and applications. The MOBI format is ideal for your Kindle, while the ePUB can be utilized on a variety of mobile devices.

Go to www.apress.com/promo/tendollars to purchase your companion eBook.

Apress®
THE EXPERT'S VOICE™

All Apress eBooks are subject to copyright. All rights are reserved by the Publisher, whether the whole or part of the material is concerned, specifically the rights of translation, reprinting, reuse of illustrations, recitation, broadcasting, reproduction on microfilms or in any other physical way, and transmission or information storage and retrieval, electronic adaptation, computer software, or by similar or dissimilar methodology now known or hereafter developed. Exempted from this legal reservation are brief excerpts in connection with reviews or scholarly analysis or material supplied specifically for the purpose of being entered and executed on a computer system, for exclusive use by the purchaser of the work. Duplication of this publication or parts thereof is permitted only under the provisions of the Copyright Law of the Publisher's location, in its current version, and permission for use must always be obtained from Springer. Permissions for use may be obtained through RightsLink at the Copyright Clearance Center. Violations are liable to prosecution under the respective Copyright Law.